# FOREWORD

It has been widely recognised among researchers that speciation data are essential for proper and reliable modelling of radionuclide behaviour in nuclear reprocessing systems, in developing improved, more specific separation agents and in the migration of radionuclides in the near and far field areas of nuclear waste repository systems.

For this reason, the NEA Nuclear Science Committee decided to hold a workshop with the aim of preparing a report on various speciation methods. The Workshop on Evaluation of Speciation Technology was successfully held at Japan Atomic Energy Research Institute (JAERI), from 26-28 October 1999, with 80 participants from 10 countries. Eleven invited papers and 27 posters were presented on recent progress and application in each of the following five main topics:

- Methods for trace concentration speciation ($<10^{-6}$ M).

- Methods for macro concentration speciation ($>10^{-6}$ M).

- Methods for empirical formula and molecular structure determination.

- Methods for redox speciation.

- Predictive approach to speciation.

On the second and third days of the workshop, subgroups according to the categories mentioned above were set up to evaluate speciation technology. The aims of these subgroups were to examine speciation data that are needed, the advantages and limitations of the different methods that can be applied to obtain this data and suggestions for future R&D to improve present methods or to develop new methods of speciation. The report prepared through each subgroup discussion would serve to guide researchers in choosing the most useful technique.

It was recommended that a web site should be established by OECD/NEA to make information on speciation technology easily available (see Annex).

These proceedings contain the abstracts and full papers presented at the workshop and the subgroup reports prepared by the participants. The editors, Professor Gregory R. Choppin, Professor Jean Fuger and Dr. Zenko Yoshida, and other anonymous referees reviewed the full papers. The editors wish to acknowledge and thank these reviewers.

## OPENING ADDRESS – Satoshi Sakurai, OECD/NEA

It is a great pleasure for me to welcome you to this workshop on behalf of the OECD Nuclear Energy Agency. The OECD Nuclear Energy Agency is an international organisation based in Paris, France, with the basic mission to contribute to the viability of the nuclear power option. The Agency pursues this mission, for example by bringing together expertise in Member countries in co-operative projects, by disseminating important information, by developing consensus opinions and by arranging meetings and workshops such as this one.

The NEA work programme addresses all the key issues in the nuclear energy area, such as nuclear safety, radioactive waste management, radiation protection, legal aspects, economics and nuclear science. In addition, the NEA Data Bank provides a direct cost-free service on nuclear data and computer programs to scientists in Member countries. For those of you who would like to know more about the NEA, I have brought with me a number of leaflets both concerning the NEA and its Data Bank. The information is also available on the NEA's web page, www.nea.fr.

A large part of the NEA's scientific programme is devoted to reactor and fuel cycle physics. However, in recent years, this programme has also been extended to cover nuclear fuel cycle chemistry issues. One preoccupation in the nuclear fuel cycle is the separation of radioactive elements from various materials. For this reason, the NEA established a small group of experts to review the different techniques and chemical processes used in the separation of actinides. The group, with the leadership of Dr. Wymer, produced a state-of-the-art report entitled *Actinide Separation Chemistry in Nuclear Waste Streams and Materials*. The report was published in 1997, and a limited number of copies are still available.

As a follow up to this state-of-the-art report, the NEA arranged the Workshop on Long-lived Radionuclide Chemistry in Nuclear Waste Treatment. This was organised by the CEA, France, in June 1997 at Villeneuve-lès-Avignon, France. The objective of this workshop was to provide up-to-date information on the chemistry of radionuclides and to provide guidance on future activities that could be undertaken in the framework of the NEA. The outcome was a recommendation that the NEA organise two workshops, one on the Evaluation of Speciation Technology (the current workshop), and the other on Speciation, Techniques and Facilities for Characterisation of Radioactive Materials at Synchrotron Light Source, which was held in October 1998 at Grenoble, France. The proceedings of the workshop was recently published and I have here one copy that will be made available for consultation. The second meeting is planned to be held, again at Grenoble, in September 2000.

Concerning the organisation of the present workshop, I would first of all like to thank Professors Choppin and Fuger for the important role they have played in the arrangement of this meeting, both as initiator and also as an active member of the organising committee. I know that organising a meeting such as this is very hard work, and I particularly want to thank Dr. Yoshida and his colleagues, who have laid the basis for a very successful and stimulating meeting. I am also grateful to Japan Atomic Energy Research Institute for hosting this workshop. Many thanks go also to Dr. Maeda, General Chairperson, and to International Scientific and Local Organising Committee members. I am sure that you will spend an interesting, instructive and profitable three days here in Tokai-mura. I wish you all a very successful meeting.

# TABLE OF CONTENTS

# EXECUTIVE SUMMARY

The Workshop on Evaluation of Speciation Technology was successfully held at Japan Atomic Energy Research Institute (JAERI) from 26-28 October 1999. This workshop was devoted to:

- Exchanging recent progress in the development of speciation technology and its application in the various fields of nuclear fuel cycle.

- Discussing the advantages, disadvantages and limitations of each speciation method and preparing a report.

- Proposing or recommending possible co-operative works to be performed within the framework or under the auspices of the OECD/NEA/NSC.

There were 80 participants gathered together from ten countries. Eleven invited papers were presented on recent progress and applications concerning the following five main topics:

- Methods for trace concentration speciation ($<10^{-6}$ M).

- Methods for macro concentration speciation ($>10^{-6}$ M).

- Methods for empirical formula and molecular structure determination.

- Methods for redox speciation.

- Predictive approach to speciation.

The keynote presentation addressed speciation requirements in the context of problems produced by nuclear energy and nuclear weapons production. This talk was followed by two invited presentations for each of the topics mentioned above. These presentations covered:

- Radiochemical procedures for actinides in the environment.

- LASER spectroscopy nanoscopic speciation.

- LASER-based analytical techniques (TRLIF, LA-OES and ES-MS) for direct studies on solids, liquids, etc.

- Luminescence lifetime measurement for determination of hydration number.

- X-ray absorption fine structure spectroscopy for Hf(IV) sorbed onto silica.

- Applications of magnetic resonance methods in the speciation of *f* elements.

- Actinide redox speciation methods, other than chemical separation techniques.

- Chemical separation methods for redox speciation of actinides.

- Estimation of thermodynamic data in assessing geological disposal.

- Chemical analogy in the case of hydrolysis species of $f$-elements.

In the poster session, 27 posters were presented on recent progress and application in each of the five main topics mentioned above. For the speciation methods with regard to trace concentration, eight posters covered the following methods: capillary electrophoresis, mass spectrometry, filtration, X-ray and low-energy $\gamma$-ray spectroscopy, and extraction.

Concerning speciation methods for macro concentration, reports were given on the following topics: LASER techniques, X-ray application, photospectrometry, vibrational spectroscopy, Mössbauer spectroscopy and NMR. The development of actinide sequestering agents was also addressed.

For empirical formula and molecular structure determination, all the posters were related to X-ray techniques: EXANES, EXAFS, XPS and XRD.

Electrochemical methods, such as coulometry, polarography, voltammetory and potentiometry with ion-selective electrode were mostly applied for the purpose of redox speciation. Other methods reported were high-performance extraction chromatography and UV-vis-NIR spectroscopy.

As posters related to predictive approach to speciation, works on calorimetry for hydroxycarbon species of lanthanides and actinides, data processing for the determination of stability constants, chemometric and computer-intensive techniques for U(VI) and spectrophotometry to obtain data for PUREX process models were presented.

After the oral presentation of invited papers and the poster session, five subgroups according to the categories mentioned above were set up. The members of each group discussed needs concerning speciation data, the advantages and limitations of the different speciation methods, and suggestions for future R&D. Resulting from these discussions, subgroup reports were prepared, which are included in these proceedings. The methods evaluated by the subgroups are summarised in the table below. The reports are also available on the NEA web site at: http://www.nea.fr/html/science/spec.

Through the comparison of various speciation methods, the participants recognised that while a great deal has been accomplished with regard to speciation techniques and methodologies, much still remains to be done. This is in part due to the difficulty and complexity of the task, caused by the ever-increasing speciation challenges. On the other hand, researchers require more information concerning recent progress in speciation technology. Therefore, it was recommended that a working group would be set up in order to review the reports and integrate them into a site devoted to speciation within the NEA web site. It was agreed among the participants that Dr. Wymer would examine the scope and content of the web site and draft a proposal.

# Speciation methods evaluated in the subgroup reports

| Methods | Category | | | | |
|---|---|---|---|---|---|
| | A | B | C | D | E |
| Electrochemical methods (voltammetry, potentiometry) | | | | *E* | |
| Electron paramagnetic resonance (EPR) | | | *E* | | |
| Gas/liquid chromatography | *E* | | | | |
| HREM micrography | | | *E* | | |
| LASER spectrometry | | | | | |
|     Photothermal spectrometry (PAS, PDS, TLS) | *E* | | *E* | *E* | |
|     Luminescence spectrometry | *E* | | | | |
|     Breakdown spectrometry | *E* | | *E* | | |
|     Time-resolved LASER fluorescence spectroscopy (TRLFS) | | | *E* | | |
| Mass spectrometry (ICP, TOF, FAB, etc.) | *E* | *E* | | | |
| Mössbauer spectroscopy | | *E* | *E* | | |
| Nuclear magnetic resonance (NMR) | | *E* | *E* | *E* | |
| Photospectrometry | | | | | |
|     Electronic spectroscopy (UV-Vis-NIR) | | *E* | *E* | *E* | |
|     Fluorescence spectroscopy | | *E* | | *E* | |
|     Light scattering | | | *E* | | |
| Luminescence | *E* | | | | |
| Radiochemical trace analysis | *E* | | *E* | *E* | |
| Thermal analysis (DTA/TG) | | | *E* | | |
| Vibrational spectroscopy (Raman, FT-IR) | | *E* | *E* | *E* | |
| X-ray technique | | | | | |
|     Scanning transmission X-ray microscopes (STXM) | | | *E* | | |
|     X-ray absorption spectroscopy (XANES, EXAFS) | | *E* | *E* | *E* | *E* |
|     X-ray photoelectron spectroscopy (XPS) | | *E* | *E* | | |
|     X-ray and neutron diffraction (XRD) | | *E* | *E* | | |

*E*: evaluated in the report

# OPENING SESSION

# SPECIATION IMPERATIVES FOR WASTE
# MANAGEMENT AND ENVIRONMENTAL POLLUTION

**Raymond G. Wymer**
188-A Outer Drive
Oak Ridge, TN 37830

## Abstract

This paper addresses speciation requirements in the context of problems produced by nuclear energy and nuclear weapons production. These problems are primarily in the areas of waste management, material contamination, and environmental pollution. They pose difficult and important measurement and speciation challenges.

Examples of speciation requirements in the context of national and international regulations are presented to exemplify and make quantitative the types of problems posed by waste management, material contamination, and environmental pollution.

The importance of identifying species present in the natural environment as well as in wastes from chemical and physical processing and from waste management activities and accidental releases is addressed. Differing speciation requirements for macro and micro concentrations of species are discussed.

The role of speciation in modelling studies is discussed.

## Introduction

Waste management and environmental pollution problems are widespread. These problems are nowhere more important and more subject to critical review and comment than when they involve radioactivity. The scope of the problems includes finding and determining the amounts and types of radioactive materials in structures, equipment, soil, water and gases to aid in deciding what treatment may be needed and in establishing whether or not regulatory requirements have been met. Solutions to the problems usually involve chemical operations that rely upon knowledge of the chemical species present. The radionuclides and other chemical elements whose species are to be determined include a large fraction of the elements on the periodic table. The elements and their compounds may be found in solids, liquids and gases in an extremely wide range of concentrations and in a great variety of compositions and mixtures. In addition, there are many toxic organic compounds in wastes whose species must be known. Thus, availability of a broad spectrum of chemical speciation methods to aid in selecting methods of treatment of the wastes and pollutants is central to solution of the waste and environmental problems.

## The regulatory setting

There are international and national bodies that set limits on doses of radiation that nuclear industry workers and the public may be exposed to. These dose limits may be related to the types of radiation and ultimately to concentrations of radioisotopes[1] through the dose analysis models employed. Conversion of annual dose limits to permissible concentrations is species dependent. It also depends on models used for human uptake and for radiation effects of individual radionuclides. The best-known and most widely followed international permissible dose recommendations are in ICRP Publication 60, a report of the International Commission on Radiological Protection. The International Atomic Energy Agency (IAEA) also provides guidance on dose limits. In the United States the comparable organisation is the National Committee on Radiation Protection (NCRP). In addition, the U.S. Environmental Protection Agency (EPA) and the U.S. Nuclear Regulatory Commission (NRC) promulgate regulations on permissible doses to the public, and the EPA provides limits on non-radioactive, toxic chemicals. These dose limits often pose very stringent measurement requirements on the analyst, requiring that measurements be made well below the level of background radiation, sometimes less than 10% of background. The EPA-proposed dose limit of 4 mrem/yr from groundwater is an extraordinarily low limit, requiring very sensitive measurement and speciation techniques. In general, 100 mrem/yr from all sources of radiation is considered to be a safe dose limit.

In the US, many of the states have also adopted concentrations and dose limits for radionuclides and toxic materials.

## Speciation matrices and special situations

Radioactive and toxic chemical elements and their compounds and toxic organic compounds are found in a wide variety of matrices, including soil, groundwater, solutions, sludges[2], sediments, solids in storage tanks, gases and vapors, as well as on and in structures and equipment. This wide range of

---

[1]   The dose limits also apply to machine-produced radiation, but in this paper we are concerned only with radioactivity.

[2]   As used here, the word sludges refers to mixtures, often heterogeneous, of semi-solids, usually, but not always, hydrous or hydroxides of metal ions such as iron, chromium, uranium. They are commonly found in tanks of neutralised wastes from nuclear reactor fuel reprocessing and in wastes from waste treatment facilities.

matrices poses difficult challenges for the analyst attempting to determine the species of a material of interest. Separation of the matrix material is often required before speciation to reduce interferences.

Although speciation of material in samples is most common, there are situations related to accessibility and cost where it is desirable to speciate *in situ*. Often this type of speciation must be done remotely because it is either very difficult to take a representative sample, or it is dangerous to do so. In other cases it may be necessary to determine species present in a gas stream, as from an incinerator or in a radioactive liquid waste stream, in "real time" to prevent prolonged accidental release of a toxic material while sample analyses are underway.

## Origins of radionuclide contaminants

Producing and processing nuclear reactor fuels, enriched uranium, and targets for production of weapons plutonium and heavy elements[3] are major source of radioactive wastes. These operations also lead to situations where significant decontamination is required. There are a variety of nuclear reactor fuel types, and two possible practical fuel cycles. The most important fuel cycle is the uranium-plutonium fuel cycle, which is the only one currently in use. However, the possibility exists for a thorium-$^{233}$U fuel cycle. These two fuel cycles present somewhat different speciation requirements. The chemical elements of principle concern from the point of view of chemical speciation and waste management in the uranium-plutonium fuel cycle are uranium, tritium, plutonium, neptunium, technetium, and iodine. A uranium isotope that does not occur in nature must be added to the list of radio elements of concern when considering the thorium-$^{233}$U fuel cycle. This isotope, $^{232}$U, and more specifically its decay daughters, is a potential source of a significant radiation hazard. Protoactinium is another element of concern in the thorium-$^{233}$U fuel cycle.

Although heavy element production is a minor activity compared to nuclear fuel processing, the production of heavy elements by neutron or charged particle irradiation of specially prepared targets produces relatively large amounts of very toxic alpha and neutron emitters. The elements of concern include americium, curium, and in special circumstances, californium, and berkelium, all of which may occur in a wide variety of chemical species. The chemical species likely to be present is strongly dependent upon the type of chemical processing the elements have undergone and on the environment in which they are present.

In addition to processing nuclear reactor fuel, enriched uranium, and heavy element production targets, wastes have been produced by the recycle of weapons plutonium and by processing scrap from various recycle operations. In the case of weapons plutonium recycle, a major element of concern is americium, in particular, the isotope $^{241}$Am. Recycle of weapons plutonium has been carried out in non-aqueous media. Consequently, speciation of elements contained in media such as halide salts may be required.

Re-enrichment of uranium that has been irradiated in a nuclear reactor has led to release of volatile technetium compounds at very low but measurable concentrations to the soil around the enrichment plant. Technetium can have a large number of valence states, and forms stable chemical species in several of the more important of them, for example, $TcO_4^-$ and Tc(IV). Recently in the US concern has been expressed about plutonium in the environment around enrichment plants that have enriched recycled uranium.

---

[3] Heavy elements are defined here as those elements with atomic numbers above 94, i.e. above plutonium in the periodic table.

In addition to the sources of large amounts of chemical elements and compounds of concern in waste management and pollution control, there are smaller but important sources of radioisotopes. These arise from medical and research activities. For the most part, the medical radioisotopes are short-lived, but this is not always the case. Examples of important medical radioisotopes are $^{131}$I, $^{99m}$Tc and $^{60}$Co. Radioisotopes used in research may be any of a large number of types, and may have long half-lives. Radioisotopes released from research and industrial activities and considered to be below regulatory concern pose a special problem for the analyst. Because they are to be released in an uncontrolled manner, their species and amounts must be well known. They may be very difficult to speciate because they are often at very low concentrations and in matrices that may cause troublesome interferences.

In general, most of the elements mentioned above may be found at a wide range of concentrations, from extremely low, requiring sophisticated methods of speciation, to concentrations high enough for the use of more conventional chemical speciation techniques. Speciation at both micro and macro concentrations is important. Although development of ultra-sensitive speciation techniques is attractive to workers in the field, it should be recognised that simpler, faster and less costly speciation methods for macro concentrations of contaminants and pollutants is also important.

## Characteristics of radioactive samples

Radioisotopes are likely to be at their lowest concentrations in soils, groundwater and gases, and at their highest concentrations in waste storage facilities, e.g., waste tanks. Equipment and facilities in spent fuel reprocessing plants may also be highly contaminated. Despite their relatively higher concentrations in tanks, the radionuclides found in the waste matrices in tanks pose severe challenges to the speciation analyst. These challenges are due to the difficulty of obtaining representative samples, the usually very high radiation fields, and the extraordinarily high cost of obtaining samples. The tanks often contain solutions that have very high concentrations of salts such as sodium nitrate and nitrite and may have very high pHs. In some cases the tanks contain high concentrations of nitric acid. These situations combine to drive the analyst to *in situ* measurements whenever possible, as discussed below.

Concentrations of radioisotopes in soil and groundwater are typically very low. Because of the complex nature of the natural environment, the possibility exists for a large number of compounds and complexes to form. Examples are carbonate, sulphate, chloride, silicate, humate and fulvate complexes. The combination of very low concentrations and complexity of the matrices makes speciation especially difficult for these radioelements. Colloids and pseudo-colloids[4] of the actinide elements and of many lanthanides and transition elements may be found in groundwater.

Only a few radioisotopes are commonly found in vapours and gases at significant concentrations. These typically arise from fuel recycle or uranium enrichment operations, most notably from spent fuel dissolver off-gas, and from incinerator operations. Iodine, and oxides of technetium and ruthenium are commonly found in the off-gas streams. In these instances it is very desirable to have on-line instrumentation to measure and speciate the elements and their compounds. Matrices, in these cases, are likely to be air, water or carbon dioxide. In some cases, the matrices may be oxides of nitrogen or halogens.

---

[4]  Pseudo-colloids are formed when an ionic or molecular species of interest attaches itself to a colloidal material, e.g. to a clay. Thus the species behaves as though it were a colloid.

Table 1 lists some of the more common radioisotopes of interest and the concentrations in water that will produce an effective dose equivalent of 100 mrem per year.

Table 2 lists concentrations of some radioisotopes found around several nuclear sites in the United States.

## Toxic elements

Toxic elements, for example beryllium, cadmium, lead, mercury and nickel, are common at most sites involved in nuclear fuel cycle operations as well as in other related operations. These elements have relatively high vapour pressures at elevated temperatures and are found in waste streams. Therefore they comprise part of an especially dangerous class of toxic materials. Because of the unusually high volatility, high toxicity and ubiquitous nature of mercury at nuclear sites it is of special interest. The very toxic methyl mercury may be formed when mercury is present in soils and sediments. Speciation to determine if mercury is elemental or is present as other species is important. Lead is also of special concern. It may be vaporised (presumably as PbO) during construction of radiation shielding. Table 3 gives examples of some toxic elements found around several nuclear sites in the United States and their maximum concentration limits in the US.

Many of the toxic elements listed in Table 3 have complex chemistries, forming a wide variety of compounds and complexes. This makes speciation especially important as well as difficult. An example is nickel, which finds use in plating applications. Nickel not only forms a large number of complexes but may form polymers and the highly toxic nickel carbonyl. Most toxic elements exist in several valence states and form a large number of compounds and complexes. In some cases, e.g. chromium, the element is much more toxic in one valence state [Cr (VI)] than in its other valence states. In order to devise chemical processes to deal with the toxic elements it is essential to know the chemical species present.

## Toxic organic compounds

Many organic compounds are toxic. In this report only a few of the organic compounds found in the nuclear industry are mentioned. Many organic compounds have found use in separations and cleaning operations in nuclear plants, for example, chlorinated hydrocarbons. They are commonly found in soils and ground waters in and around the plants as well as in waste storage facilities such as tanks. Because of the extremely low dose limits for many of these compounds and because they are often present in low concentrations in soil and groundwater, their speciation poses an especially difficult challenge for the speciation analyst. Polychlorobiphenyls (PCBs), which were widely used in the past in electrical transformer coolants and some other applications, are a good example of toxic organic compounds. They may be found in and around uranium enrichment gaseous diffusion plants at unacceptable concentrations and are a toxic organic waste.

Biological processes that change the chemical composition of the compounds are often present and active at the time of speciation. Incineration of contaminated organic wastes at plants handling radioisotopes produces off gases that may contain a variety of radioelements of concern. These and other toxic organic materials are subject to thermal degradation leading to a range of products, some of which are themselves, toxic. Table 4 gives illustrative examples of some organic compounds found around some nuclear sites in the United States.

## Mixed wastes

Mixtures of radioisotopes and toxic elements and/or organic compounds are common at nuclear sites. These wastes are called mixed *wastes*, and form a special class of problems for the speciation analyst because both radioactive and toxic materials must be analysed for, often in complex matrices. Mixed wastes may be in waste storage facilities such as tanks, and in soils, groundwater, and sediments, and in structures and equipment. They may be present in a very wide range of concentrations. The baseline treatment for organically contaminated mixed wastes is incineration. However, incineration can produce volatile dioxins and furans and release volatile toxic metals and radionuclides in the off-gases. Consequently, off-gas monitoring is required. Current monitoring processes are time consuming and costly. Continuous-emission monitors for speciation and measurements could potentially overcome these shortcomings.

More than 100 000 cubic meters of mixed low level and transuranic waste are stored at more than 20 sites in the US Department of Energy (DOE) complex. Mercury has been identified in more than 50 000 cubic meters of waste at 19 DOE sites. In some cases it may be possible to separate radioactive contaminants from non-radioactive toxic materials, thereby changing both the amounts of various types of waste and the waste disposal options available for them. It will be necessary to know the chemical species present in order to devise separations processes.

## Sample treatment considerations

It is not uncommon for special circumstances to exist that require sample pre-treatment, or special sample handling. The most common situation is that requiring pre-concentration of the sample if it is so dilute that it is impractical to analyse it directly. The usual method of concentrating aqueous samples is by evaporation. A potential problem with this approach is that either an elevated temperature or a change in concentration of chemicals in the sample may change the nature of the species. Equilibria may be shifted, or slowly reversible reactions may occur. This is especially true of species that form colloids, such as tetravalent plutonium, which forms a very stable polymeric or colloidal solution. Other methods of sample concentration include ion exchange, solvent extraction, precipitation, membrane processes, and electrically driven processes.

## Handling interferences

A sample of a waste containing only the material to be speciated is rare. The usual situation is one where several, and often many, additional substances are present. If these are in large enough concentrations, or are similar enough to the material being speciated that they interfere with speciation, they must be removed. A potential problem is that the removal operations may alter the species being analysed for. The methods available for removal of interferences are ion exchange, solvent extraction, precipitation, membrane processes and electrical processes, as mentioned for sample concentration.

## Speciation requirements for modelling

Modelling is used extensively to demonstrate that regulatory requirements have been met in treating hazardous wastes entering or already present in the environment and in determining that the decontamination operations have been effective. It is in modelling that waste speciation finds one of its most important applications. Computer models are often at the heart of demonstrating regulatory compliance.

In the complex biosphere there are many coupled chemical processes. That is, there are systems where a chemical or physical change in one part of the system causes changes in another part of the system. These coupled processes may be very complex, and their proper introduction into the model requires that the species be known at each step, and after each system change. This requirement poses a very challenging speciation task.

## Conclusion

Speciation of chemical elements and compounds is essential to carry out the measurements necessary to predict and control movement or releases of radioactive and toxic materials from sites, soil, ground water, surface water, equipment and facilities contaminated with those materials. Speciation is also essential for devising processes for remediating contaminated and polluted facilities, sites, and the environment. Preservation of human health and the natural environment depend upon acquisition of the knowledge speciation enables. In addition, the information provided by speciation points the way to determination of the need for and cost effective methods of containment and remediation of radioactive and toxic materials. New, improved, and less costly methods of speciation at the micro and macro concentration levels are needed and will continue to be needed for the foreseeable future.

**Table 1. Derived concentration guides for radiation protection for common radionuclides in water***

| Radionuclide | Concentration, $\mu$Ci/L | Molarity |
|:---:|:---:|:---:|
| $^{3}$H | $2 \times 10^{0}$ | $6.9 \times 10^{-11}$ |
| $^{54}$Mn | $5 \times 10^{-2}$ | $4.4 \times 10^{-11}$ |
| $^{60}$Co | $5 \times 10^{-3}$ | $7.4 \times 10^{-14}$ |
| $^{90}$Sr | $1 \times 10^{-3}$ | $8.0 \times 10^{-14}$ |
| $^{99}$Tc | $2 \times 10^{0}$ | $1.2 \times 10^{-6}$ |
| $^{106}$Ru | $6 \times 10^{-3}$ | $1.7 \times 10^{-14}$ |
| $^{125}$Sb | $5 \times 10^{-2}$ | $3.9 \times 10^{-13}$ |
| $^{129}$I | $5 \times 10^{-4}$ | $2.2 \times 10^{-08}$ |
| $^{137}$Cs | $3 \times 10^{-3}$ | $2.5 \times 10^{-13}$ |
| $^{238}$U** | $6 \times 10^{-7}$ | $7.5 \times 10^{-6}$ |
| $^{237}$Np | $4 \times 10^{-8}$ | $2.39 \times 10^{-10}$ |
| $^{239}$Pu | $3 \times 10^{-5}$ | $2.0 \times 10^{-12}$ |
| $^{241}$Am | $3 \times 10^{-5}$ | $3.6 \times 10^{-14}$ |

* Derived guidelines from DOE Order 5400.5.
** Estimated.

**Table 2. Typical concentrations of some radioisotopes around US nuclear sites**

| Radioisotope | Concentration |
|---|---|
| $^{137}$Cs | 1 pCi/kg |
| $^{60}$Co | 1 pCi/kg |
| $^{99}$Tc | 51 Bq/kg |
| $^{232}$Th | 51 Bq/kg |
| $^{3}$H | 2 pCi/L |
| $^{234}$U | 4 pCi/L |
| $^{238}$U | 4 pCi/L |
| $^{238}$Pu | 2 pCi/L |
| $^{239,240}$Pu | 2 pCi/L |

**Table 3. Concentrations of some toxic elements around US nuclear sites**

| Element | Typical site concentration in water | EPA limit in water, mg/L |
|---|---|---|
| Arsenic | 8 mg/L | 0.05 |
| Barium | 13 mg/L | 2 |
| Beryllium | 13 mg/L | 0.004 |
| Cadmium | 13 mg/L | 0.005 |
| Cadmium | 75 mg/L* | – |
| Chromium | 13 mg/L | 0.1 |
| Copper | 13 mg/L | 1.3* |
| Lead | 8 mg/L | 0.015** |
| Lead | 78 mg/kg*** | – |
| Mercury (inorganic) | 8 mg/L | 0.002 |
| Mercury (inorganic) | 210 µg/g*** | – |
| Nickel | 13 mg/L | 0.14 |

\*    Action level: 1.3 mg/L; "action level" is the concentration of lead or copper in water which determines, in some cases, the treatment requirements that a water system is required to complete.
\*\*   Action level: 0.015 mg/L.
\*\*\*  Values in soil.

**Table 4. Representative concentrations of organic compounds common in nuclear operations**

| Compound | Typical site concentration | EPA limit |
|---|---|---|
| Carbon tetrachloride | 135 µg/L | 5 µg/L |
| Chloroform | 135 µg/L | 5 µg/L |
| Nitrobenzene | 51 µg/L | – |
| Polychlorobiphenyls | 394 µg/L | 0.5 µg/L |
| Tetrachloroethene | 1 µg/L | 5 µg/L |
| Trichloroethene | 135 µg/L | 5 µg/L |
| Vinyl chloride | 135 µg/L | 2 µg/L |

# SESSION A

*Methods for Trace
Concentration Speciation ($<10^{-6}$ M)*

# RADIOCHEMICAL PROCEDURES FOR SPECIATION OF ACTINIDES IN THE ENVIRONMENT: METHODOLOGY AND DATA OBTAINED IN CONTAMINATED REGIONS OF RUSSIA BY RADIONUCLIDES

**B.F. Myasoedov and A.P. Novikov**
V.I. Vernadsky Institute of Geochemistry and Analytical Chemistry
Russian Academy of Sciences
Kosygin str.19, Moscow, Russia

## Abstract

The migration ability of radionuclides in biosphere mainly depends on their occurrence forms, which are ultimately responsible for the potential hazards to human beings. In the paper, the radiochemical methods for actinides speciation in aqueous and solid environmental samples are discussed. They include membrane fractionation, ion exchange, solvent extraction, coprecipitation, adsorption, electrophoresis for water samples and selective leaching, organic substances fractionation, liquid and membrane extraction, partition chromatography, membrane retention with water-soluble polymers, etc. for solid samples. These techniques were used during radiomonitoring of zones near Production Association "Mayak" and Krasnoyarsk Mining-Chemical Plant. The data on oxidation states of plutonium in aqueous samples and in soil and its leached solution was obtained. The organic matter fractionating of soils allowed to explain experimental values of migration coefficients of $^{137}$Cs, $^{90}$Sr, $^{239}$Pu, and $^{241}$Am. This data was used for long-term prediction of radionuclides migration behaviour in zone under investigation.

## Introduction

Experimental tests of nuclear and thermonuclear weapon in atmosphere and under ground, nuclear power engineering and numerous accidents that took place at the nuclear power plants (NPP), unauthorised dump of radioactive materials in various places of the ocean, and pouring off the strong dump of radioactive wastes from ships and submarines equipped with nuclear power engines made artificial radionuclides a constant and unretrievable component of the modern biosphere, becoming an additional unfavourable ecological factor.

To control environmental pollution with radionuclides and to predict their behaviour in ecosystems, we should not only know the radioisotopes content and interphase distribution, but also their occurrence forms. The latter determine the migration ability in different systems, viz. soil-plant, soil-ground water, water-suspension-sediments and so on. Such migration is ultimately a potential hazard for population. However, this important problem has not been thoroughly studied, first of all because of the difficulties of a methodological character. It is known that for actinides the direct methods of the complex structure study in the environment practically cannot be applied due to low content (excepting U). For example, plutonium content in natural waters is about $10^{-14}$-$10^{-18}$M, and in soils – about $10^{-9}$-$10^{-13}$ g/Kg. Therefore, indirect methods of actinides speciation is now developed, which are based on modern achievements of separation chemistry.

In many natural systems, actinides mainly exist in the form of Th(IV), Pa(V), U(VI), Np(V), Pu(IV) and Am(III). Stability of Th(IV), Pa(V) and Am(III) in the known heterogeneous systems is high. Redox form of uranium and neptunium may change due to particular conditions. According to many authors, plutonium exists in various oxidation states from +3 to +6 [1]. However, it is the actinide oxidation state that defines the character of interaction of these radionuclides with the natural matrices and, correspondingly, their occurrence forms in nature. Among the main factors influencing the oxidation states of actinides, apart from pH and Eh, hydrolysis, hydrolytic polymerisation and colloid formation, adsorption, complex formation with the water-soluble organic and non-organic substances, as well as the reactions of disproportionation, particularly for neptunium and plutonium, are important. Radiolytic decomposition of matrix may also affect the equilibrium of the redox reactions [2]. In solid phase, actinides strongly interact with the humus matter being introduced into the soil organic-mineral complexes formed by humic and fulvic acids with the soil metals (Fe, Al, Ca, Mg). Continuous reproduction of the soil humus provides stability and importance of the indicated factor for actinide occurrence forms. In the presence of soil solution organo-mineral complexes form hetero-phase systems, where all components (including actinide elements) have definite distribution between the mobile forms (in solution) and insoluble forms (solid and/or gel phase). In case of a strong fixation of actinides by the organo-mineral complex, the character of the inter-phase distribution will depend upon the solubility of the organo-mineral complex, whereas the solubility is dependent upon the concentration of humus acids and soil metals in the complex, hydro-chemical composition of the soil solution, and the nature of pH-defining cations.

## Sampling

When investigating actinides interaction with environment matrices, it is necessary to consider the peculiarity of sampling and classification, storage and pre-treatment of samples. It explains it by possible modifying of actinides species during mechanical and thermo-photochemical effects.

## Aqueous samples

Water samples should be analysed either immediately after sampling without conservation or using the known methods of conservation. Selective pre-concentration of actinides from natural waters remains rather complicated and this is a weakly developed step of analysis. The ultrafiltration method is most widespread for this purpose [3].

## Solid samples

Besides the time dependent factors mentioned above, the problem of the solid sample analysis is related to its considerable heterogeneity. Therefore, the study of the radionuclide occurrence forms in soil needs not only mechanical homogenisation, but also preliminary extraction of the artificial inclusions. The latter may be highly radioactive (so called hot particles), and without the extraction procedure, they may considerably change the equilibrium between the organo-mineral soil complexes after the sample treatment by mineral acids. Hot particles are of particular interest for radio monitoring of the zone in the vicinity of the Chernobyl NPP, severely affected by the accident. However hot particles with considerable radioactivity due to [137]Cs were found by us together with the scientists from the Institute of Biophysics on some islands of the Yenisei river close to the Krasnoyarsk Mining Chemical Combine (KMCC) (table 1, [4]).

Presented data show that despite a much lower concentration of plutonium in the hot particles as compared with radiocesium and radiostrontium, its contribution to the total alpha activity of the soil containing hot particles is rather high. In fact, plutonium concentration in alluvial soils of this zone ranges from 0.1 to 5.0 Bq/kg. The weight of the analysed sample does not exceed 10 to 50 g, which means that 50% to 100% of the total activity of the sample is due to plutonium present in one hot particle.

Another problem of radiomonitoring in the impact zone close to KMCC is the high contribution of the technogenic magnetisable fractions with the enhanced concentration of plutonium (Table 2). To clean such soils, it is necessary to start with magnetic separation.

## Methods of pre-treatment of natural water samples

In general, to pre-concentrate actinides from aqueous media without chemical destruction of their occurrence forms, one can use methods of inverse osmose, piezodialisis, or thermodiffusion. However, since we are not aware of the changes in equilibrium of the occurrence forms of actinides due to the increase of macrocomponents' concentration and to a complicated analytic technique that can't be used in field conditions, these methods are not spread widely.

## Radiogeochemical methods of selective leaching of actinides binding by various groups of organic and inorganic substances from solid samples

In contrast to samples of natural waters and soil solutions, solid part of soils and bottom sediments can not be under investigation by radiochemical methods without pre-treatment. However, for purposes of actinides speciation radiochemical methods well combine with usual technique of the organic matter fractionating and determination of the forms of geochemical mobility (Figures 1 and 2). These methods are based on different solubility of binding radionuclides with various humus acids and other organic acids of low molecular mass. Apart from that, saturation with sesquialteral trivalent metal (iron, aluminium) hydroxides typical for soils changes the solubility of the fulvic acids complexes. As a result, fulvic acid exists in soil in two states; as an insoluble precipitant of these hydroxides fulvates and in the form of the soluble fulvates less saturated with these metals. This process is accompanied by an active heavy metals fixation by the insoluble humic acids. These two dynamically linked forms define the presence of the migrated active complexes.

After the selective leaching, usual radiochemical procedures are used to determine the actinides. Alpha-spectrometric, electrochemical, spectroscopic and luminescence methods are developed to measure the content and to carry out for speciation of actinides.

Despite some conventionalism of the selective leaching and some imperfections due to the possible changes of the forms during the sample processing, the obtained data enable to compare the mobility of radionuclides in different biogeocenoses and provide with some quantitative data which can be used in modelling radionuclide migration.

Some results of such investigations which were carried out at industrial basin B-10 (PA "Mayak") are given in Tables 3 and 4. Partial fractionation of organic compounds of the bottom sediments showed that practically all organic fractions contain radionuclides. Analysis of the relations in radionuclide distribution between various organic fractions gives insight into the nature of the difference in speciation. For example, high mobility in the bottom sediments of radiostrontium, americium, and curium compared to raidiocesium and plutonium correlates with primary localisation of the first group of the radionuclides in sum of the decalcinate (Fraction 1a) and the first fraction of fulvic acids. Fraction 1a of fulvic acids contains free fulvic acids, non-specific organic compounds (low-molecular-weight acids, amino acids, polysaccharides, etc.) and their complexes with mobile iron and aluminium hydroxides, calcium, and other chemical elements. Since fulvic and low molecular weight acids are strong complex forming agents, their interactions with chemical elements influence the sorption of radionuclides by the solid phase and activity of their migration in ecosystems [5].

## Methods of actinides speciation in solutions

After selective leaching several techniques to isolate and concentrate actinides in different oxidation states can be used. They are co-precipitation, ion exchange, micro and ultra filtration, liquid extraction, membrane complex forming retention, membrane extraction and some kinds of chromatography. Most of these methods have some disadvantages connected with complicated biochemical and physico-chemical processes proceeding in samples under investigation, possible effect on the actinide's oxidation states by used chemical reagents, necessity for correction of chemical composition of the aqueous samples, etc.

Elaboration of the technique depends upon the objectives and tasks. They can be the determination of the oxidation forms of actinides, the degree of their complexation, the charge of the soluble species and particles, the nature and amount of the strongest ligands of actinides complexes, etc. Conditions for separation of various actinide elements are established usually in experiments with micro-concentrations or the stable analogous elements. This helps to increase reliability of the obtained results. Unfortunately, it does not guarantee from errors since the velocity of chemical processes that actinides participate in is rather high [1].

## Co-precipitation

Co-precipitation is applied more often as it is a simple and convenient technique for identification of the actinide oxidation form. Fluoride deposition of Ac(III,IV) in presence of $LaF_3$ is believed to be a classic technique [2]. The most attractive aspect of this technique is its favourable combination with alpha spectrometry after filtration of the pseudo-deposition on lavsan membranes with a pore diameter about 0.1-0.2 micrometers. In optimum conditions, the resolution of the alpha spectra peaks is not lower then 50 keV.

## Ion exchange and membrane complex forming retention

Actinide distribution in systems with resins of various nature are well studied. Ion exchange can be used for identification of the oxidation states as well as for the study of complexation and hydrolysis processes. However, the necessity to correct pH values of the solution and relatively low velocity of the ion-exchange processes significantly limits application of this method. In recent years, a more effective analogue Membrane Complex Forming Retention (ultra filtration with water-soluble metal binding polymers) was developing. The main advantage of this method is the homogenous phase in which all chemical reactions are going, and therefore their velocity is higher. Moreover, currently there have been synthesised polymers which are able to extract effectively definite actinide forms in neutral media, e.g. [6]. We have developed a technique of producing membranes saturated with polyethelenimine-ethylendiamintetraacetic acids (or its sodium salt) to extract Ac(III,IV). Unfortunately, this technique is not applicable to identification of the oxidation state of neptunium and plutonium, as EDTA is an agent with ability of a rather strong reducing and stabilising the form 4+.

## Electrophoresis and ultrafiltration

Electrophoresis is used to determine the charge and molecular weight of suspended particles, whereas micro and ultra filtration helps to evaluate the size of complexes and particles that contain actinides. These methods should be used prior to the others because they enable to optimise elaboration of systems and to exclude the non-soluble coarse inclusions.

Table 5 shows the data on radionuclide occurrence forms in ground water sample taken from an observation well 2 km away from the lake Karachai (PA "Mayak") and in the same sample after its durable contact with the aluminophosphate matrix used by PA "Mayak" in waste solidification.

Obtained data show that 30% to 60% of plutonium and americium are present as anions and the leaching products of the aluminophosphate matrix (condensed phosphates) considerably increase the portion of anion forms. Although only 40% of plutonium and 60% of americium of initial ground waters are not detached by a membrane of 10 kDa, concentration of these elements in filtrate after the contact with the aluminophosphate matrix increases up to 70-80%. Therefore the products of decomposition of the aluminophosphate matrix enhance actinide migration in hydraulic network.

## Sorption

Sorption on various collectors ($MnO_2$, $CaCO_3$, $TiO_2$, silica gel, etc.) is used for selective concentration of the actinide forms from a large volume of sea and fresh water. In contrast to ultrafiltration with water-soluble metal binding polymers, this technique needs a long contact of the solid phase with the water, the usage of buffering solutions and regulation of pH value of the aqueous phase under study. Moreover, there still exists a problem of the possible changes of the actinide occurrence forms on the sorbing surfaces during long contact of the phases.

In recent years, methods based on radionuclide sorption by complex-forming agents were widely developing. In GEOKHI RAS they developed sorbents POLYORGS applied in the form of modules of fibre type [7]. The main advantage of such sorbents is in high dynamics of the process and the possibility of selective extraction of particular actinide forms without correction of the aqueous phase. POLYORGS-VIIM may be used to separate Ac(III), POLYORGS-32 for Ac(III)+Ac(VI), POLYORGS-XXIV for Ac(IV)+Ac(III). However, a weak knowledge of the kinetics of the redox reactions for the adsorbed actinide forms remains a shortcoming of such systems.

### Solvent extraction, membrane extraction and extraction chromatography

Extraction methods are most widely used to determine actinide forms in solutions of different nature. This is due to a considerable amount of data on kinetics and thermodynamics of actinide extraction by various organic reagents. In more simple model systems, extraction methods help to determine rather rigorously the composition and stability of the actinide aqueous complexes. In studies of the environmental components, these methods are useful for identification of the oxidation states and the general degree of complexation of particular actinide species. These methods have some evident advantages, although they have common drawbacks due to the necessity in some cases to correct composition of the aqueous phase and the known disproportionation reactions that take place in organic phase, for example, for neptunium and plutonium. The advantages are: high dynamics of the process, possibility of a fast phase separation, availability of tabled values of the equilibrium constants of extraction and that of diffusion coefficients for the extracted forms. The latter parameter is important while kinetic method for identification of the reactions of actinide reduction and disproportionation are used. Now kinetic curves for mass migration of particular actinide forms can be calculated rigorously.

Classical works by G.R. Choppin who devoted to the determination of actinide occurrence forms in a number of aqueous systems, have been performed with the help of β-diketone, dibenzoylmethane (DBM) [8,9]. This agent extracts actinides present in neutral media in different oxidation forms. DBM helps to separate effectively Pu(IV), Pu(V) and Pu(VI), with spectroscopic studies showing comparable results. Extraction chromatography technique helps to analyse considerable volumes of solution and to increase factor of separation of the studied forms.

We consider membrane extraction to be one of the promising trends in the development of the technique for actinide separation. Among the main advantages of this technique there are: relatively short presence of the studied element in the organic phase (not longer than 10 sec); weak action on the composition of the analysed aqueous phase (ratio between the aqueous and membrane liquid bulk is approximately $10^4$); possibility of selection of a wide range of re-extracting agents, selectivity for particular non-organic actinide forms; achievement of the necessary values of the local distribution coefficients on the boundary between membrane and the analysed solution indifferent to the intensity of the hydrolysis and complexation reactions; possibility to use kinetic factors of separation and identification of the reduction reactions; favourable combination of the membrane extraction and luminescent determination of actinides [10].

Table 6 shows some results of the study of plutonium occurrence forms in aqueous, acetate (exchanging forms), and hydrochloric acid (acid-soluble forms) leach from the alluvial soil of the Yenisei river basin (the island of Atamanovsky). Obtained data showed that almost 80% of the plutonium extracted by aqueous solution corresponds to Pu(V), but one should note that the portion of this plutonium in its total concentration in the soil does not exceed 1,3%. In acetate leach the fraction of Pu(V) decreases to 55%, but this amount equals to 3,4% of the total plutonium found in the soil. In hydrochloric acid leach the portion of plutonium other than Pu(IV) is lowed more. This leads to the conclusion that the most mobile plutonium forms in soils really depend upon the ability of plutonium to exist in the environment in the higher oxidation degrees.

## Conclusion

The ability of actinides to easily change the oxidation forms, even at minor hydro-chemical disturbances, stands in the way of the study of its forms occurring in the environment. By now, there are no radiochemical methods that have rigorous theoretic grounding for the problem of the true picture of actinide behaviour in the environment. Development of such methods should be related to the further accumulation of any information on actinide occurrence forms, to the new achievements in the development of the new selective reagents, and to the development of new separation technologies based on membrane and chromatographic techniques. It is our opinion that the most promising research field is a combination of the radiochemical methods using selective extraction and concentration of particular actinides forms and direct methods of detecting these species.

## REFERENCES

[1]    Choppin, G.R., "Proceedings of International Trace Analysis Symposium", Sendai, Japan, 1990, pp. 123-132.

[2]    Choppin, G.R., Bond, A.H., "Russian Analytical Chemistry", 1996, 51, 1240.

[3]    Myasoedov, B.F., Novikov, A.P., Pavlotskaya, F.I., *Russian Analytical Chemistry*, 1996, 51, 124.

[4]    Bolsunovsky, A.Y, Goryachenkova, T.A, Charkizyan, V.O, Myasoedov B.F., "Russian Radiochemistry", 1998, 40, 271.

[5]    Myasoedov, B.F., Novikov, A.P, Pavlotskaya, F.I., "Russian Radiochemistry", 1998, 40, 461.

[6]    Novikov, A.P, Shkinev, V.M., Spivakov, B.Ya., *et al., Radiochemica Acta.* 1989, 45, 35.

[7]    Molochnikova, N.M., Scherbinina, N.I., Myasoedova, G.V., Myasoedov, B.F. "Russian Radiochemistry", 1997, 39, 280.

[8]    Saito, A., Choppin G.R., *Analytical Chemistry*, 1983, 55, 2454.

[9]    Choppin G.R., *J. Radioanal. Nucl. Chem.*, 1991, 222, 109.

[10]   Myasoedov, B.F., Novikov, A.P., *J. Radioanal. Nucl. Chem.*, 1998, 229, 33.

## Table 1. Data on radiochemical analyses of "hot particles"

| Particle N | Weight, mg | $^{137}$Cs, MBq | $^{90}$Sr, Bq | $^{238-240}$Pu, Bq |
|---|---|---|---|---|
| 1544 | 0.45 | 6.85 | 1 350 | 0.14 |
| NR8/1 | 0.50 | 0.18 | 474 | 0.80 |
| NR7 | – | 0.70 | 54.4 | 5.7 |

## Table 2. Relative weight of magnetised fraction and its percentage of total plutonium content in soil sample. Soil profile on island Atamanovski (Enisei River).

| Depth of sampling, cm | 0 | 5 | 20 | 35 |
|---|---|---|---|---|
| Weight, g per 100g of soil | 3.8 | 2.7 | 1.8 | 1.7 |
| Content of plutonium, % | 42.3 | 21.6 | 5.2 | 2.8 |

## Table 3. Speciation of radionuclides in bottom sediments of basin B-10, % of the total content

| Form | Sample 4 (0-3 cm) | | | | | Sample 4a (20-23 cm) | | | | |
|---|---|---|---|---|---|---|---|---|---|---|
| | $^{137}$Cs | $^{90}$Sr | $^{239}$Pu | $^{241}$Am | $^{244}$Cm | $^{137}$Cs | $^{90}$Sr | $^{239}$Pu | $^{241}$Am | $^{244}$Cm |
| Exchangeable | 8 | 33 | 1 | 5 | 7 | 10 | 28 | 2 | 4 | 5 |
| Mobile | 6 | 29 | 16 | 68 | 65 | 10 | 40 | 14 | 39 | 40 |
| Acid-soluble | 24 | 29 | 53 | 20 | 20 | 42 | 26 | 79 | 36 | 34 |
| Poorly soluble | 62 | 9 | 30 | 7 | 8 | 38 | 6 | 5 | 21 | 21 |

## Table 4. Distribution of radionuclides between fractions of humus acids of bottom sediments of industrial basin B-10, % of the total content in the group

| Sample | Radionuclide | Fraction of fulvic acid | | | Fraction of humic acid | |
|---|---|---|---|---|---|---|
| | | 1a | 1 | 2 | 1 | 2 |
| 4 | $^{90}$Sr | 84 | 16 | – | 100 | – |
| | $^{137}$Cs | 58 | 17 | 25 | 38 | 62 |
| | $^{239,240}$Pu | 50 | 20 | 30 | 8 | 92 |
| | $^{241}$Am | 94 | 6 | – | 40 | 60 |
| 4a | $^{90}$Sr | 89 | 11 | – | 100 | – |
| | $^{137}$Cs | – | 17 | 83 | 81 | 19 |
| | $^{239,240}$Pu | 16 | 45 | 39 | 86 | 14 |
| | $^{241}$Am | 86 | 14 | – | 100 | – |

**Table 5. Speciation of radionuclides in groundwaters (1) before and (2) after interaction with aluminophosphate glass, % of the total content in the group**

| Form of occurrence | $^{137}$Cs | | $^{90}$Sr | | $^{237}$Np | | $^{239-240}$Pu | | $^{241}$Am | |
|---|---|---|---|---|---|---|---|---|---|---|
| | 1 | 2 | 1 | 2 | 1 | 2 | 1 | 2 | 1 | 2 |
| Cationic | 62-96 | 89 | 99-100 | 28 | – | 21 | 43-69 | <1 | 40-69 | 7 |
| Anionic | 3-24 | 11 | <1 | 62 | – | 46 | 30-56 | 78 | 31-60 | 69 |
| Neutral | 1-14 | – | – | 10 | – | 33 | 1-12 | 21 | – | 24 |

**Table 6. Speciation of plutonium in various soil leach**

| Water leach | | | | | | |
|---|---|---|---|---|---|---|
| Fraction | After 0.2 µm filter | | | After 10 kDa filter | | |
| Oxidation state | IV | V | III+VI | IV | V | III+VI |
| Pu, % of total in fraction | 14 | 79 | 7 | 12 | 82 | 6 |
| Pu, % of total in soil sample | 0.23 | 1.3 | 0.12 | 0.08 | 0.6 | 0.04 |
| 1 M CH3COO leach | | | | | | |
| Fraction | After 0.2 µm filter | | | After 10 kDa filter | | |
| Oxidation state | IV | V | III+VI | IV | V | III+VI |
| Pu, % of total in fraction | 39 | 55 | 6 | 30 | 61 | 9 |
| Pu, % of total in soil sample | 2.4 | 3.4 | 0.4 | 1.3 | 2.6 | 0.4 |
| 1 M HCl leach after 0.2 µm filter | | | | | | |
| | Pu | | | Am | | |
| Oxidation state | IV | III+V+VI | | IV | III+V+VI | |
| Ac, % of total in fraction | 73 | 27 | | – | 100 | |
| Ac, % of total in soil sample | 25.0 | 9.3 | | | 49.4 | |

**Figure 1. Flowsheet of isolation of particular forms of radionuclides from (a) water and (b) bottom sediments**

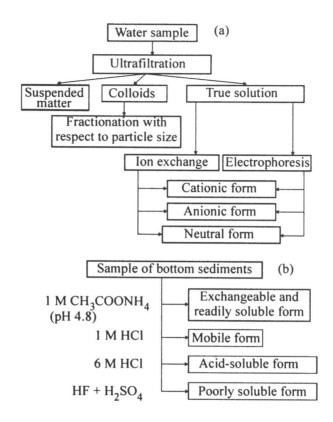

**Figure 2. Flowsheet of partial fractionation of
organic compounds from soils and bottom sediments**

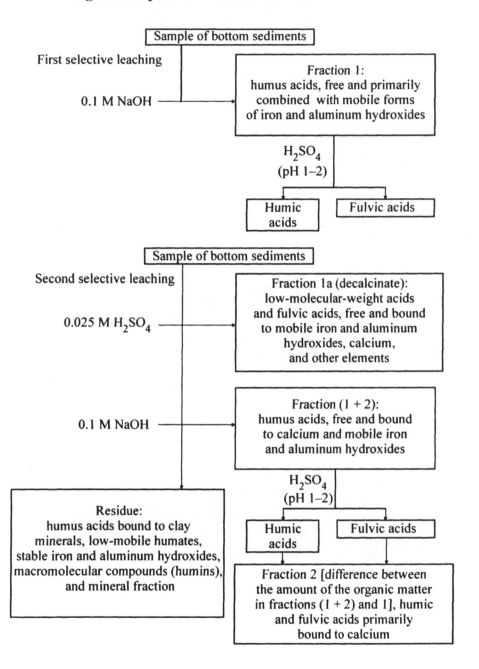

# NANOSCOPIC SPECIATION OF AQUATIC ACTINIDE IONS BY LASER SPECTROSCOPY

**J.I. Kim**
Forschungszentrum Karlsruhe, Institut für Nukleare Entsorgungstechnik
76021 Karlsruhe, Germany

## Abstract

Speciation of actinides is fundamental for a detailed appraisal of their chemical behaviour in the geological environment and hence entails straightforward methods of high sensitivity. Using laser as a light source, various spectroscopic methods have been developed and applied for the study of aquatic chemical reactions in the nano-mole range, i.e. complexation reactions, solid-water interface interactions and colloid generation of actinide ions of different oxidation states. A brief overview on the laser spectroscopy is presented for the general principle and the fields of application.

## Introduction

Appraisal of actinide chemical reactions in a dynamic natural aquatic system requires direct speciation methods that allow the characterisation of their physical and chemical states without perturbation. The solubility of actinides under natural aquatic conditions is sparingly low in the absence of strongly complexing ligands. As a result, the geochemical behaviour of aquatic actinide ions is constrained by nano-mole or subnano-mole scale reactions. The appraisal of such chemical reactions demands, therefore, non-conventional speciation methods that are operational without perturbation or with minimal perturbation of the system under investigation.

Mobile and immobile actinide species possibly present in aquatic systems [1] are schematically illustrated in Figure 1.

**Figure 1. Mobile and immobile actinide species in natural aquatic system**

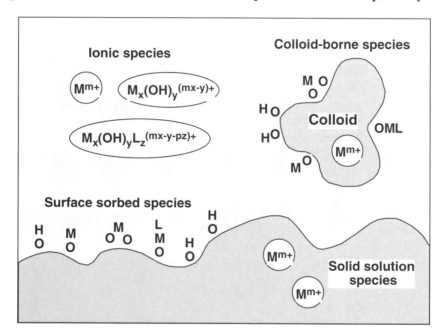

Ionic actinide species present in aqueous phase are hydrated metal ions, hydrolysed ions and binary or ternary complexes. Immobile species are present as pure homogeneous solids, as surface sorbed species or as species incorporated into mineral phases. Additionally, actinides may form "real-colloids" by homogeneous nucleation of hydrolysed species or may be sorbed onto natural colloids in groundwater and thus present as "pseudo-colloids" [2]. The migration behaviour of ionic and colloid-borne species depends on the extent of their interactions with available solid matrices.

Spectroscopic speciation of actinides in various physical and chemical states, as shown in Figure 1, can be performed without perturbation of a given system. Absorption spectroscopy on the transitions of the 5f-electron shell is a useful speciation method for the characterisation of oxidation and complexation states. The spectroscopic resolution of actinide transition bands is in general very high but their intensity is relatively weak (molar absorbance < 500 L/mole cm) [3]. Therefore, the application of conventional absorption spectroscopy to the speciation of actinide ions under natural conditions is much limited, because of the poor spectroscopic sensitivity. The low solubility of actinides under given conditions (usually below $10^{-6}$ mole/L) necessitates a speciation method of high sensitivity.

A number of spectroscopic methods have been developed in the recent past based on laser as a light source, which then facilitate the direct speciation of individual chemical reactions of actinides under natural aquatic conditions. Various laser spectroscopic techniques appropriate for the speciation of trace level actinides are illustrated schematically in Figure 2. The enhancement of sensitivity is attained by a direct measurement of different relaxation processes of photon excited states, instead of transmission measurement as practised in conventional absorption spectroscopy [4]. The methods shown in Figure 2 can be classified into the following four categories: photo-thermal spectroscopy, luminescence spectroscopy, light scattering and plasma generation. Along with this classification, the speciation capability of each method is discussed in the following.

**Figure 2. Schematic illustration of the basic principle of highly sensitive laser spectroscopic methods and compilation of various laser spectroscopic methods**

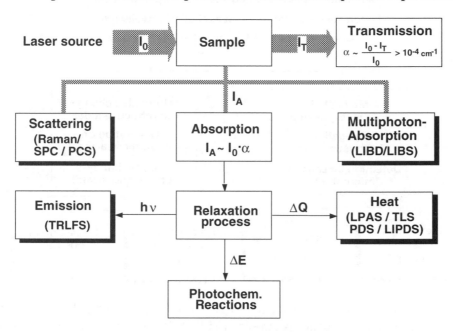

The names of individual spectroscopy are summarised as follows:

| Photo-thermal spectroscopy | |
|---|---|
| LPAS: | Laser-induced photo-acoustic spectroscopy |
| LIPDS: | Laser-induced photo-thermal displacement spectroscopy |
| TLS: | Thermal lensing spectroscopy |
| PDS: | Photo-thermal deflection spectroscopy |
| **Luminescence spectroscopy** | |
| TRLFS: | Time resolved laser fluorescence spectroscopy |
| **Light scattering** | |
| SPC: | Single particle counting |
| PCS: | Photon correlation spectroscopy |
| **Plasma generation** | |
| LIBD: | Laser-induced breakdown detection |
| LIBS: | Laser-induced breakdown spectroscopy |

**Photo-thermal spectroscopy**

The thermal effect induced by non-radiative relaxation of a given excited light absorber can be measured by different detection procedures [4], for which different names of spectroscopy are given. There are two different detection procedures distinguished in principle, direct or indirect detection. A schematic illustration of photo-thermal spectroscopic processes is shown in Figure 3.

**Figure 3. Schematic illustration of photo-thermal spectroscopic process**

## Laser-induced Photothermal Spectroscopy

Modulated light source pulsed or chopped cw laser
Ion specific absorption at excitation wavelength
Generation of heat by nonradiative relaxation

| Modulated volume expansion | Modulated change in refractive index |
|---|---|
| Propagation of acoustic wave | Generation of photothermal lens |
| Detection by sensitive "microphone": | Detection by probe laser beam: |

| Piezoelectric transducer LPAS | Interferometer LIPDS | Defocussing of probe beam TLS | Deflection of probe beam PDS |

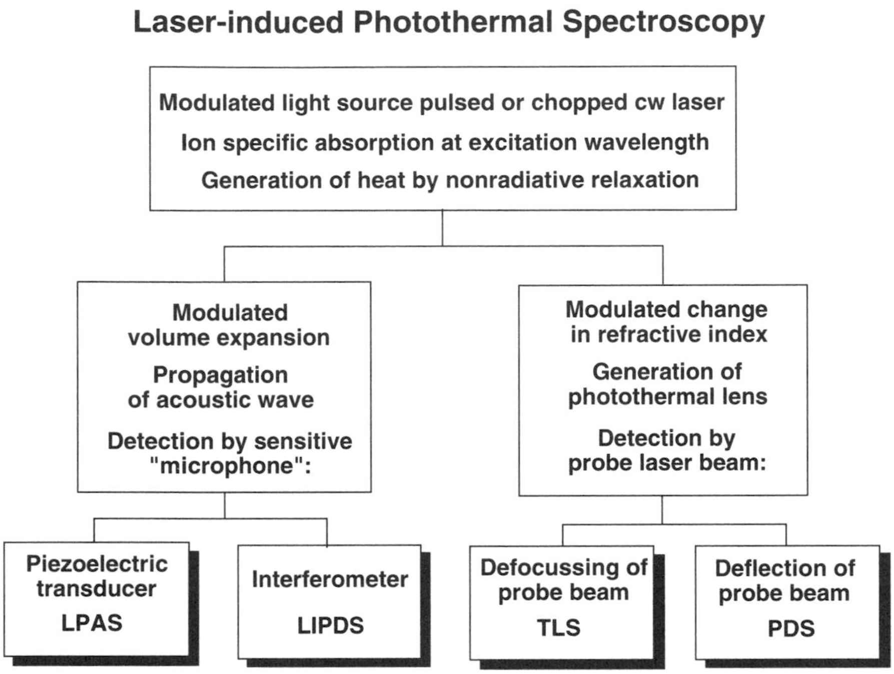

Acoustic signals generated by thermal expansion can be detected either by a direct contact mode known as laser-induced photo-acoustic spectroscopy (LPAS) [5,6,7] or by a remote contact mode via interferometer known as laser-induced photo-thermal displacement spectroscopy (LIPDS) [8]. Difference in the two methods is simply whether the acoustic detector is contacted directly to a sample cuvette or not. Another mode of measurement is deflection or de-focusing of a probe laser beam by the change of the refractive index within the absorbing sample, which are known as thermal lensing spectroscopy (TLS) [9] or photon deflection spectroscopy (PDS) [10], respectively. All four spectroscopy with different names are the same for their function, i.e. they all function as conventional absorption spectroscopy but with higher sensitivity. The detection of the relaxation process as executed in the given spectroscopy is advantageous for the sensitivity over the measurement of light transmission as done in conventional absorption spectroscopy.

A typical application of LPAS is shown in Figure 4 for the speciation of Np(IV). The concentration of Np(IV) at $1.7 \times 10^{-7}$ mole/L can be well speciated. The speciation sensitivity is found to be $10^{-8}$ mole/L. For this purpose, the aqueous background is depressed by using deuterium water.

LPAS provides the sensitivity of about $10^{-6}$ cm$^{-1}$ in absorbance for different oxidation states of actinides [7]. The limiting factor is the water absorbance for the aqueous speciation. Other laser-induced photo-thermal spectroscopic methods have the same limiting factor and comparable sensitivity.

**Figure 4. LPAS and UV-Vis spectra of Np(IV) in 1 M DClO₄/D₂0**

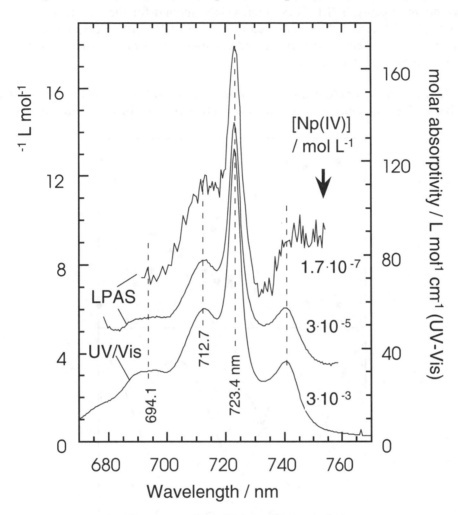

**Luminescence spectroscopy**

The most applied spectroscopy in this category is time resolved laser fluorescence spectroscopy (TRLFS), which is based on the measurement of radiative relaxation of a light induced excited state. The characteristic of this method is the selective excitation of ground state combined with the selective measurement of emission. Additionally, a time resolved measurement could be applied as well. All these provide a high sensitivity in the speciation. However, the method is limited to the elements with fluorescence emission, to which belong Cm(III) [11,12], Am(III) [13]and U(VI) [14,15] as actinides. With these elements, the chemical behaviour of trivalent and hexavalent actinides in aqueous phase can be investigated in the nano-mole concentration range.

Figure 5 shows the application of TRLFS for the carbonate complexation study of Cm(III) [16]. Individual carbonate species of Cm (III) in aqueous phase are speciated in the concentration range of $10^{-8}$ mole/L by varying the carbonate ion concentration via pH change. While measuring the fluorescence of radiative relaxation, the lifetime ($\tau$) of each excited state is also determined and thereby the number of hydrated water is calculated as illustrated in this figure. A decrease in the number of hydrated water is observed as a function of the carbonate ion number involved in the complexation. The example shown in Figure 5 demonstrates the speciation capability of TRLFS, which provides detailed information on the chemical state of Cm(III) in the course of complexation. For such well-known reasons, TRLFS has been widely applied for the chemical behaviour study of Cm(III) [17], U(VI) [14,15,18] and some rare earth elements (REE(III)) in aqueous phase. Solid-water interface interactions have been also investigated for these elements in trace concentrations for the appraisal of their migration behaviour [19].

**Figure 5. Life time and hydration numbers (upper part) and the Cm(III) species distribution derived by TRLFS speciation (lower part) as a function of the carbonate concentration**

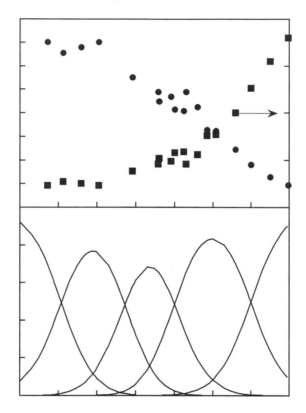

## Plasma generation

The speciation comprises not only the chemical state characterisation but also the characterisation of physical state for actinides in aqueous phase, i.e. identification of colloidal species. Actinides in natural water undergo, besides a variety of complexation, also the colloid generation, either "real colloids" by homogeneous nucleation of actinide ions themselves or "pseudo colloids" by heterogeneous nucleation, i.e. sorption of actinide ions onto groundwater colloids [2]. In groundwater, the generation of pseudo colloids is prevalent, because of the low actinide solubility. The chemical state of colloidal actinide species can be speciated to a certain extent by TRLFS, as aforementioned. The quantification of a colloid number density in groundwater requires a sensitive detection method, since colloids present in groundwater are in general small in average size (<50 nm) and concentration (ppb range or less) [2].

A noble method has been recently developed for the quantification of aquatic colloids, which is called laser-induced breakdown detection (LIBD) [20]. The method is based on the plasma generation on colloidal particles by intense laser light absorption. The event takes place, upon multi-photon absorption, through the excitation and ionisation avalanche of de-localised electrons within a colloidal composite and hence it is called breakdown. The breakdown event can be detected by measuring plasma emission light or photo-acoustic signal. As the threshold energy density required for solid is much lower than for liquid, the plasma generation occurs with colloids selectively in their suspension medium. A sensitivity comparison of LIBD is shown in Figure 6 with that of light scattering methods, i.e., static light scattering and photon correlation spectroscopy (PCS).

**Figure 6. Detection limit of LIBD in comparison to conventional light scattering**

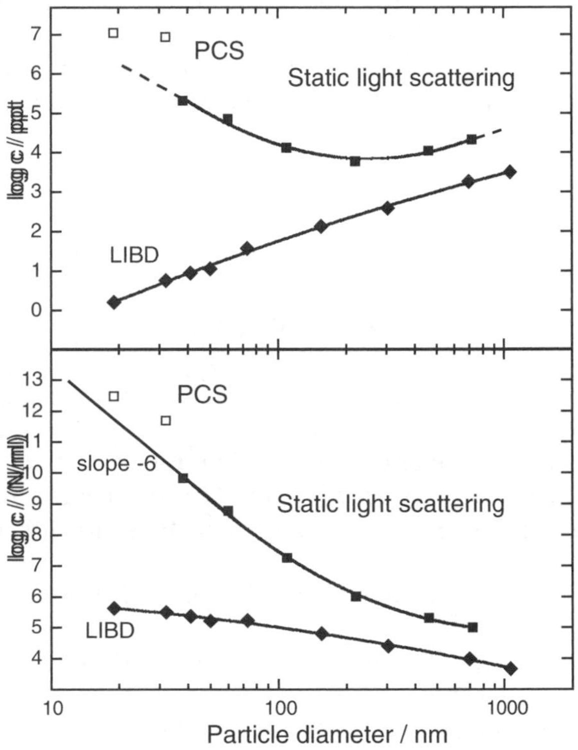

In the low part of nm particle size, LIBD appears distinctively more sensitive than light scattering over many orders of magnitude. Therefore, with LIBD the colloid number density and their average predominant size in natural groundwater can be quantified without perturbation. Such a colloid quantification allows the investigation on the pseudo colloid generation of actinides in aquatic systems, i.e. chemical interactions of actinide ions with groundwater colloids.

In the course of plasma relaxation, the discrete plasma or atomic emission takes place after the recombination of electrons. The process can be used, as laser induced breakdown spectroscopy (LIBS), for the characterisation of colloid borne elements selectively in solution [21]. Accordingly, actinides present as ionic and colloid borne species in aquatic systems can be differentiated by LIBS, since the colloid borne actinides are better excited than their aqueous ions by the same laser energy intensity. This is a particular speciation method to detect pseudo-colloids of actinides in groundwater.

## Light scattering

There are different kinds of colloid detection methods available. As non-disturbing methods, two light scattering methods are widely used: single particle counting method (SPC) and photon correlation spectroscopy (PCS). As they are based on the light scattering, their sensitivity is much inferior to LIBD, as shown in Figure 6, particularly in the colloid size range of < 50 nm, in which the Rayleigh scattering prevails proportionally to $d^{-6}$ (d: particle diameter). For this reason, the two methods are of limited use for the study of groundwater colloids, which are in general small in size and concentration as mentioned already.

## Conclusions

The hitherto available laser spectroscopy provides the possibility to investigate actinide chemical reactions in the nano-mole concentration range under natural aquatic conditions. This possibility cannot be attained by conventional methods. Further development in this subject area is necessary to broaden our knowledge on the aquatic chemistry of actinides, which is an essential prerequisite for understanding the geochemical behaviour of actinides in the nuclear waste repository environment.

## REFERENCES

[1]     J.I. Kim, "The Chemical Behaviour of Transuranium Elements and Barrier Functions in Natural Aquifer Systems, in Scientific Basis for Nuclear Waste Management XVI, C.G. Interante and R.T. Pablan (eds.), Mater. Res. Soc. Symp. Proc. 294, Pittsburg, PA, p. 3 (1993).

[2]     J.I. Kim, "Actinide Colloids in Natural Aquifer Systems", MRS Bull. 19, 47 (1994).

[3]     J.V. Beitz, "Similarities and differences in trivalent lanthanide- and actinide molecules", in Handbook of Physics and Chemistry of the Actinides, Vol. 18, K.A. Gschneider, Jr., L. Eyring, G.R. Choppin, and G.H. Lander (Eds.), North-Holland Phys. Publ., Amsterdam, p. 120 (1994).

[4]     D.S. Kliger, (ed), "Ultrasensitive Laser Spectroscopy", Academic Press, New York (1983).

[5]     W. Schrepp, R. Stumpe, J.I. Kim, H. Walther, "Oxidation-State-Specific Detection of Uranium in Aqueous Solution by Photoacoustic Spectroscopy", *Appl. Phys.*, B32, 207 (1983).

[6]     R. Stumpe, J.I. Kim, W. Schrepp,, H. Walther, "Speciation of Actinide Ions in Aqueous Solution by Laser-Induced Pulsed Photoacoustic Spectroscopy", *Appl. Phys.* B34, 203 (1984).

[7]     J.I. Kim, R. Stumpe, R. Klenze, "Laser-Induced Photoacoustic Spectroscopy for the Speciation of Transuranic Elements in Natural Aquatic Systems", *Topics in Current Chemistry*, No. 157, Chemical Applications of Nuclear Probes, Yoshihara, K. (ed), Berlin, Springer, p. 183 (1990).

[8]     T. Kimura, Y. Kato, Z. Yoshida, "Laser-Induced Photothermal Displacement Spectroscopy (LIPDS) for Speciation of Lanthanides and Actinides", *Radiochim. Acta* 82, 141 (1998).

[9]     H.L. Fang, R.L. Swafford, "The Thermal Lens in Absorption Spectroscopy", in [4], p. 175.

[10]   W.B Jackson, N.M. Amer, A.C. Boccara,, D. Fournier, "Photothermal Deflection Spectroscopy and Detection", *Appl. Opt.* 20, 1333 (1981).

[11]  J.V. Beitz,, J.P. Hessler, "Oxidation State Specific Detection of Transuranic Ions in Solution", *Nucl. Tech.* 51, 169 (1980).

[12]  J.I. Kim, R. Klenze, H. Wimmer, "Fluorescence Spectroscopy of Curium(III) and Application", Eur. J. Solid State Inorg. Chem., 28 (Suppl.), 347 (1991).

[13]  A.B. Yusov, "Photoluminescence of Americium(III) in Aqueous and Organic Solution, *J. Radioanal. Nucl. Chem.*, Articles 143, 287 (1990).

[14]  Y. Kato, G. Meinrath,T. Kimura, Z. Yoshida, "A study of U(VI) Hydrolysis and Carbonate Complexation by Time-Resolved Laser-Induced Fluorescence Spectroscopy (TRLFS)", *Radiochim. Acta* 64, 107 (1994).

[15]  Ch. Moulin, P. Decambox, V. Moulin, J.G. Decaillon, "Uranium Speciation in Solution by Time-Resolved Laser-Induced Fluorescence", *Anal. Chem.* 67, 348 (1995).

[16]  Th. Fanghänel,, H.T. Weger, Th. Könnecke, V. Neck, P. Paviet-Hartmann, E. Steinle, J.I. Kim, "Thermodynamics of Cm(III) in Concentrated Electrolyte Solutions: Carbonate Complexation at Constant Ionic Strength (1 m NaCl), *Radiochim. Acta* 82, 47 (1998).

[17]  Th. Fanghänel, J.I. Kim, "Spectroscopic Evaluation of Thermodynamics of Trivalent Actinides in Brines", *J. Alloys and Compounds* 271-273, 728 (1998).

[18]  G. Bernhard, G. Geipel, V. Brendler, H. Nitsche, "Uranium Speciation in Waters of Different Uranium Mining Areas", *J. Alloys and Compounds* 271-273, 201 (1998).

[19]  K.H. Chung, R. Klenze, K.K. Park, P. Paviet-Hartmann, J.I. Kim, "A Study of the Surface Sorption Process of Cm(III) on Silica by Time-Resolved Laser Fluorescenec Spectroscopy (I), *Radiochim. Acta* 82, 215 (1998).

[20]  F.J. Scherbaum, R. Knopp, J.I. Kim, "Counting of Particles in Aqueous Solutions by Laser-Induced Photoacoustic Breakdown Detection", *Appl. Phys.* B: Lasers Opt. B63, 299 (1996).

[21]  R. Knopp, F.J. Scherbaum, J.I. Kim, "Laser Induced Breakdown Spectroscopy (LIBS) as an Analytical Tool for the Detection of Metal Ions in Aqueous Solutions", Fresenius' *J. Anal. Chem.* 355, 16 (1996).

# SESSION B

## *Methods for Macro Concentration Speciation ($>10^{-6}$ M)*

# SPECIATION FROM PHOTON TO ION DETECTION

**C. Moulin**
CEA Saclay
Direction du Cycle du Combustible-Département des Procédés d'Enrichissement
Service de Physique et de Chimie des Procédés
91191 Gif-sur-Yvette, France
E-mail: cmoulin1@cea.fr

## Abstract

New analytical techniques allowing to perform speciation in the framework of the nuclear fuel cycle are more and more needed. Among them, several laser-based analytical techniques present several advantages (non intrusive). Hence, Thermal Lensing (TL)/Photoacoustic (LIPAS), Time Resolved selective, sensitive Laser-Induced Fluorescence (TRLIF) have been used for actinides/lanthanides interaction and speciation studies in inorganic and organic matrices, Laser Ablation-Optical Emission Spectroscopy (LA-OES or LIBS) for direct studies on solids, liquids, ... where *in situ* measurements (elemental or isotopic) are mandatory. In complementary to these photon-based methods, new ion detection methods such as ElectroSpray-Mass Spectrometry (ES-MS) seems promising for speciation studies. Principle, advantages and limitations as well as results obtained and trends for these different methods will be presented.

# Introduction

The search for analytical techniques that are more sensitive in order to detect very low level of radionuclides, more selective in order to be able to deal with very complex solutions of interest is always of great concern in the nuclear industry. Moreover, in the framework of speciation studies, this technique should also be the less intrusive possible in order to keep the real image of the initial solution. In this framework it seems interesting to describe the evolution of the following three techniques in the CEA: Time-Resolved Laser-Induced Fluorescence (TRLIF), Laser Ablation-Optical Emission Spectroscopy (LA-OES) and ElectroSpray-Mass Spectrometry (ES-MS)

TRLIF first started in the CEA in the beginning of the 80s by a PhD thesis [1] on the use of a pulsed UV laser compared to classical UV lamp for a better detection of uranium. At that time, the improvement, in terms of LOD (20 ng/l) was closed to $10^2$ compared to classical fluorimetry which is limited to few µg/l. Then, the technique was adapted in gloves box (1984) and extended to the determination of lanthanides, americium and curium in very complex matrices [2-6]. In 1986, TRLIF was adapted in shielded cell with fiber optics and optode [7] at CEA-Marcoule for analysis in highly radioactive solutions. Finally, in order to respond to numerous demand from CEA and partners (COGEMA, ANDRA), a commercial apparatus (under CEA licence), the FLUO 2001 [8] (DILOR company) was constructed (1988) where the classical photomultiplier was replaced by a multichannel photodiodes array. During all these years of TRLIF developments and analysis, the LOD for uranium has improved from 20 ng/l to 0.1 ng/l [9] and more than hundred different matrices have been studied from sodium to grass going through organic solvents or highly radioactive samples [10-12]. It should also be noted that this technique was also implemented with vocal recognition and synthesis for easier manipulations in gloves box. From that point (1989) and till now, TRLIF has been used for interaction and speciation studies beneficying of the very good limit of detection obtained for classical analysis. Nowadays, 10 FLUO 2001 are used for routine analysis and 4 experimental set-up (with more versatility in terms of laser source and detection system) are used for speciation studies. The main results obtained with this technique will be presented in results and discussion.

LA-OES or Laser Induced Breakdown Spectroscopy (LIBS) was first developed in the mid 80s in the CEA for the analysis of metals in cell. But at that time, little was known for analytical applications of laser based plasma. Then, in the framework of numerous thesis and UE contract, the technique has evolved toward a better understanding of the interaction for analytical applications. It was shown that UV ablation was much more efficient in terms of ablated mass as well as a better reproducibility due to a less thermal interaction (inverse Bremstrhalung). Thanks to all these improvements, a prototype (1992) for µ-analysis was developed and this technique has been implemented in gloves box (1995) for plutonium analysis. Various types of samples have been analysed from solid (metals, ceramics, minerals, powder, fishes,…) to liquids (oil, colloid,…). Sensitivity reached is in the ppm range for classical emission spectroscopy but can reach ppb level with fluorescence detection. It is also possible to reach uranium isotopy with such an optical technique.

ES-MS was directly implemented in the 1996 in the CEA for speciation purposes, this rather new technique, the principle has been quoted in the mid 80s by Fenn and Yamashita and the first commercial apparatus was put together in 1992. This technique seems very promising since, it is the first time that it is possible to directly couple a liquid at atmospheric pressure to a mass detection working at reduced pressure with a soft mode of ionisation that should allow to give informations on chemical species present. A lot of work have been carried out in the beginning of the 90s on metal speciation but they were more directed towards mass spectrometry. Our purpose in the CEA is to use this technique for chemistry purposes, with at start relying on known data (OECD) and mature speciation techniques such as TRLIF (on element that fluoresce) to first prove the capacity of the technique for speciation (i.e. the non intrusivity) and then extend the palette of element such as fission products, non fluorescent actinides, etc. It is planned to have this technique in radioactive laboratories in Marcoule in the years 2000.

## Experimental

### *Time-resolved laser-induced fluorescence*

A laser (Nd-YAG, dye, OPO or microchip) was used as the excitation source, wavelengths varies from UV (266 nm) to visible depending on the element under study, frequency ranged from 10 Hz to 5 kHz and energy from few mJ to few µJ. The beam is directly directed into a 4 ml quartz cell or in a fiber optics (case of remote sensing). The radiation coming from the cell is focused on the entrance slit of the polychromator. The detection is performed by an intensified photodiodes array cooled by Peltier effect and positioned at the polychromator exit. Logic circuits, synchronised with the laser shot, allow the intensifier to be active with determined time delay and aperture time.

### *Laser ablation optical emission spectroscopy*

Pulses from laser (Nd-YAG, excimer) were focused on a target placed on a microdriven platform (µ-analysis) or in a gloves box (radioactive samples) with a lens to generate the plasma. The pulse energy incident on the sample ranges from few mJ to hundred giving a power density from few MW to hundred GW/cm$^2$. The plasma light was collected directly or via fibre optics by a multichannel analyser (ICCD with its controller). A programmable pulse-delay generator was used to gate the ICCD.

### *MS apparatus and conditions*

Electrospray ionisation mass spectrometric detection was performed on the mass spectrometer Quattro II. The sample in introduced into the source with a syringe pump. Initial gas flow rate was 250 l/h and source temperature 80°C. For most experiments, the cone voltage was set to 25 V and the sample solution flow rate set to 10 µl/min. Full scan modes by scanning the quadrupole from 50 to 1 000 m/z with a scan duration of 4 s and an acquisition time of 2 min. was used. For MS/MS measurements, collision induced dissociation was performed with Ar ($10^{-3}$ mbar) and the collision energy set to 25 eV.

## Results and discussion

In the framework of this presentation, it seemed important to only highlight the main performances and results obtained with the previously quoted techniques. It should also be emphasis on the fact that these studies have been made possible thanks to collaboration with other teams from the CEA, and partners COGEMA, ANDRA as well as European (in the framework of UE contracts) and international (see acknowledgements).

Concerning TRLIF, Table 1 lists the figure of merits in terms of LOD for both analysis and speciation. In analysis you "dominate" the medium by the addition of a suitable complexing reagent such as phosphoric acid for uranium determination; in speciation, you have to stick with the matrix

**Table 1. Limit of detection and speciation limit for the main element of interest by TRLIF**

| Elements | Limit of detection (M) | Speciation limit (M) |
|---|---|---|
| Uranium | $5.10^{-13}$ | $5.10^{-9}$ |
| Curium | $5.10^{-13}$ | $10^{-8}$ |
| Americium | $10^{-9}$ | $10^{-6}$ |
| Europium | $5.10^{-12}$ | $10^{-8}$ |

and therefore, speciation limits are much higher (roughly 3 to 4 order of magnitude). Figure 1 presents, the fluorescence spectra of the uranium hydroxo complexes together with their lifetime that were obtained both by varying pH and uranium concentration. It can be seen as important changes both on spectra and lifetime. All the different transitions from the two excited states to the different ground states are present. A complete spectral database has been obtained for hydroxo [13-14] but also nitrate [15], phosphate [16], citrate, etc. By the same token, it is possible to perform speciation studies on the other fluorescent actinides such as americium and curium but also on lanthanides such as europium [17]. Figure 2 presents the evolution of the europium fluorescence spectra from free species ($Eu^{3+}$) to the one obtained in presence of carbonate and humic substances with the possible presence of mixed europium-carbonate humic substances complexes. It is interesting to note the spectral modifications where a symmetrical spectrum is obtained for free europium surrounded by 9 water molecules and europium – humic substances which seems to be an outer complex since the water molecules are still present (same lifetime). Aside this speciation capability, it is also possible to obtain interaction constant by spectral titrationn [18-19] where the principle consists in the study of fluorescence in a non complexing medium, then progressive addition of complexant (humic acids, micelles, etc.), increase or decrease of the fluorescence signal until saturation, determination of the complexing capacity (intersection of the two slopes) and the interaction constant $\beta$ by non linear regression. It has been among other results that the interaction constant varies with metal concentration.

**Figure 1. TRLIF spectra and lifetime of uranium-hydroxo complexes together with free uranium**

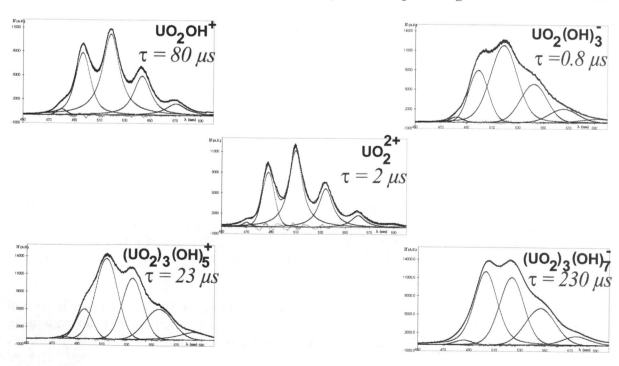

# Figure 2. TRLIF spectra and lifetimes of europium carbonate and humic complexes in synthetic waters

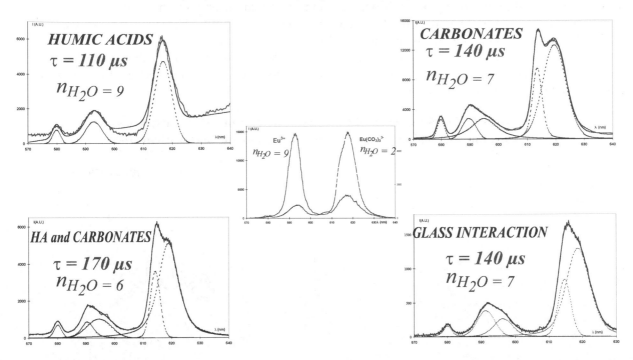

Concerning AL-SEO or LIBS, this technique has allowed to perform fast 2-D μ-analysis at atmospheric pressure as shown on Figure 3 with a resolution of 3 μm. Other striking results are its capability to perform uranium isotopic analysis as well as direct analysis of plutonium impurities [20] in powder or *in situ* elemental analysis in dirty solutions (oils, precipitates, colloids, etc.).

# Figure 3. AL-SEO 2D mapping on copper alloy and comparison with Castaing probe

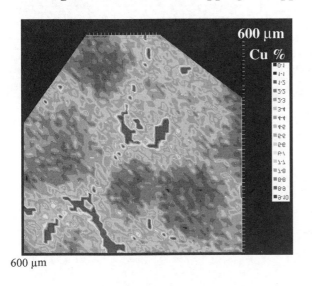

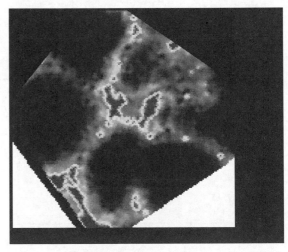

Resolution 5 μm, time 30 min
depth 2-3 μm

*With permission from CEA-DCC/DPE/SPCP/LSLA*

Concerning ES-MS, this technique has allowed to perform speciation on calixarene-caesium complexes [21] as shown on Figure 4. Hence, when there is twice less caesium than calixarene, it is possible to visualise the complex with one caesium at 961 amu, and when there is twice as more caesium than calixarene, it is possible to visualise a complex (doubly charged) with two caesium at 547 amu, together with the monocharged and free caesium. It has also been possible to perform uranium speciation [22] as shown on Figure 5 where free uranium and the first hydroxo complexes have been perfectly identified (adducts) as well as one of the polymeric uranium species. Results obtained have been corroborated by TRLIF studies.

**Figure 4. ES-MS spectra of Cs-calixarene complexes : a/ twice less caesium than calixarene, b) twice more caesium than calixarene**

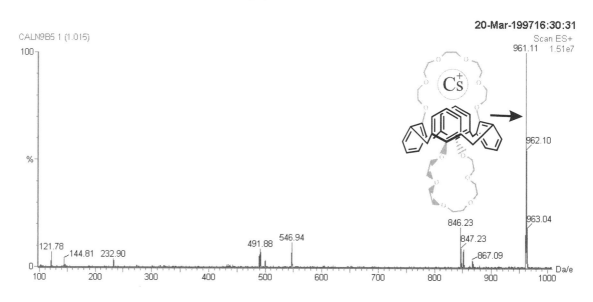

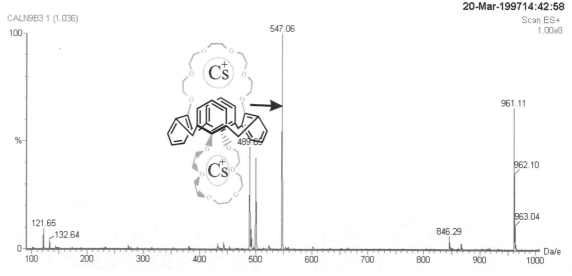

**Figure 5. ES-MS of uranium hydroxo complexes**

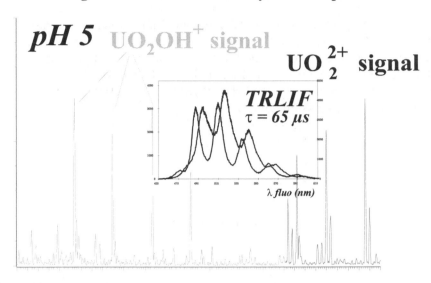

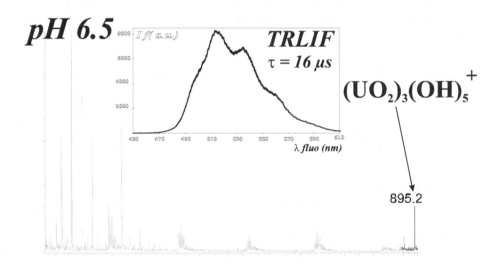

## Conclusions and trends

These different techniques are very powerful for speciation studies they have advantages and drawbacks and are complimentary. As spectroscopic technique, they all lead to spectral data banks that are very useful when one starts from simple to complex matrices Concerning trends, for TRLIF aside, more complex studies with the use of spectral data bank, the evolution is toward in situ speciation with miniaturised system [23]. For LIBS, it is toward the use of multichannel analyser with high resolution, for ES-MS to a better understanding of the mechanisms taking place in the source as well as its evolution. Finally, great care should be taken in terms of sample preparation as well as results interpretation. Inter-comparison studies are needed for a better use of these techniques as well as the use of thermodynamical database and molecular modelling.

*Acknowledgements*

CEA – Laboratoire de Spectroscopie Laser Analytique, Laboratoire de Migration et de Géochimie du Solide, Laboratoire d'Analyse et de Synthèse Organique, Laboratoire de Chimie Analytique Appliquée, Laboratoire Nouveaux Extractants, Laboratoire Instrumentation, Laboratoire des Procédés de Traitement des Effluents – IPN/Groupe de Radiochimie – COGEMA La Hague (SPR), COGEMA Marcoule (LABM), ANDRA Soulaines, IPSN/DPHD/SDOS Pierrelatte, Université Pierre et Marie Curie, SCK Mol, European Commission, OECD.

# REFERENCES

[1]     "Dosage de l'Uranium par Spectrofluorimétrie à Source d'Excitation Laser", PhD thesis P. Mauchien CNAM Paris (1983).

[2]     "Direct Uranium Trace Analysis in Plutonium Solutions by Time-Resolved Laser-Induced Spectrofluorometry" T. Berthoud, P. Decambox, P. Mauchien, C. Moulin, Anal. Chem., 60, 1296 (1988).

[3]     "Time Resolved Laser Induced Fluorescence for Curium Trace Determination", C. Moulin, P. Decambox, P. Mauchien, Radiochimica Acta, 48, 23 (1989).

[4]     "Direct Determination of Traces of Lanthanide Ions in Aqueous Solutions by TRLIF", P. Decambox, T. Berthoud, B. Kirsch, P. Mauchien, C. Moulin, Analytica Chimica Acta, 220, 235 (1989).

[5]     "Determination of Curium at 0.1 ng/L Levels in a Micellar Medium by TRLIF", C. Moulin, P. Decambox, P. Mauchien, Analytica Chimica Acta, 254, 145 (1991).

[6]     "Americium Trace Determination in Aqueous and Solid Matrices by TRLIF", S. Hubert P. Thouvenot, C. Moulin, P. Mauchien, P. Decambox, Radiochimica Acta, 61, 15 (1993).

[7]     "Uranium Determination by Remote Time-Resolved Laser-Induced Fluorescence", C. Moulin, S. Rougeault, D. Hamon, P. Mauchien, Applied Spectroscopy, 47, 2007 (1993).

[8]     "FLUO 2001-Nouveau Spectrofluorimètre avec Détection par Barrette de Photodiodes Intensifiée et Pulsée", C. Moulin, P. Mauchien, P. Decambox, B. Kirsch, Proceedings OPTO 88, ESI Pub. 357 (1988).

[9]     "Determination of Uranium in Solution at the ng/l Level by TRLIF: Application to Geological Survey", C. Moulin, C. Beaucaire, P. Decambox, P. Mauchien, Analytica Chimica Acta 238, 291 (1990).

[10]    "Direct and Fast Determination of Uranium in Human Urines by Time-Resolved Laser-Induced Spectrofluorometry", C. Moulin, P. Decambox, P. Mauchien, Applied Spectroscopy, 45, 116 (1991).

[11] "Dual use of Micellar Enhanced Ultrafiltration and TRLIF for the Study of Uranyl Exchange at the Surface of Alkylsulfate Micelles", P. Reiller, D. Lemordant, C. Moulin, C. Beaucaire, JCIS 162 (1994).

[12] Direct and Fast Uranium Determination in Zirconium by TRLIF", C. Moulin, P. Decambox, L. Trecani, Analytica Chimica Acta. 321, 121 (1996).

[13] "Uranium Speciation in Solution Studied by Time-Resolved Laser-Induced Fluorescence", C. Moulin, P. Decambox, V. Moulin, J.G Decaillon, Analytical Chemistry. 34, 348 (1995).

[14] "TRLIF as a Unique Tool for Uranium Speciation at low Level", C. Moulin, I. Laszak, V. Moulin, C. Tondre, Applied Spectroscopy 52, 528 (1998).

[15] "Direct Uranium(VI) and Nitrate Determinations in Nuclear Reprocessing by TRLIF", C. Moulin, P. Decambox, P. Mauchien, D. Pouyat, L. Couston, Analytical Chemistry. 68, 3204 (1996).

[16] "Investigations by TRLIF and CE of the Uranium Phosphate System: Application to Blood Serum", S. Scapollan, E. Ansoborlo, C. Moulin, C.J. Madic, Alloys and Compounds 271, 106 (1988).

[17] "Europium Complexes Investigations in Natural Waters by TRLIF", C. Moulin, J. Wei, P. Van Iseghem, I. Laszak, G. Plancque, V. Moulin, Anal. Chim. Acta 253, 396 (1999).

[18] "On the use of TRLIF for Interaction Studies Between Organic Matters and Actinides: application to Cm", C. Moulin, P. Decambox, P. Mauchien, V. Moulin, M. Theyssier, Radiochim. Acta 52/53, 119 (1991).

[19] "Complexation Behaviour of Humic Substances Towards Actinides and Lanthanides Studied by TRLIF", V. Moulin, J. Tits, C. Moulin, P. Decambox, P. Mauchien, O. De Ruty, Radiochim. Acta 58/59, 121 (1992).

[20] "Determination of Impurities in Uranium and Plutonium Dioxides by Laser Induced Breakdown Spectroscopy", P. Fichet, P. Mauchien, C. Moulin, Appl. Spec. 111, 53 (1999).

[21] Behaviour of Calix[4]Arene-Biscrown-6 Under Irradiation. ES-MS Investigations and Molecular Dynamics. Simulations on Cs and Na Complexes of Degradation Compound, V. Lamare, J.F. Dozol, F. Allain, H. Virelizier, C. Moulin, C. Jankowski, J.C. Tabet, ACS 1999.

[22] Speciation of U by ES-MS. Comparison with TRLIF. Appl. Spec. Submitted.

[23] "Pulsed Microchip LIF for In Situ Tracer Experiments", C. Moulin, X. Vitart, J. Fresenius, Anal. Chem, 361, 85 (1998).

# DETERMINATION OF THE HYDRATION NUMBER OF ACTINIDES(III) AND LANTHANIDES(III) BY LUMINESCENCE LIFETIME MEASUREMENT AND ITS APPLICATION TO THE SPECIATION STUDY

**Takaumi Kimura, Yoshiharu Kato and Hideyo Takeishi**
Japan Atomic Energy Research Institute, Tokai-mura, Ibaraki 319-1195, Japan

**Yoshio Takahashi**
Hiroshima University, Higashi-Hiroshima, Hiroshima 739-8526, Japan

**Yoshitaka Minai**
Musashi University and Nezu Institute of Chemistry, Nerima, Tokyo 176-8534, Japan

**Gregory R. Choppin**
Florida State University, Tallahassee, Florida 32306-3006, USA

## Abstract

The inner-sphere hydration number $N_{H2O}$ (i.e., the number of $H_2O$ molecules in the first hydration sphere of a metal ion) provides significant information regarding the co-ordination environment of the chemical species in solution, in solid and at their interface. A correlation between the luminescence decay constant $k_{obs}$ (i.e., the reciprocal of the excited state lifetime) and the $N_{H2O}$ of actinides(III) [An=Am, Cm] and lanthanides(III) [Ln=Nd, Sm, Eu, Tb, Dy] was investigated systematically by time-resolved laser-induced fluorescence spectroscopy. The calibration relations of the $k_{obs}$ vs. $N_{H2O}$ were proposed on the basis of the linear correlation of the $k_{obs}$ vs. molar fraction of $H_2O$ in $D_2O$-$H_2O$ solutions and the $N_{H2O}$ in $H_2O$, i.e., nine for Am(III), Cm(III), Nd(III), Sm(III) and Eu(III) and eight for Tb(III) and Dy(III). The relationships were applied for studies on hydration states of An(III) and Ln(III) in the polyaminopolycarboxylate complexes, separation behaviour of Cm(III) and Eu(III) into a cation exchange resin, and sorption behaviour of Eu(III) species onto silica and clay minerals. It was demonstrated that the determination of the $N_{H2O}$ of An(III) and Ln(III) by the luminescence lifetime measurement is an effective method for the speciation of these ions in various circumstances.

# Introduction

The solution chemistry of actinides plays an important and basic role in the field of nuclear technology, especially in the back end of the nuclear fuel cycle. For future advances in the technology of the nuclear fuel cycle, accurate and microscopic elucidation of phenomena encountered in various conditions will be required to ensure further developments in the separation chemistry and in the environmental modelling of the actinides. Therefore, highly sensitive and selective speciation methods of the actinides are indispensable for in-depth understanding of the mechanisms based on the identification and characterisation of the species at the molecular level.

The hydration of a metal ion is an important factor in the structural and chemical behaviour of the species, since the hydrated ion is a fundamental unit of the chemical reaction in aqueous solution. Knowledge of the structure and properties of hydrated ions is necessary but not sufficient to understand the chemistry of the ions in aqueous solutions, as the electron configuration and associated characteristics of the metal ion must also be understood well. The techniques for studying the size and/or structure of the hydration sphere can be classified as direct or indirect methods. The direct methods include X-ray and neutron diffraction, extended X-ray absorption fine structure (EXAFS), luminescence and nuclear magnetic resonance (NMR) relaxation measurements, while the indirect methods involve compressibility, NMR exchange and absorption spectrum measurements. These measurements, except luminescence spectroscopy, require aqueous solutions of relatively high metal concentrations ($>10^{-4}$ M), but the luminescence method allows determination of the hydration number of some metal ions in very dilute solutions ($<10^{-4}$ M metal concentrations) and is the most promising method for studying the hydration of radioactive actinide ions.

By using time-resolved laser-induced fluorescence spectroscopy in addition to conventional fluorescence spectrometry, the excitation and emission spectra and the lifetimes have been measured for f-elements. The luminescence of the f-elements was studied extensively for lanthanides(III) but not thoroughly for actinides(III) [1-5]. Examples of such studies include determination of the f-element concentration, speciation of the metal complexes, elucidation of factors affecting non-radiative decay, and measurement of binding in fluorescence immunoassays. The wavelength shift and intensity variation has been employed for the purpose with less use of lifetime information. It has been shown that the measurement of luminescence lifetimes in both $H_2O$ and $D_2O$ solutions is a useful method to determine quantitatively the number of water molecules co-ordinated to Eu(III) or Tb(III) in aqueous solutions[6-8]. However, this method was not extended to other lanthanides(III) or to actinides(III). Recently, a correlation between the luminescence decay constant $k_{obs}$ (i.e., the reciprocal of the excited state lifetime) and the inner-sphere hydration number $N_{H2O}$ (i.e., the number of $H_2O$ molecules in the first hydration sphere) has been investigated for Am(III) and Cm(III) ions[9-11] and additional lanthanide(III) ions[Ln = Nd, Sm, Eu, Tb, Dy][9-13]. There have also been attempts to evaluate the relations of the $k_{obs}$ vs. $N_{H2O}$[10, 14-18].

In the present paper, systematic study of the luminescence properties of An(III) and Ln(III) was carried out to establish a method for determining the values of $N_{H2O}$ of the ions from the measurements of the luminescence lifetime. The applications to the co-ordination chemistry, separation chemistry, and environmental chemistry of the f-elements are discussed from the viewpoint of the speciation in aqueous solution, in solid containing water and at solid-water interface, respectively.

# Experimental

## Materials and sample preparation

Am(III) and Cm(III) stock solutions in 0.01 M HClO$_4$ were prepared from [241]Am(t$_{1/2}$ = 432 y) and [244]Cm(t$_{1/2}$ = 18.1 y) oxides supplied by Amersham, UK and by CEA, France, respectively. [244]Cm was purified previously from its daughter nuclide [240]Pu by passage through anion exchange resin (AG 1X8) with elution by 7 M HNO$_3$. To avoid the use of highly radioactive [244]Cm as solid phase, [248]Cm (t$_{1/2}$ = 3.5 × 10$^5$ y) purified from its spontaneous fission products by cation exchange resin[9] was also used for preparation of Cm(III) co-precipitated with lanthanum bromate. The concentration of the Am(III) and Cm(III) stock solutions was determined by α-ray spectrometry and liquid scintillation counting. Ln(III) stock solutions in 0.01 M HClO$_4$ were prepared by dissolving an appropriate amount of the oxides (Wako Pure Chem. Ind., Ltd.) in perchloric acid. D$_2$O (99.9 at.%) was obtained from Merck, Canada and used to prepare the D$_2$O-H$_2$O solutions and the various samples in D$_2$O.

Nitrilotriacetic acid (NTA), N-(2-hydroxyethyl)ethylenediamine-N,N',N'-triacetic acid (HEDTA), ethylenediamine-N,N'-diacetic-N,N'-dipropionic acid (ENDADP), ethylenediamine-tetraacetic acid (EDTA), 1,2-diaminopropane-N,N,N',N'-tetraacetic acid (PDTA), trans-1,2-diaminocyclohexane-N,N,N',N'-tetraacetic acid (CDTA), diethylenetriaminepentaacetic acid (DTPA), glycoletherdiamine-tetraacetic acid (EGTA) and triethylenetetraamine-N,N,N',N'',N''',N'''-hexaacetic acid (TTHA) were used as received from Tokyo Kasei Kogyo Co., Ltd. Solutions of the polyaminopolycarboxylate complexes were prepared by mixing stoichiometric amounts of metal(III) and ligand stock solutions of known concentration. The solution pH was adjusted by the addition of standard NaOH(D) or H(D)ClO$_4$ and was checked before and after the measurement of the luminescence lifetime. The supporting electrolyte concentration was adjusted with NaClO$_4$.

AG 50WX8 (functional group, R-SO$_3^-$; matrix, styrene divinylbenzene; proton exchange capacity, 2.1 meq/g) purchased from Bio-Rad Lab., Inc. was used as the cation exchange resin. Montmorillonite and kaolinite purchased from Nacalai Tesque, Inc. and Wako Pure Chemical Ind. Ltd., respectively, were characterised by powder X-ray diffraction and IR spectroscopy. The surface areas of montmorillonite and kaolinite were 55 and 7.5 m$^2$/g, respectively, determined by BET analysis (N$_2$) and cation exchange capacities (CEC) were determined by modified Schofield method (pH 8, NH$_4$Cl 2.0 mM) as 30 and 2.9 meq/100g, respectively. Nonporous silica (Aerosil 200: surface area, 200 m$^2$/g) and porous silica (Fuji Silica Gel AB: surface area, 590 m$^2$/g; CEC, 15 meq/100g; porous diameter, 40 Å) were purchased from Japan Aerosil Co. Ltd. and Fuji Division Chemicals Ltd., respectively. All other chemicals used were of analytical reagent grade and aqueous solutions were prepared from doubly distilled deionised water.

The distribution of Cm(III) and Eu(III) between the solid and aqueous phases was measured with [244]Cm and [152]Eu tracers. From the radiometric analysis of aqueous phase, the distribution coefficient K$_D$(cm$^3$/g), as defined by K$_D$ = {(C$_0$-C)/C}•(V/M), was calculated, where C$_0$ is the initial concentration of a metal ion in the aqueous phase; C, the equilibrium concentration of the metal ion in the aqueous phase; V, the volume of the aqueous phase (cm$^3$); M, the mass of the solid phase (g).

## Luminescence lifetime measurement

The lifetimes of An(III) and Ln(III) samples were measured using a 5 mm-i.d. quartz cell and a standard 1 cm fluorescence spectrometry cell, respectively. The suspended solid water solution was continuously stirred in the quartz cell and directly exposed to a pulsed laser beam. When the percentage of the metal sorbed was less than 90%, the suspension was placed into a quartz tube and centrifuged. The separated aqueous phase was removed and the slurry at the bottom of the quartz tube was exposed to the laser beam. The metal (III) ion in the samples was excited to the excited states by the pulsed laser beam and the emission from the lowest luminescent level to the ground state manifold was measured subsequently to obtain the luminescence lifetime. The excitation and emission wavelengths used for each ion[19-23] are summarised in Table 1. The 355 nm (third harmonic) laser beam was obtained directly from a pulsed (10 Hz) output of a Spectron SL-803 Nd:YAG laser. The 390-410, 504 or 593 nm laser beam was obtained with a pulsed (10 Hz) 308 nm output of a Lambda Physik COMPex201 XeCl excimer laser pumping PBBO, Coumarin 307 or Rhodamine 6G (Lambda Physik), respectively, in dioxane or methanol solution in a Lambda Physik SCANmate2 dye laser head. The pulse power was typically 3-7 mJ per pulse and the pulse width was about 15 ns. The emission light was collected at 90° into an Oriel 77257 monochromator using an optical fiber and detected by a Hamamatsu R3896 or R928 photomultiplier tube. The whole emission wavelength range (12.8 nm width) was observed. Sharp cut optical filters (Y, O or R series, Toshiba Glass Co.) were used to minimise the scattering laser light. The signal was fed into a Hewlett Packard 54510A digitising oscilloscope, which was connected to a NEC PC-9801RX computer through a GP-IB interface. The luminescence decay curves observed in this work were fitted to single-exponential curves with correlation coefficients 0.99-0.999.

## Results and discussion

### *Determination of the inner-sphere hydration number by luminescence lifetime measurement*

The inner-sphere hydration number $N_{H2O}$ of Eu(III) and Tb(III) has been obtained by using the difference in the decay constants of the species in $H_2O$ and $D_2O$ solutions [7]. The luminescence decay constants $k_{obs}(s^{-1})$ of $An^{3+}$ and $Ln^{3+}$ were, therefore, measured in $D_2O$-$H_2O$ solutions as a function of molar fraction of $H_2O$. As shown in Figure 1, the $k_{obs}$ of these ions increased linearly with increasing molar fraction of $H_2O$ in the solutions with correlation coefficients above 0.99 for $Nd^{3+}$ and $Sm^{3+}$ and 0.999 for the others [9,11,13], indicating that the quenching behaviour of the ions is attributable primarily to energy transfer from the excited states of the metal ions to OH vibrations of the hydrated $H_2O$ and that the OH vibrators in the water molecule act independently in the de-excitation process.

The lifetimes of the ions observed in $H_2O$ and $D_2O$, i.e. $\tau_{H2O}$ and $\tau_{D2O}$, are summarised in Table 1. The $\tau_{H2O}$ and $\tau_{D2O}$ decreased with decreasing energy gap, $\Delta E$, defined as the difference in energy between the emitting state and the next lower lying state[5,24]. The decay constant $k_{obs}$ of the ions in aqueous solution can be expressed by $k_{obs} = 1/\tau_{obs} = k_F + \Sigma k_i^{nr} + \chi_{H2O} \cdot k_{H2O}$, where $k_F$ is the constant for the radiative process, the second term represents the rate constants for non-radiative de-excitation processes not involving water, and $\chi_{H2O}$ is the volume percentage of $H_2O$ in the $D_2O$-$H_2O$ solution. Non-radiative decay rate $k_{H2O}$ due to $H_2O$ is estimated from the $\tau_{H2O}$ and $\tau_{D2O}$ by $k_{obs}(H_2O)-k_{obs}(D_2O)$ for each ion. Figure 2 shows different trends of the $k_{H2O}$ values between $An^{3+}$ and $Ln^{3+}$ as a function of $\Delta E$; i.e., if the $\Delta E$ values are similar, the $k_{H2O}$ of $An^{3+}$ is larger than that of $Ln^{3+}$, especially at higher $\Delta E$. This suggests that the interaction of $An^{3+}$ with co-ordinated waters is much stronger than that of $Ln^{3+}$ perhaps reflecting that the 5f-orbitals of $An^{3+}$ are less well shielded than the 4f-orbitals of $Ln^{3+}$. The lifetime ratios $\tau_{D2O}/\tau_{H2O}$ of $Sm^{3+}$, $Eu^{3+}$, $Tb^{3+}$ and $Dy^{3+}$ in Table 1 are very similar to the intensity ratios $I_{D2O}/I_{H2O}$ in the literature [24,26] and seem to be independent of the electron configuration and energy gap.

It is known that the water exchange rates between the inner co-ordination sphere and bulk water are about $10^8$ $s^{-1}$ for ions of the lanthanide series, which have been measured using both [17]O-NMR and ultrasound techniques [27]. Even for the case of $Am^{3+}$ in $H_2O$, water molecules in the inner co-ordination sphere would exchange on a time scale short compared to the luminescence lifetime and, therefore, the luminescence decay curves observed in this work were fitted well to single-exponential curves.

To determine the $N_{H2O}$ of $Cm^{3+}$ and $Ln^{3+}$ in $H_2O$, the $k_{obs}$ of the bromate salts, $M(BrO_3)_3 \cdot 9H_2O$, were measured. The bromates prepared were confirmed to have the expected composition because the $k_{obs}$ of the our Eu(III) and Tb(III) bromates were very similar to those reported for these bromates in the literature[7,28]. By using the $k_{obs}$ of M(III) in $H_2O$ and of the bromate, the $N_{H2O}$ of $Cm^{3+}$, $Sm^{3+}$, $Eu^{3+}$, $Tb^{3+}$ and $Dy^{3+}$ in $H_2O$ were calculated to be $9.1 \pm 0.5$, $9.0 \pm 0.5$, $9.1 \pm 0.5$, $8.3 \pm 0.4$ and $8.4 \pm 0.4$, respectively[9,12]. Considering the information from X-ray and neutron diffraction, luminescence, Raman and visible spectroscopy, etc., Rizkalla and Choppin [27] concluded that the light lanthanides($La^{3+}$-$Nd^{3+}$) form a series with the hydration number of nine, whereas the heavier elements($Tb^{3+}$-$Lu^{3+}$) apparently form octahydrates. The middle-series lanthanides($Nd^{3+}$-$Tb^{3+}$) either assume intermediate structures or a rapid exchange equilibrium exists between both hydrate structures. Our results of the $N_{H2O}$ with those obtained by X-ray diffraction [29] are shown in Figure 3 as a function of ionic radius. The comparison of Cm(III) and Ln(III) indicates that the hydration number is independent on the electronic configuration of the ion but is governed by the ionic size. The $N_{H2O}$ of nine for $Am^{3+}$, $Cm^{3+}$, $Nd^{3+}$, $Sm^{3+}$ and $Eu^{3+}$ and of eight for $Tb^{3+}$ and $Dy^{3+}$ in $H_2O$ were employed for the derivation of the calibration relations of $k_{obs}$ vs. $N_{H2O}$, since the trend of our results agreed with the conclusion of Rizkalla and Choppin [27]. From the $N_{H2O}$ in $H_2O$ and the linear correlation obtained in $D_2O$-$H_2O$ solutions (Figure 1), the following correlations were proposed for the determination of the hydration numbers of An(III) and Ln(III) [11,13].

$$N_{H2O} = 2.56 \times 10^{-7} \, k_{obs}(Am) - 1.43 \tag{1}$$

$$N_{H2O} = 6.12 \times 10^{-4} \, k_{obs}(Cm) - 0.48 \tag{2}$$

$$N_{H2O} = 3.58 \times 10^{-7} \, k_{obs}(Nd) - 1.97 \tag{3}$$

$$N_{H2O} = 2.54 \times 10^{-5} \, k_{obs}(Sm) - 0.37 \tag{4}$$

$$N_{H2O} = 1.05 \times 10^{-3} \, k_{obs}(Eu) - 0.44 \tag{5}$$

$$N_{H2O} = 4.03 \times 10^{-3} \, k_{obs}(Tb) - 0.87 \tag{6}$$

$$N_{H2O} = 2.11 \times 10^{-5} \, k_{obs}(Dy) - 0.60 \tag{7}$$

If there is no contribution from the ligand to the de-excitation of the excited state, the $N_{H2O}$ of a metal ion in the different complexes can be obtained from the $k_{obs}$ measured in $H_2O$. The slopes of Eqs. (5) and (6) for Eu(III) and Tb(III), respectively, agreed with those in the literature [7] within $\pm 5\%$. Eqs. (5) and (6) gave results consistent with those obtained by the procedure of Horrocks and Sudnick[7] within the uncertainty of the luminescence method, $\pm 0.5$ water molecules. This suggests that Eqs. (5) and (6) are reliable for determination of the $N_{H2O}$ of Eu(III) and Tb(III) and, presumably, so are the Equations (1)-(4) and (7) for Am(III), Cm(III), Nd(III), Sm(III) and Dy(III), respectively, within the uncertainty $\pm 0.5$ $H_2O$ molecules. Using Cm(III) and Eu(III) co-precipitated with lanthanum compounds of known hydration structure, we have reported the calibration relations of the $k_{obs}$ versus $N_{H2O}$ for the ions [9]. These relationships were utilised successfully to determine the $N_{H2O}$ of various species [10]. Eqs. (2) and (5) are in acceptable agreement with those in the previous paper and give the same $N_{H2O}$ values within $\pm 0.3$ or $\pm 0.1$ $H_2O$ molecules for Cm(III) and Eu(III), respectively.

For the hydrated ions without any complexing media, the detection limits for determination of the $N_{H2O}$ are about $10^{-8}$-$10^{-9}$ and $10^{-5}$-$10^{-6}$ M for $An^{3+}$ and $Ln^{3+}$, respectively. For analytical purpose of An(III) and Ln(III), the detection sensitivity of the ions can be improved by enhancing the luminescence by dehydration via inner-sphere complexation and/or energy transfer from ligand to metal ion, i.e. sensitisation.

### *Speciation of actinide(III) and lanthanide(III) complexes with polyaminopolycarboxylate ligands*

The residual hydration of the polyaminopolycarboxylate complexes of Eu(III) and Tb(III) has been determined by luminescence measurements [30-32]. By using Eu(III) and Tb(III) as reference ions, the hydration states of An(III) and Ln(III) complexes with polyaminopolycarboxylate ligands were studied in order to validate the applicability of the calibration relations (1)-(7) [9,11,13].

The $k_{obs}$ of An(III) and Ln(III) complexed with NTA, EDTA, DTPA and TTHA in $D_2O$ were measured at pH 4.0-5.0 and the $N_{H2O}$ were evaluated to be 0±0.4, indicating that these ligands were not effective in causing non-radiative de-excitation of the excited states of the ions. The quenching effect of CH vibration on the excited state $Am^{3+}$ which has the smallest energy gap with the ground state (Table 1) was estimated using Eq. (1) from the lifetime data in $(CH_3)_2SO$ and $(CD_3)_2SO$ [33]. The difference of the effect in both solvents is equivalent to 0.25 $H_2O$ molecules although the calculated values of $N_{H2O}$ are negative in these solvents, indicating that the quenching by $(CH_3)_2SO$ is much less than that by $D_2O$. This also supports the assumption that the high-frequency ligand vibrations such as CH and CN have less contribution and the OH vibration of the residual $H_2O$ molecules in the complex is the primary quencher for the excited states of Am(III).

The pH dependence of the hydration of An(III) and Ln(III) in the presence of these ligands in $H_2O$ was studied over the pH range of 1-12. The hydration behaviour of the complexes is similar qualitatively to the pH dependence of Eu(III) and Tb(III) complexes in the literature[30-32]. As an example of the complexes, the hydration states of An(III)- and Ln(III)-EDTA complex systems are shown as a function of pH in Figure 4. The co-ordination environment of the metal ion varies successively with pH, that is, free hydrated ions at lower pH, 1:1 complexes in the pH range of 4-8, and ternary hydroxo complexes at higher pH (ca. 9-11). In the case of the NTA system, the first plateau around pH 3-5 is caused by the formation of the $M(nta)^0$ complexes and the second plateau at higher pH is due to the formation of the $M(nta)_2^{3-}$ and $M(OH)_2^+$ complexes, according to the speciation calculation [9,11,13]. The complexation of Am(III) with CDTA was not sufficient and that with ENDADP was shifted to the higher pH, compared with those of the other ions [11]. The luminescence decay curves of the Am(III) complexes in the plateaux region of the $N_{H2O}$ were fitted well to single-exchange curves, suggesting that the formation rate constant of the polyaminopolycarboxylate complexes would be similar order with those of oxalate [34] and murexide [35,36] complexes of Nd(III) and that the rate of water release was the controlling step in these reactions.

The average numbers of co-ordinated water molecules $N_{H2O}$, ligand co-ordination numbers $CN_L$, and total co-ordination numbers $CN_T$ of the 1:1 metal-polyaminopolycarboxylate complexes are shown in Table 2. The $CN_T$ were calculated as the sum of the $CN_L$ and the $N_{H2O}$ measured, where $CN_L(NTA) = 4$, $CN_L(HEDTA) = 5.5$, $CN_L(ENDADP) = CN_L(EDTA) = CN_L(PDTA) = CN_L(CDTA) = 6$, $CN_L(DTPA) = 7.5$, $CN_L(EGTA)= 8$ and $CN_L(TTHA)= 8.5$ were employed as the number of the ligand donor groups [37]. The $N_{H2O}$ in each 1:1 complex and the $CN_T$ in the complexes, i.e. $CN_T(Eu) = 8.8 \pm 0.5$ and $CN_T(Tb)=8.5 \pm 0.5$, of Eu (III) and Tb (III) agreed well with those in the literature [27]. The $N_{H2O}$ of Am(III), Cm(III) and Nd(III) in these complexes were apparently larger than those of

Eu(III) and Tb(III). The energy gaps of Sm(III) and Dy(III) are quite similar, which indicates a similar quenching behaviour of the ions in aqueous solution. Nevertheless, the $N_{H2O}$ of Sm(III) were larger than those of Dy(III) in the complexes. For the Ln(III)-polyaminopolycarboxylate complexes, the results of $CN_T(Nd) = 9.9 \pm 0.3$, $CN_T(Sm) = 10.2n \pm 0.5$, $CN_T(Eu) = 8.9 \pm 0.2$, $CN_T(Tb) = 8.6 \pm 0.4$ and $CN_T(Dy) = 9.1 \pm 0.2$ are consistent with the trend observed in Figure 3, i.e. the inner-sphere hydration or co-ordination number decreases with decreasing the ionic radius. The results also indicate that the $CN_T$ is about ten for the light lanthanides(III) and around nine for the heavy lanthanides(III) and that the change of the $CN_T$ occurs between Sm(III) and Eu(III). The result $CN_T(Am) = 10.7 \pm 0.3$ is appreciably larger than $CN_T(Cm) = 9.8 \pm 0.3$, suggesting that the $CN_T$ of the An(III) complexes may be in the range between ten and eleven.

In summary, our results suggest that the calibration relations proposed as Eqs. (1)-(7) for An(III) and Ln(III) are valid to determine the $N_{H2O}$ by measurements of the luminescence lifetime and that the total co-ordination numbers of Am(III), Cm(III), Nd(III) and Sm(III) are possibly unity or above larger than those of Eu(III), Tb(III) and Dy(III) in the polyaminopolycarboxylate complexes.

## Speciation of Cm(III) and Eu(III) in cation exchange resin-hydrochloric acid solution system

Since ion exchange resins are opaque to most spectroscopic investigations, experimental information on the sorbed species has been limited to techniques such as Mössbauer spectroscopy, electron spin resonance, neutron diffraction, etc. The $N_{H2O}$ values of Cm(III) and Eu(III) in HCl solutions and in the cation exchange resin phases were measured to clarify the relationship between the separation behaviour and the hydration structure of the ions [15].

The distribution coefficients $K_D$ of Cm(III) and Eu(III) in the AG 50WX8 cation exchange resin-hydrochloric acid solution system are shown in Figure 5. Up to about 5 M HCl there was very little difference in the sorption behaviour of Cm(III) and Eu(III) and both ions behaved qualitatively in a similar manner to those in $HNO_3$ or $HClO_4$ [15], indicating that below 5 M the dominant effect of HCl variation was the common ion effect caused by varying $H^+$ ion concentration. An appreciable divergence occurred above 5 M HCl, with $K_D$ (Eu) increasing and $K_D$(Cm) continuing to decrease, which is quite similar to the behaviour of An(III) and Ln(III) with Dowex 50 resin (similar to AG 50W) in the literature [38, 39]. This difference in behaviour has been assigned to stronger inner-sphere chloro complexing of An(III) than Ln(III), possibly due to a greater extent of covalent bonding in the An(III) than Ln (III) [38,40].

Figure 5 also shows the $N_{H2O}$ of Cm(III) and Eu(III) in HCl solutions as a function of the concentration. There is very little difference in the hydration structure of Cm(III) and Eu(III) up to 5 M HCl in agreement with the trends in $K_D$ to ca. 5 M HCl. Again, a marked divergence is seen in the $N_{H2O}$ values above 5 M HCl, i.e. $N_{H2O}(Eu) > N_{H2O}(Cm)$, due presumably to the increased inner-sphere complexation of Cm(III) relative to that of Eu(III) as mentioned in the previous paragraph. Figure 5 also shows the $N_{H2O}$ of Cm(III) and Eu(III) in the AG 50WX8 resin phases as a function of HCl concentration. In contrast with the $K_D$ and $N_{H2O}$ in HCl, almost no difference was found in the hydration structure of sorbed Cm(III) and Eu(III) even in higher HCl concentration and the $N_{H2O}$ of both ions decreased gradually with increasing HCl concentration from ca. 8 at 0.1 M to ca. 5 at 11 M. In the previous study [14], from comparison of the $N_{H2O}$ of sorbed Eu(III) on AG 50WX8 resin with that of Eu(III) complexed with poly(4-styrenesulfonic) acid as linear polyelectrolyte analogue of the resin, it was shown that the cross-linked structure of the resin affected the hydration structure of Eu(III) to some extent. The $N_{H2O}$, ca. 7.5, of the sorbed Eu(III) in pH region [14] is in agreement with those of sorbed Cm(III) and Eu(III) at lower acid concentration within the uncertainty of ±0.5 $H_2O$ molecules. The similarity in the $N_{H2O}$ for Cm(III) and Eu(III) in the resin phases in HCl solutions suggests that the hydrated metal ions are sorbed in the resin and that at higher HCl concentration, which results in lower water content in the resin phase, an average of three sulfonate groups enter the inner hydration sphere of the ions.

It has been well known that the separation of An(III) from Ln(III) can be performed by the cation exchange from concentrated chloride media but not from nitrate and perchlorate media. As mentioned above, the separation behaviour of Cm(III) and Eu(III) in these media was well correlated with the hydration structure of the ions in the solutions, but is almost independent of that in the resin phases. The difference in the $N_{H2O}$ between both ions observed in concentrated HCl was not seen in concentrated $HNO_3$ and $HClO_4$ [15]. The present results on the hydration structure of An(III) and Ln(III) provides spectroscopic evidence that the chloride ion is a softer donor and forms inner-sphere complexes with An(III) at lower chloride concentrations than do the analogous Ln(III) cations, although the nitrate and perchlorate ions are strong bases which favour ionicity.

### *Speciation of Eu(III) species sorbed onto silica and clay minerals*

The direct spectroscopic analysis to clarify the structure of a metal ion sorbed on solid surface is very limited, despite its necessity for the understanding of the metal sorption reaction at the solid-water interface. The determination of $N_{H2O}$ was applied to characterise Eu(III) species sorbed onto silica and clay minerals suspended in the aqueous solution [17].

In our partitioning experiment without adsorbents, Eu(III) began to precipitate above pH 6.5. Speciation calculation for the dissolved Eu(III) species in equilibrium with air was conducted, taking account hydrolysis and carbonate complexation [27,41,42]. The result shows that $EuOH^{2+}$ and $EuCO_3^+$ begin to form at pH 6, while $Eu(CO_3)_2^-$ is most stable above pH 8 at a supporting electrolyte ($NaClO_4$) concentration of $C_s = 0.02$ M. The $N_{H2O}$ of Eu(III) was ca. 9 below pH 6 in the present system showing that Eu(III) was dissolved as hydrated ion (Figure 6). Between pH 6 and 7.5, the $N_{H2O}$ decreased to 6-7 due to formation of dissolved inorganic complex. These results were consistent with the speciation calculation.

Above pH 7.5, in the absence of adsorbents, the $k_{obs}$ increased dramatically and the $N_{H2O}$ exceeds 9. In the presence of Gd(III) or La(III) in the Eu(III) solution at the higher pH region, such abrupt increase of $k_{obs}$ did not occur, because of the large energy gap of Gd(III) and the absence of an f-electron in La(III). These phenomena are explained by energy transfer in the solid among Eu(III) ions which are close enough to interact with neighbouring Eu(III) via phonon-assisted process [43-47]. Such energy transfer effect above pH 7.5 in Eu(III) colloid was also observed when $D_2O$ was used as solvent. In $Eu_2O_3$ powder suspended in water, a similar short lifetime due to Eu(III) colloid was observed, while the lifetime in $Eu_2(CO_3)_3$ was rather long (250 μs). The result implies that the colloid may mainly consist of polynuclear Eu(III) hydroxides.

More than 80% of Eu(III) was removed from aqueous phase in the nonporous silica system above pH 6.5 at $C_s = 0.02$ M. The pH dependence of the $N_{H2O}$ in the system is shown in Figure 6 for various ageing times. The $N_{H2O}$ decreased with increasing pH and reached $N_{H2O}$ below 1 above pH 10 for one day ageing of the sample. This value is extraordinarily low, showing that only very little water remains bound to the Eu(III). This suggests that Eu(III) was incorporated into silica matrix, though the sample was not heated nor dried. The time dependence also supports the speculation that the $N_{H2O}$ decreased with increasing ageing time due to increasing incorporation of Eu(III) into the bulk phase.

The dissolution of Si into the aqueous phase was determined by silicomolybdic acid method [48]. The result was that ca 0.1% of silica was dissolved below pH 8.5 and the percentage increased up to 1% from pH 8.5 to 11 after one day. This may imply that the Eu(III) was taken into the silica in the course of dynamic equilibrium of silica dissolution-precipitation process. The energy transfer does not occur in the silica system estimated from its lifetime in $D_2O$ system, in contrast to much shorter lifetime of Eu(III) hydroxide colloids and $Eu_2O_3$ particles. The absence of energy transfer in the present study indicates that Eu(III) species in the silica seems to be rather dispersed compared with Eu(III) in the polynuclear hydroxides.

In employing silica gel (porous silica) as adsorbents, the result was generally similar. The $N_{H2O}$ decreased with pH and the energy transfer among Eu(III) seemed to be absent (Figure 6). However, the $N_{H2O}$ in the silica gel system was larger than that of nonporous silica system between pH 6 and 10. This might be due to the trapping of Eu(III) complexes in the pores of the silica gel.

More than 80% of the Eu(III) was removed from the aqueous phase in the montmorillonite system above pH 2 at $C_s = 0.02$ M (Figure 7). At $C_s = 0.1$ M, a smaller amount of Eu(III) was adsorbed on montmorillonite below pH 7. This indicates that a nonspecific ion exchange reaction, due to the permanent charge induced by isomorphic substitution, was responsible for the adsorption of Eu(III) by montmorillonite in the lower pH region [49,50]. Above pH 8, the effect of $C_s$ was not observed. The pH dependence of $k_{obs}(H_2O)$ or $k_{obs}(D_2O)$ is shown in Figure 8. The $k_{obs}(H_2O)$ of Eu(III) in the montmorillonite system was about 1.5 times larger than that of hydrated Eu(III), suggesting a quenching effect by montmorillonite. Therefore, the difference in the rate constants ($k_{obs}(H_2O) - k_{obs}(D_2O)$), which corresponds to one hydrated water molecule, is also noted in Figure 8. The $N_{H2O}$ of Eu(III) was ca. 9 in the montmorillonite system below pH 6, showing that the outer-sphere complex of hydrated Eu(III) with montmorillonite was formed. This is consistent with the data for lutetium(III) using XAS analysis [51].

On the contrary, Eu(III) sorbed on kaolinite seems to show somewhat different results. Above pH 5, the $N_{H2O}$ in the kaolinite system is lower than 9 showing that some water molecules were removed from Eu(III) in contact with kaolinite; the $N_{H2O}$ is ca. 9 below pH 4 suggesting the outer sphere complex is important in the pH region. In Figure 7, the independence on $C_s$ of Eu(III) sorbed to kaolinite shows that the sorption is mainly due to the interaction with specific sites of kaolinite compared with non-specific ion exchange.

Above pH 8, the luminescence decay curve indicates contributions of two species. The lifetime of the faster component decreased with an increase in pH similar to that of Eu(III) hydroxide colloids. This component was also seen in the kaolinite system. The $k_{obs}(H_2O)$ and $k_{obs}(D_2O)$ for the slower component, of minor, is plotted against pH in Figure 8 for the montmorillonite and kaolinite systems. The reproducibility was not so good probably due to the fitting by double-exponential function and contributions of various species involved in the slower component. The $N_{H2O}$ of the fraction above pH 7 was 2-3 for montmorillonite system, showing that a large fraction of the boundwater was released by adsorption (Figure 8). Further studies are needed to identify the species, while the present study suggests the sorption of the Eu(III) species as (1) the species do not show the energy transfer and (2) more than six water molecules were removed from the co-ordination sphere of Eu(III). For example, sorption of an inorganic complex or the formation of polynuclear Eu(III) species different from Eu(III) hydroxides can be postulated as responsible for this behaviour.

## Conclusion

The luminescence properties, especially the lifetimes, of Am(III), Cm(III), Nd(III), Sm(III), Eu(III), Tb(III) and Dy(III) in aqueous solution have been investigated systematically to establish the method for determining the $N_{H2O}$ of the species. From measurements of the lifetimes of An(III) and Ln(III) in $D_2O-H_2O$ solutions and of the bromate salts, the calibration relations of the $k_{obs}$ vs. $N_{H2O}$ were developed and applied to the speciation of the f-elements in the polyaminopolycarboxylate complexes, in the cation exchange resin, and on silica and clay minerals in aqueous solutions.

The results confirm that the relationships proposed are valid for determining the $N_{H2O}$ from the measured lifetime and that the determination of the $N_{H2O}$ is a useful method for studying the speciation of the f-elements in various circumstances. The sensitivity of this method makes it possible to compare directly the $N_{H2O}$ between An(III) and Ln(III) by the same experimental procedure. Therefore, the method can be used to study the co-ordination structure and bonding and species involved in separation methods of An(III) and Ln(III) in aqueous solution. In this paper, we have emphasised mainly the importance of the $N_{H2O}$ determination for the speciation study, but the use of lifetime and spectral luminescence data can be expected to play an importance in expanding our knowledge of the co-ordination chemistry of the f-element families.

## REFERENCES

[1]   Carnall, W.T., "Handbook on the Physics and Chemistry of Rare Earths",, Vol. 3, Chap. 24, (Gschneidner, Jr., K.A., Eyring, L., eds.), North-Holland, Amsterdam, pp. 171-208 (1979).

[2]   Carnall, W.T., Beitz, J.V., Crosswhite, H., Rajnak, K., Mann, J.B., "Systematics and the Properties of the Lanthanides", Ch. 9, (Sinha, S.P., ed.), Reidel, Dordrecht, pp. 389-450 (1983).

[3]   Sinha, S.P.,"Systematics and the Properties of the Lanthanides",, Ch. 10, (Sinha, S.P., ed.), Reidel, Dordrecht, pp. 451-500 (1983).

[4]   Horrocks, Jr., W.D., Albin, M., "Progress in Inorganic Chemistry", Vol. 31, (Lippard, S.J., ed.), Wiley and Sons, New York, pp. 1-104 (1984).

[5]   Beitz, J.V., "Handbook on Physics and Chemistry of Rare Earths", Vol. 18, Chap. 120, (Gschneidner, Jr., K.A., Eyring, L., Choppin, G.R., Lander, G.H., eds.), Elsevier, Amsterdam, pp. 159-195 (1994).

[6]   Horrocks, Jr., W.D., Schmidt, G.F., Sudnick, D.R., Kittrell, C., Bernheim, R.A., *J. Am. Chem. Soc.*, 99, 2378 (1977).

[7]   Horrocks, Jr., W.D., Sudnick, D.R., *J. Am. Chem. Soc.*, 101, 334 (1979).

[8]   Horrocks, Jr., W.D., Sudnick, D.R., *Acc. Chem. Res.*, 14, 384 (1981).

[9]    Kimura, T., Choppin, G.R., *J. Alloys Comp.*, 213/214, 313 (1994).

[10]   Kimura, T., Choppin, G.R., Kato, Y., Yoshida, Z., *Radiochim. Acta*, 72, 61 (1996).

[11]   Kimura, T., Kato, Y., *J. Alloys Comp.*, 271-273, 867 (1998).

[12]   Kimura, T., Kato, Y., *J. Alloys Comp.*, 225, 284 (1995).

[13]   Kimura, T., Kato, Y., *J. Alloys Comp.*, 275-277, 806 (1998).

[14]   Takahashi, Y., Kimura, T., Kato, Y., Minai, Y., Tominaga, T., *Chem. Commun.*, 223 (1997).

[15]   Kimura, T., Kato, Y., Takeishi, H., Choppin, G.R., *J. Alloys Comp.*, 271-273, 719 (1998).

[16]   Takahashi, Y., Kimura, T., Kato, Y., Minai, Y., Makide, Y., Tominaga, T., *J. Radioanal. Nucl. Chem.*, 239, 335 (1999).

[17]   Takahashi, Y., Kimura, T., Kato, Y., Minai, Y., Tominaga, T., *Radiochim. Acta*, 82, 227 (1998).

[18]   Kimura, T., Kato, Y., *J. Alloys Comp.*, 278, 92 (1998).

[19]   Conway, J.G., *J. Chem. Phys.*, 40, 2504 (1964).

[20]   Carnall, W.T., Rajnak, K., *J. Chem. Phys.*, 63, 3510 (1975).

[21]   Carnall, W.T., Fields, P.R., Rajnak, K., *J. Chem. Phys.*, 49, 4424 (1968).

[22]   Carnall, W.T., Fields, P.R., Rajnak, K., *J. Chem. Phys.*, 49, 4447 (1968).

[23]   Carnall, W.T., Fields, P.R., Rajnak, K., *J. Chem. Phys.*, 49, 4450 (1968).

[24]   Stein, G., Würzberg, E., *J. Chem. Phys.*, 62, 208 (1975).

[25]   Riseberg, L.A., Moos, H.W., *Phys. Rev.*, 174, 429 (1968).

[26]   Heller, A., *J. Am. Chem. Soc.*, 88, 2058, (1966).

[27]   Rizkalla, E.N., Choppin, G.R., "Handbook on Physics and Chemistry of Rare Earths", Vol. 15, Chap. 103, (Gschneidner, Jr., K.A., Eyring, L. eds.), Elsevier, Amsterdam, pp. 393-442 (1991).

[28]   Barthelemy, P.P., Choppin, G.R., *Inorg. Chem.*, 28, 3354 (1989).

[29]   Habenschuss, A., Spedding, F.H., *J. Chem. Phys.*, 73, 442 (1980).

[30]   Brittain, H.G., Jasinski, J.P., *J. Coord. Chem.*, 18, 279 (1988).

[31]   Brittain, H.G., Choppin, G.R., Barthelemy, P.P., *J. Coord. Chem.*, 26, 143 (1992).

[32]   Chang, C.A., Brittain, H.G., Telser, J., Tweedle, M.F., *Inorg. Chem.*, 29, 4468 (1990).

[33]   Yusov, A.B., *J. Radioanal. Nucl. Chem.*, 143, 287 (1990).

[34]   Graffeo, A.J., Bear, J.L., *J. Inorg. Nucl. Chem.*, 30, 1577 (1968).

[35]   Geier, G., *Ber. Bunsenges. Physik. Chem.*, 69, 617 (1965).

[36]   Geier, G., *Helv. Chim. Acta*, 51, 94 (1968).

[37]   Choppin, G.R., *J. Alloys Comp.*, 192, 256 (1993).

[38]   Diamond, R.M., Street, K., Seaborg, G.T., *J. Am. Chem. Soc.*, 76, 1461 (1954).

[39]   Choppin, G.R., Chatham-Strode, A., *J. Inorg. Nucl. Chem.*, 15, 377 (1960).

[40]   Street, K., Seaborg, G.T., *J. Am. Chem. Soc.*, 72, 2790 (1950).

[41]   Maes, A., Brabandere, J.D., Cremers, A.: *Radiochim. Acta*, 44/45, 51 (1988).

[42]   Lee, J.H., Byrne, R.H., *Geochim. Cosmochim. Acta*, 57, 295 (1993).

[43]   Lochhead, M.J., Bra, K.L., *Chem. Mater.*, 7, 572 (1995).

[44]   Yen, M.W., Selzer, P.M., *"Laser Spectroscopy of Solids"*, Springer-Verlag, Berlin (1981).

[45]   Uitert, L.G.V., *J. Electrochem. Soc.*, 107, 803 (1960).

[46]   Buijs, M., Meyerink, A., Blasse, G., *J. Lumin.*, 37, 9 (1987).

[47]   Fan, X., Wang, M., Xiong, G., *J. Mat. Sci. Lett.*, 12, 1552 (1993).

[48]   Nisimura, M., Tsunogai, S., Noriki, S., *"Kaiyo-kagaku"*, Sangyo-Tosho, Tokyo, p. 258, (1983), [in Japanese].

[49]   Nagasaki, S., Tanaka, S., Suzuki, A., *J. Nucl. Mat.*, 244, 29 (1997).

[50]   Papelis, C., Hayes, K.F., *Colloids Surf.*, A 107, 89 (1996).

[51]   Muñoz-Páez, A., Alba, M.D., Castro, M.A., Alvero, R., Trillo, J.M., *J. Phys. Chem.*, 98, 9850 (1994).

# Table 1. Luminescence properties of hydrated actinide (III) and lanthanide(III) ions in $H_2O$ and $D_2O$

| Ion | $nf^a$ | $\lambda_{ex}$ (nm) | $\lambda_{em}$ (nm) | $\Delta E$ (cm$^{-1}$) [a] | $\tau_{H2O}$ (μs) [b] | $\tau_{D2O}$ (μs) [b] | $\tau_{D2O}/\tau_{H2O}$ |
|---|---|---|---|---|---|---|---|
| $Cm^{3+}$ | $5f^7$ | 397 ($^8S_{7/2} \to {}^6I_{17/2,11/2}$) | 594 ($^6D_{7/2} \to {}^8S_{7/2}$) | 16 943 | 65 | 1 200 | 18 |
| $Tb^{3+}$ | $4f^8$ | 355 ($^7F_6 \to {}^5D_2$) | 543 ($^5D_4 \to {}^7F_5$) | 14 804 | 430 | 4 100 | 9.5 |
| $Eu^{3+}$ | $4f^6$ | 394 ($^7F_0 \to {}^5L_6$) | 592 ($^5D_0 \to {}^7F_1$) | 12 255 | 110 | 3 900 | 35 |
| $Sm^{3+}$ | $4f^5$ | 401 ($^6H_{5/2} \to {}^6P_{3/2}$) | 594 ($^4G_{5/2} \to {}^6H_{7/2}$) | 7 438 | 2.7 | 64 | 24 |
| $Dy^{3+}$ | $4f^9$ | 355 ($^6H_{15/2} \to {}^4I_{11/2}$) | 479 ($^4F_{9/2} \to {}^6H_{15/2}$) | 7 374 | 2.5 | 43 | 17 |
| $Nd^{3+}$ | $4f^3$ | 593 ($^4I_{9/2} \to {}^4G_{5/2},{}^2G_{7/2}$) | 890 ($^4F_{3/2} \to {}^4I_{9/2}$) | 5 473 | 0.032 | 0.17 | 5.3 |
| $Am^{3+}$ | $5f^6$ | 504 ($^7F_0 \to {}^5L_6$) | 691 ($^5D_1 \to {}^7F_1$) | 4 942 | 0.025 | 0.16 | 6.4 |

a) Difference in energy between emitting "free-ion" state and next lower "free-ion" state.
b) The experimental errors of the lifetime values are about 2.5%

# Table 2. Average numbers of co-ordinated water molecules $N_{H2O}$, ligand co-ordination numbers $CN_L$ and total co-ordination numbers $CN_T$ for actinide(III) and lanthanide(III) polyaminopolycarboxylate complexes

| Ligand ($CN_L$ [37]) | Am(III) $N_{H2O}$ | $CN_T$ | Cm(III) $N_{H2O}$ | $CN_T$ | Nd(III) $N_{H2O}$ | $CN_T$ | Sm(III) $N_{H2O}$ | $CN_T$ | Eu(III) $N_{H2O}$ | $CN_T$ | Tb(III) $N_{H2O}$ | $CN_T$ | Dy(III) $N_{H2O}$ | $CN_T$ |
|---|---|---|---|---|---|---|---|---|---|---|---|---|---|---|
| NTA (4) | 6.5 | 10.5 | 6.3 | 10.3 | 5.6 | 9. | 5.8 | 9.8 | 4.7 | 8.7 | 4.2 | 8.2 | 4.7 | 8.7 |
| HEDTA (5.5) | 5.1 | 10.6 | 4.3 | 9.8 | 4.5 | 10.0 | 4.2 | 9.7 | 3.4 | 8.9 | 2.9 | 8.4 | 3.4 | 8.9 |
| ENDADP (6) | 4.9 | 10.9 | – | – | 3.9 | 9.9 | 5.0 | 11.0 | 3.3 | 9.30 | 3.0 | 9.0 | 3.2 | 9.2 |
| EDTA (6) | 4.8 | 10.8 | 3.9 | 9.9 | 4.0 | 10.0 | 4.2 | 10.2 | 2.8 | 8.8 | 2.4 | 8.4 | 3.1 | 9.1 |
| PDTA (6) | 4.8 | 10.8 | – | – | 3.9 | 9.9 | 4.3 | 10.3 | 2.8 | 8.8 | 2.3 | 8.3 | 3.1 | 9.1 |
| CDTA (6) | 7.3 | 13.3* | 3.9 | 9.9 | 4.5 | 10.5 | 5.0 | 11.0 | 2.6 | 8.6 | 2.1 | 8.1 | 3.3 | 9.3 |
| DTPA (7.5) | 3.1 | 10.6 | 1.9 | 9.4 | 2.6 | 10.1 | 2.4 | 9.9 | 1.1 | 8.6 | 1.1 | 8.6 | 1.5 | 9.0 |
| EGTA (8) | 3.0 | 11.0 | – | – | 2.0 | 10.0 | 2.3 | 10.3 | 1.2 | 9.2 | 1.2 | 9.2 | 1.5 | 9.5 |
| TTHA (8.5) | 1.6 | 10.1 | 0.9 | 9.4 | 0.7 | 9.2 | 1.1 | 9.6 | 0.4 | 8.9 | 0.6 | 9.1 | 0.5 | 9.0 |
| av. $CN_T$ | | 10.7 | | 9.8 | | 9.9 | | 10.2 | | 8.9 | | 8.6 | | 9.1 |

The $CN_T$ were calculated as the sum of the $CN_L$ and the $N_{H2O}$ obtained at pH 4-5.
The error of the $N_{H2O}$ values was estimated to be within ±0.4.
* This value was omitted for the calculation of the average $CN_T$.

**Figure 1. Luminescence decay constants $k_{obs}$ for actinide(III) and lanthanide(III) ions as a function of molar fraction of $H_2O$ in $D_2O$-$H_2O$ solutions: 0.01 M H(D)ClO$_4$; [Am] = $1.2 \times 10^{-5}$ M; [Cm] = $6.3 \times 10^{-7}$ M; [Nd] = [Sm] = [Eu]= [Tb] = [Dy] = $1.0 \times 10^{-2}$ M**

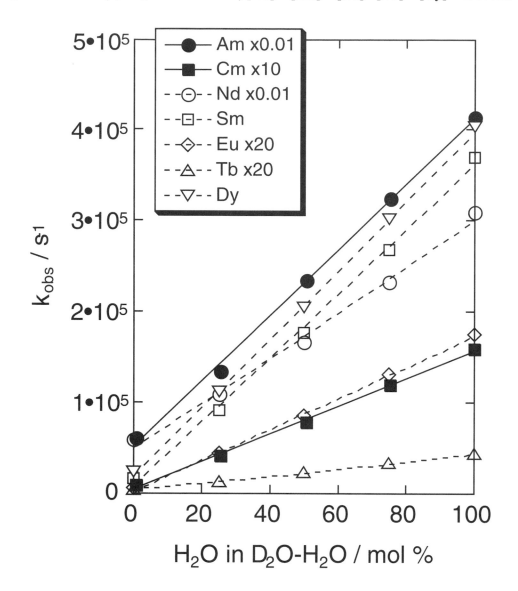

**Figure 2.** Non radiative decay rate $k_{H2O}$ due to $H_2O$ for actinide(III) and lanthanide(III) ions in $H_2O$ solution as a function of energy gap $\Delta E$. Data points were fitted to an energy gap law expression [25], $k_{H2O} = C \exp[\alpha \Delta E]$, where $C$ and $\alpha$ are characteristic constants in a particular medium.

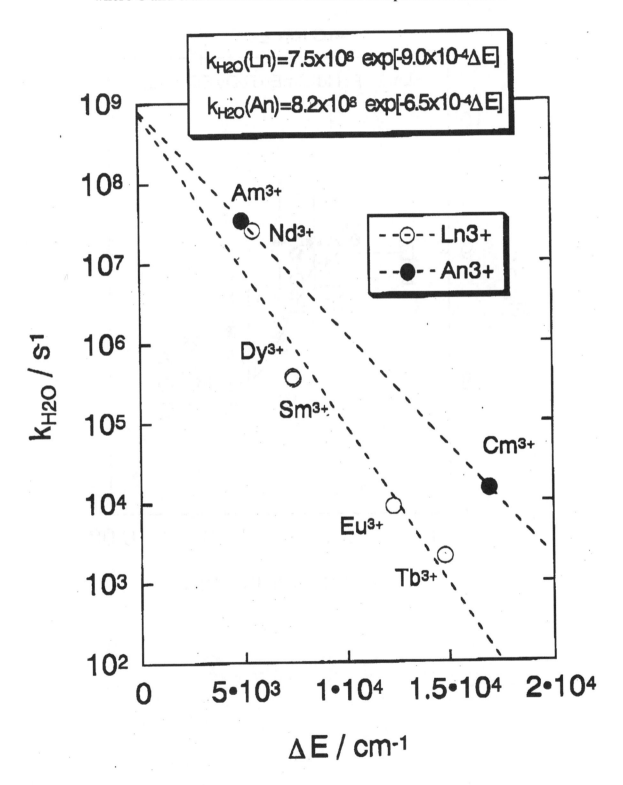

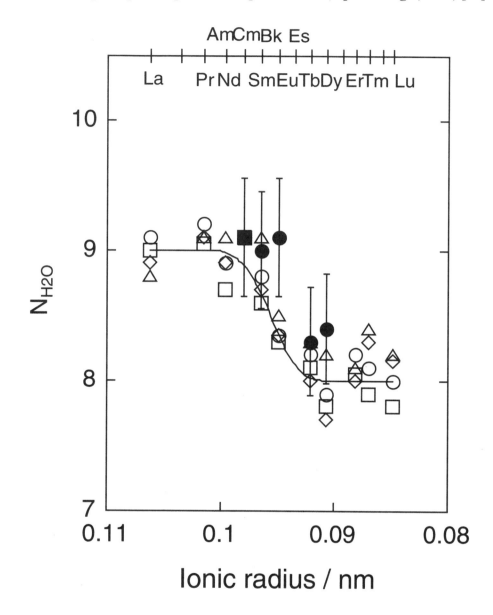

Figure 3. Plot of the inner-sphere hydration number $N_{H2O}$ as a function of ionic radius of actinide(III) and lanthanide(III) ions: luminescence lifetime (closed square, $Cm^{3+}$; closed circle, $Ln^{3+}$); X-ray diffraction using different constants (open square, open circle, open diamond, open triangle; $Ln^{3+}$) [29]

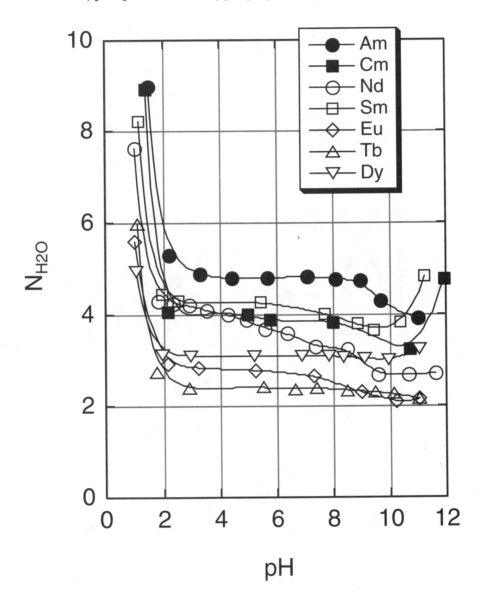

**Figure 4.** Inner-sphere hydration numbers $N_{H2O}$ for actinide(III) and lanthanide(III) complexes with EDTA as a function of pH: 0.1 M $NaClO_4$; $[Am] = 1.2 \times 10^{-5}$ M; $[Cm] = 6.3 \times 10^{-7}$ M; $[Nd] = [Sm] = [Eu] = [Tb] = [Dy] = 1.0 \times 10^{-2}$ M

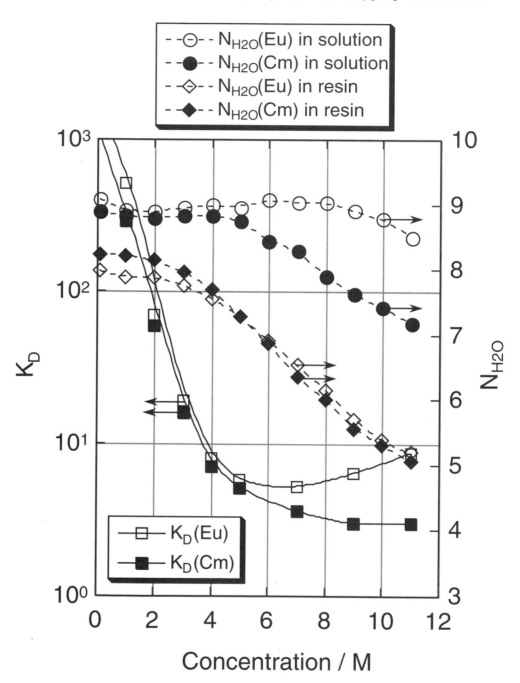

**Figure 5.** Distribution coefficients $K_D$ and inner-sphere hydration numbers $N_{H2O}$ of Cm(III) and Eu(III) in the AG 50WX8 resin-hydrochloric acid system as a function of HCl concentration: $[Cm] = 6.3 \times 10^{-7}$ M; $[Eu] = 1.0 \times 10^{-3}$ M

**Figure 6.** Inner-sphere hydration numbers $N_{H2O}$ of Eu(III) in the silica system for various aging time as a function of pH: 0.02 M NaClO$_4$; [Eu] = $2.0 \times 10^{-4}$ M; unporous silica, 8.0 g/dm$^3$; silica gel, 20 g/dm$^3$. Aqueous phase was separated by filtration (450 nm).

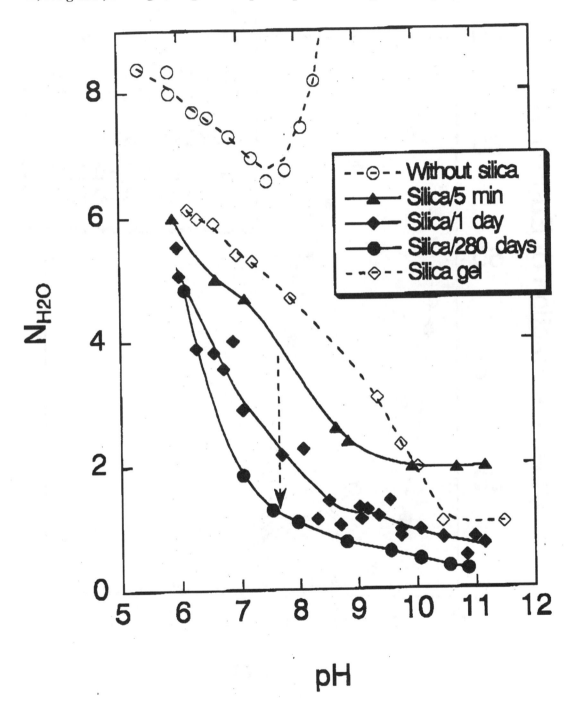

**Figure 7.** Dissolved fraction of Eu(III) in the montmorillonite and kaolinite systems at $C_s = 0.02$ and $0.1$ M as a function of pH: Montmorillonite, $4.0$ g/dm$^3$; Kaolinite, $20$ g/dm$^3$

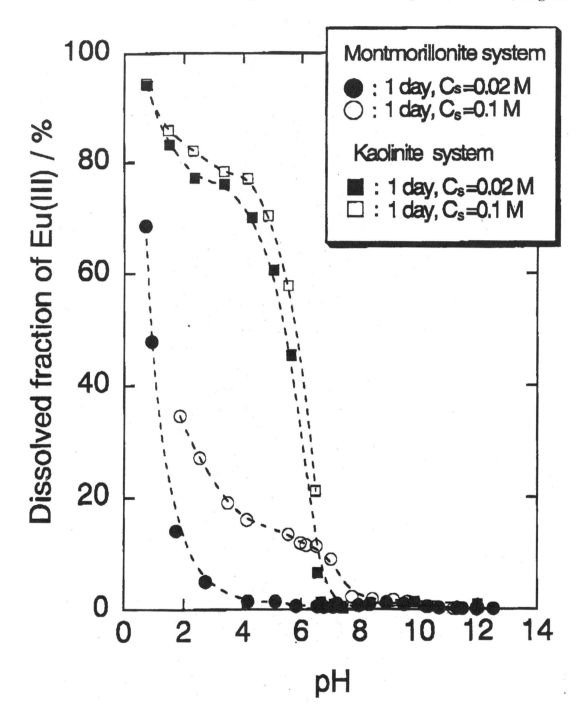

**Figure 8.** Luminescence decay constants $k_{obs}$ of Eu(III) in the montmorillonite and kaolinite systems as a function of pH: 0.02 M NaClO$_4$; [Eu(III)] = $2.0 \times 10^{-4}$ M; Montmorillonite, 4.0 g/dm$^3$; Kaolinite, 20 g/dm$^3$. The difference of rate constants ($k_{obs}(H_2O)$ - $k_{obs}(D_2O)$) corresponding to one hydrated water molecule was also noted as a measure. Above pH 8, $k_{obs}(H_2O)$ and $k_{obs}(D_2O)$ for the slower lifetime component were plotted, while two components having different lifetimes were detected in the pH region.

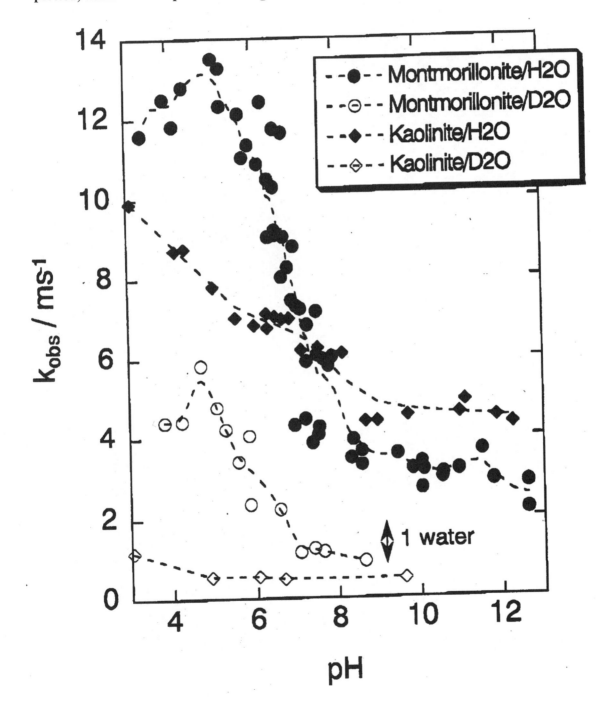

# SESSION C

*Methods for Empirical Formula and Molecular Structure Determination*

# X-RAY ABSORPTION FINE STRUCTURE SPECTROSCOPY FOR RADIONUCLIDE SPECIATION STUDIES

**Melissa A. Denecke, J. Rothe, C. Pohlmann, H. Geckeis**
Forschungszentrum Karlsruhe, Institut für Nukleare Entsorgungstechnik
PO Box 3640, 76021 Karlsruhe, Germany
E-mail: denecke@ine.fzk.de

**D. Degering**
Technische Universität Bergakademie Freiberg, Institut für Angewandte Physik
Bernhard-von-Cotta-Str. 4, 09596 Freiberg, Germany

**J. Janssen**
Physikalisches Institut der Universität Bonn
Nussallee 12, 53115 Bonn, Germany

## Abstract

Studies combining parallel time differential perturbed angular correlation, extended X-ray absorption fine structure and grazing incidence X-ray absorption fine structure (TDPAC, EXAFS and GIXAFS, respectively) experiments on Hf(IV) sorbed onto silica as a function of pH have been performed and results from these studies are presented. From these investigations, the sorbed Hf(IV) species is characterised. The combination of these techniques helps to overcome the limited sensitivity of the standard EXAFS method.

## Introduction

The application of X-ray absorption fine structure (XAFS) spectroscopy for the investigation of radionuclide speciation is becoming increasingly widespread [1]. The reasons for this are two-fold. First, XAFS is elemental specific and is often applicable without prior sample preparation, separation or dissolution steps, thereby minimising possible speciation changes. This renders metal cation speciation studies in complicated matrices possible [2]. The second reason for the increase in XAFS radionuclide speciation studies is the fact that the number of experimental facilities where XAFS experiments on radioactive samples are possible is expanding. These include the installation of experimental stations specifically dedicated to such studies such as ROBL at the ESRF and the JAERI beamline at SPring8.

Speciation information concerning valence and co-ordination geometry, as well as metrical parameters for nearest neighbours surrounding a radionuclide of interest are accessible from the analysis of XAFS data [3]. Figure 1 depicts a representative X-ray absorption spectrum, where the product of the absorption coefficient and sample thickness, $\mu d = \log(I_0/I)$, is measured as a function of incident photon energy. X-rays are absorbed by the sample, whereby electrons are promoted from core states to outlying unoccupied states. At the core state threshold energy, $E_0$, a sharp rise in absorption or edge is observed. The principle quantum number of the core electron is used to denote the absorption edge, e.g. K, L and M edges. The classical division of XAFS spectra into two energy regions is also shown in Figure 1: the lower energetic region usually referred to as the X-ray absorption near-edge structure (XANES) region, and the extended X-ray absorption fine structure (EXAFS) at higher energies. Determination of the radionuclide oxidation state in a sample is possible from the energy position of the absorption edge in XANES spectra [4]. Theoretically, XANES spectra also contain information concerning the radionuclide co-ordination geometry. The EXAFS oscillations represent an interferogram of the atomic arrangement surrounding the absorbing atom (see Figure 2). The outgoing photoelectron wave created by ionisation of a core electron of the absorbing atom interferes with a wave that is backscattered on surrounding atoms. Analysis of the interference pattern amplitude and frequency yields information concerning the type and number of backscattering atoms (N) and the bond distances (R), respectively.

The greatest advantage of XANES/EXAFS for speciation studies is that these techniques are element specific and applicable for investigating an element embedded in a complex matrix. This makes *in situ* studies possible, without separation/extraction steps that may inadvertently cause alterations in the speciation. The concentration sensitivity of XAFS, however, is limited. Transmission experiments on L2,3 edges of the actinide elements require solution concentrations on the order of 10-50 mM, depending on path length. Such high concentrations can exceed the solubility product of the species of interest, as in the case of aqueous solutions of tetravalent actinides, for example. Sensitivity can be enhanced by one to two orders of magnitude for dilute samples, for which the XAFS signal is only a small fraction of the total absorption intensity, by using an energy dispersive detector to discriminate background radiation from fluorescent radiation emitted from the sample upon relaxation processes filling the absorber atom core hole.

Surface sensitivity, and with it an inherent increase in concentration sensitivity, is possible by performing XAFS experiments at a grazing angle of incidence (GIXAFS) [5]. A schematic representation of the GIXAS set-up is shown in Figure 3. Monochromatic synchrotron radiation, SR, is defined to a small horizontal width (<300 µm) by two horizontal slits. The sample chamber is on a high precision goniometer cradle, which allows vertical and angular (within 1 mdeg) positioning. The sample surface is oriented with respect to the impinging beam at an angle slightly below the critical angle for total external reflection. This gives rise to a penetration depth limited typically to a few nm. The intensity of the incoming monochromatic beam is measured with a first ionisation chamber (IC1).

The reflected beam is monitored with IC2; the remaining fractions of the beam are cut off at the lower jaw of a third horizontal slit. Fluorescent detection is also possible with a solid state detector placed vertically above the sample. Investigations of monolayer and sub-monolayers of radionuclides on surface substrates are possible with this technique. Even with the increased concentration sensitivity with fluorescent detection and the GIXAFS technique, the concentration range for XAFS investigations remains too high to be relevant to any far field scenario speciation studies. The combination of XAFS investigations with complementary techniques having higher concentration sensitivities is essential to improve our understanding of radionuclide speciation, especially in geochemical media.

Time differential perturbed angular correlation (TDPAC) spectroscopy on [181]Hf as a Pu(IV) homologue may be a promising complementary technique to XAFS studies. No known actinide nuclides exist which can serve as probes in PAC investigations. However, due to their having the same oxidation state, similar ionic radii and comparable complexing stability constants, Hf(IV) is frequently used as chemical homologue for the tetravalent actinides [6]. The TDPAC technique is a nuclear probe method and detects the hyperfine interaction between the nuclear quadrupole moment of an excited state and the electric field gradient (EFG) at the site of the probe nuclei [7]. Analysis of TDPAC spectra yields parameters defining the quadrupole interaction, which provide information on the components of the EFG tensor and its symmetry and, thereby, images the chemical environment. Additionally, the fraction of nuclei occupying different sites, e.g. on a solution/mineral interface, can also be determined.

In the present study, initial studies combining parallel TDPAC, XAFS and GIXAFS experiments on Hf(IV) sorbed onto silica as a function of pH have been performed and results from these studies will be presented. The goal of these studies is to characterise the sorbed Hf(IV) species.

## Materials and methods

TDPAC and EXAFS sorption samples are prepared by reacting a suspension of 250 mg purified non-porous, amorphous silica (Aerosil 380, Degussa) with 0.5 ml of ~30 mM Hf stock solution (~5 mL reaction volume, I = 0.1 M) and adjusting the pH to values in the range 1-7 with 1-1.5 M aqueous NaOH solution. After 24 hours reaction time, the pH of the reaction mixtures is measured again. The reaction solution pH values drifted slightly after 24 h. The pH of reaction solutions for EXAFS samples, taken as the mean between initial and final pH, are 0.9, 1.9, 2.8, 4.2, 5.2, 6.6, 7.3. Samples are isolated by centrifugation, the supernatant liquid removed and the sample investigated by TDPAC or EXAFS, in the form of a wet-paste. Separate batches of sorption samples were prepared using a [181]Hf stock solution (see below) for TDPAC samples and an inactive Hf(IV) stock solution for EXAFS samples.

[181]Hf is produced by activating $HfOCl_2*8H_2O$ 10 h with a neutron flux density ~ $10^{13}$ n*$cm^{-2}$*$s^{-1}$ in the TRIGA-HD II reactor, Deutsches Krebsforschungszentrum, Heidelberg. A 29.5 mM, 4.1 MBq/ml [181]Hf stock solution is prepared by dissolving the irradiated material in 1 M $HClO_4$ and further dilution to 0.1 M $HClO_4$. For the TDPAC measurements, the 133-482 keV $\gamma$-$\gamma$ cascade of [181]Ta emitted after the $\beta$-decay of [181]Hf to [181]Ta is observed. Details of the method and experimental set-up are described elsewhere [8].

Samples with Hf(IV) sorbed onto the basal (0001) plane of a mica surface for the GIXAS experiments are obtained by immersing thin sheets of muscovite mica (75 mm × 25 mm, thickness 0.2 mm) in ~ 3mM Hf(IV) solution, that was adjusted to pH 1, 4 and 7 using 1-1.5 M aqueous NaOH.

The mica was left in the solution for 97, 86 and 50 hours, respectively. Prior to mounting into the sample chamber, the sheets are rinsed with Milli-Q water and adherent solution blown off with a $N_2$ stream.

## *EXAFS measurements*

Hafnium L3 edge EXAFS measurements are performed at the Hamburger Synchrotron strahlungslabor, HASYLAB, beamline A1, using either a Si(111) double-crystal monochromator detuned ~50% of the maximum incident flux. Two to four transmission spectra are recorded at room temperature and averaged. All spectra are calibrated by defining the Hf L3 absorption maximum as 9 565 eV. This energy is also arbitrarily defined as $E_0$. EXAFS oscillations are isolated using the program autobk of the UWXAS package for background subtraction [9]. Theoretical fits to raw data are performed using the WinXAS program [10] using theoretical backscattering amplitude and phase-shift functions calculated from FEFF7 [11] using a 46 atom cluster having atomic positions from $HfSiO_4$ [12].

## *GIXAFS measurements*

GIXAFS measurements are performed at the RefleXAFS endstation at beamline BN2 of the ELSA storage ring, Bonn University Institute of Physics, using a Ge(422) double-crystal monochromator. A single diode Si(Li) solid state detector, mounted above the sample, is used for the detection of the Hf $L_\alpha$ fluorescence yield as a function of photon energy. The sample surface is oriented at an angle of 150 mdeg, relative to the incident beam. Ten spectra are recorded and averaged. Further data analysis is done in a manner analogous to the Hf L3 edge EXFAS data on amorphous silica.

## Results and discussion

### *TDPAC results*

An exemplary room temperature TDPAC spectrum and its corresponding Fourier transform for the sorption sample prepared at pH = 5 is shown in Figure 4. TDPAC parameters describing the hyperfine interaction are obtained from fits to the spectra. The best fit to pH 1-2 sample spectra is obtained using a model with two binding sites, sites 1 and 2, having similar quadrupole coupling constants, $v_{Q1} \approx 1\ 300$ MHz and $v_{Q2} \approx 1\ 360$ MHz. This means that the z-component $V_{zz}$ of the EFG tensor for both sites is similar. The difference between sites 1 and 2 lies in the asymmetry parameters obtained from the fits, $\eta_1 \approx 0$ and $\eta_2 \approx 0.4$. The EFG of site 1 is axial symmetric, that of site 2 is axial asymmetric. In the range above pH 3, a third binding state (site 3, $v_{Q3} \approx 350$ MHz) appears. This site is rather unspecific because of its broad frequency distribution of about 50%. The relative fraction of sites 1+2 and site 3, normalised to 100%, is shown in Figure 5. The relative amount of [181]Hf associated with site 3 increases with increasing pH. At pH 5, one half of [181]Hf are at sites 3; at pH 7, sites 1+2 disappear and only site 3 remains.

TDPAC parameters measured for the Hf(IV) sorption samples are clearly different than those obtained for polynuclear Hf(IV) species formed in aqueous solutions (c.f. [13] and the references cited therein), so that the signals from sorption samples must be from sorbed Hf(IV) species on the silica surface.

## EXAFS results

Hf L3 edge EXAFS spectra and their corresponding Fourier transforms (FT's) for the Hf(IV) sorbed onto amorphous silica at varying pH are shown in Figure 6. There are pH dependent trends in the Hf(IV)-silica sample EXAFS, which coincide to those in the TDPAC studies. The EXAFS results indicate a change in the Hf-O co-ordination as a function of pH. At least two nearest neighbour Hf-O distances are found at low pH. With increasing pH, another species with a slightly longer average Hf-O distance appears and dominates the spectrum of the sample prepared at pH 7.

No characteristic features of a Hf-Hf interaction are present in the spectra for the Hf(IV)-silica samples, excluding the possibility of the formation of polynuclear surface species. The EXAFS for Hf(IV) species sorbed onto the amorphous silica surface at the pH values studied is dissimilar in frequency and beat pattern to that measured for a Hf(IV) solution (~30 mM pH = 1) and precipitated Hf(IV) hydrolysis product (spectra not shown). These conclusions corroborate the TDPAC finding that the Hf(IV) species in the sorption samples can be assigned to a sorbed Hf(IV) species on the silica surface. Furthermore, the sorbed Hf(IV) species is dissimilar to the Hf(IV) hydrolysis product and the Hf(IV) solution species.

The EXAFS spectra in Figure 6 are fit to a model with a single co-ordination shell of oxygen atoms as backscatterers. The metrical parameters from the fits are listed in Table 1. Although FT spectra indicate that low pH samples likely have more than one oxygen nearest neighbour shell, fit results provide an average of Hf-O distances, useful for interpretation purposes. Indeed, the large EXAFS Debye-Waller factors ($\sigma^2$) obtained for the low pH sample result from there being more than the one Hf-O distance used in the model. For theoretical fits to the Hf(IV)-silica sorption sample spectra, an average $R_{Hf-O}$ for all pH values of 2.03 Å is obtained. These short bond lengths are comparable to the Hf-O distance in crystalline $HfSiO_4$ associated with Hf atoms bound to corners of Si-tetrahedra (2.05 Å); the bidentate, edge sharing Hf-O distances in $HfSiO_4$ are considerably longer (2.44 Å) [12]. A short, average $R_{Hf-O}$ of 2.03 Å for the sorption samples suggests that the primary interaction of hafnium cations with the silica surface occurs *via* links to Si-tetrahedra corners, i.e. Hf(IV) bound to surface silanol groups. The observed differences in the EFG surrounding sorbed Hf(IV) cations as a function of pH in TDPAC spectra must result from changes in overall species geometry but not in a change in the primary Hf-silica surface interaction. If a change in the primary interaction from a monodentate to a bidentate surface bound Hf(IV) species occurs, i.e. change from Hf(IV) linked to tetrahedra corners (silanol) to tetrahedra edges (geminal), then a significant elongation of the Hf-O bond is expected. This is not the case.

## GIXAFS results

Hafnium L3 edge GIXAS spectra of the Hf(IV)-mica sorption samples prepared at varying pH, their corresponding Fourier transforms, and Fourier-filtered Hf-O shell oscillations are shown in Figure 7. A change in the GIXAFS spectra as a function of pH is evident, which is similar to that observed in the bulk EXAFS measurements of Hf(IV) sorbed onto amorphous silica. At pH 1, there is more than one Hf-O distance. This is evident in the Fourier transform but especially evident in the Fourier-filtered spectrum, where interference of oscillations from two Hf-O distances lead to a beat in the oscillatory pattern at $k \approx 7$ Å$^{-1}$. In contrast, there is only one oscillation evident in the Fourier-filtered spectrum for the sample prepared at pH 7. As in the bulk study, the Hf-O distance is shifted to somewhat longer bond length, compared to samples prepared at lower pH. The spectrum of the sample prepared at pH 4 is intermediate to the spectra from the other two samples.

That the GIXAS spectra of the Hf(IV) species sorbed on mica show similar pH trends and similar spectra as does the EXAFS of the Hf(IV) sorbed species on amorphous silica, indicates that the interaction of Hf(IV) with the surface in both systems is similar. The (0001) mica surface is composed of unreactive siloxane rings. The most reactive mica sites are edge sites, where the siloxane ring network ends. At an edge site, an oxygen atom forms a silanol group instead of acting as a bridge to a neighbouring silicon atom in the siloxane network structure. Edge sites are not exclusively located on the periphery of a mica sheet, but are also found at steps on the cleaved mica surface or as imperfections on the basal plane. It is therefore likely that similarities between the EXAFS and GIXAFS results presented here reflect the presence of Hf(IV) linked to corners of Si-tetrahedra, i.e. bound to silanol groups, on both the amorphous silica and the mica surfaces.

## Conclusion

EXAFS and TDPAC spectroscopic characterisation of the Hf(IV) species formed from reacting 5 mL ~ 3 mM Hf(IV) solution with 250 mg amorphous silica, at varying pH, for 24 h, show that no polynuclear Hf(IV) surface species are formed in a pH range 1-7. This is similar to results reported for the actinides U(VI) [14] and Th(IV) [15] sorbed onto amorphous silica, where no metal-metal interaction was observed, even when polynuclear species were in the reaction solution or for samples with high surface loadings. These findings are, nevertheless, surprising. In contrast, the formation of dimer and polynuclear species for Cu(II) and Cr(III) transition metal sorption onto silica have been reported [16 17]. According to Hf-solubility data published by Linke [18] (log $K_{sp}$ = -25), the solubility limits are exceeded at pH > 3, under the experimental conditions for sample preparation used here. Therefore, one would expect to observe polynucleation or precipitation on the silica surface [19]. TDPAC and especially EXAFS results, however, point towards a disperse Hf(IV) distribution over the silica surface. Our findings indicate that modelling surface sorption solely based on thermodynamic solubility data and on solution species present during sorption might be erroneous. The necessity for spectroscopic speciation in each system under investigation becomes clear.

All the samples of Hf(IV) sorbed onto amorphous silica exhibit a short Hf-O average bond length, which indicates that the primary interaction of Hf(IV) cations with the silica surface occurs via silanol groups in the entire pH range studied. However, TDPAC results show that there are three silica surface binding sites for Hf(IV) with different EFG's, which is indicative of differences in overall site geometries between the sites. However, the existence of different site geometries is apparently not associated with any change in the primary Hf-silica surface interaction, which is a monodentate Hf(IV) species, corner-linked to Si-tetrahedra. A variation in the Hf-O-Si bond angle in the samples could explain the observed differences in EFG's.

The similarity between GIXAFS and EXAFS spectra also supports the conclusion that the primary interaction of Hf(IV) on the silica surface is via links to Si-tetrahedra corners. Metal cation binding to silanol edge sites associated with steps or kinks on the mica surface has been previously reported for the sorption of Cu(II) [20]. Indeed, the silanol edge sites are the reactive surface sites on mica and can be expected to bind Hf(IV). The GIXAFS experiment probes only the surface, so that the silanol groups involved in binding Hf(IV) must result from steps on the (0001) cleavage plane.

The success of these investigations has inspired further research in this direction. By using carrier-free [181]Hf, TDPAC investigations of samples containing hafnium in trace amounts (~$10^{-8}$ M) will be performed in the near future. It is of particular interest to observe if the EFG varies with decreasing amounts of Hf(IV) on silica surface. Combining the concentration sensitivity of the TDPAC method with metrical parameters and speciation information from XAFS can help judge the applicability of extrapolating structural information obtained from higher concentrations to trace levels. Combination with GIXAS experiments on other model systems should also help assist interpretation of spectra. In separate experiments, high resolution Hf L3 edge XANES measurements of similar Hf sorption samples are planned. In preliminary studies, a splitting of the main XANES absorption feature, the so-called white line, in spectra of Hf(IV) species sorbed onto amorphous silica is observed. This splitting may indicate a crystal field splitting of final-state 5d orbitals [21]. It will hopefully be possible to correlate XANES results to EFG symmetry information from TDPAC measurements. Investigations of the XANES region can be used to increase XAFS sensitivity, because the absorption coefficient shows large changes in this region. This allows recording usable XANES spectra at concentrations where the signal to noise ratio of the EXAFS spectra may be too poor for analysis [22].

## Acknowledgements

For beam time allotment and experimental assistance, HASYLAB, Dr. Kathy Dardenne (INE), Professor Dr. J. Hormes (Physikalisches Institut der Universitaet Bonn), are gratefully acknowledged.

## REFERENCES

[1]    See, for example, proceedings of the XAFS IX Conference in Grenoble, France, appearing in J. Physique IV 7, (1997) and the XAFS X Conference in Chicago, IL, USA, J. Synchrotron Rad. 6 (1999).

[2]    Conradson, S.D., Appl. Spectrosc. 52 (1998) 252A.

[3]    Koningsberger, D.C., Prins, R., eds.: *X-Ray Absorption: Principles, Applications, Techniques of EXAFS, SEXAFS and XANES*. John Wiley & Sons, New York, 1988.

[4]    For an example of Tc oxidation state determination see Allen, P.G., Shuh, D.K., Bucher, J.J., Edelstein, N.M., Reich, T., Denecke, M. A., J. Physique IV 7, 2-789-92 (1997), An example for U is in Bertsch, P.M., Hunter, D.M., Sutton, S.R., Bajt, S., Sasa, Rivers, M.L., Environ. Sci. Technol. (1994), 28, 980-4. Pu oxidation state determination is reported in Conradson, S.D., Al Mahamid, I., Clark, D.L., Hess, N.J., Hudson, E.A., Neu, M.P., Palmer, P.D., Runde, W.H., Tait, C.D., Polyhedron 17, 599 (1988).

[5]    Crapper, M.D., Vacuum 45, 691-704 (1994).

[6]    Chapman, N.A., McKinley, I.G., Smellie, J.A.T., *The Potential of Natural Analogues in Assessing Systems for Deep Disposal of High-Level Radioactive Waste*, Eidgenössisches Institut für Reaktorforschung, Würenlingen, EIR-545 (1984).

[7]    Frauenfelder, H., Steffen, R.M., "Angular Correlations" in *Alpha, Beta- and Gamma-Ray Spectroscopy, Vol. 2* (K. Siegbahn, ed.) North Holland Publishing Co. Amsterdam 1968, pp.997-1198.

[8]    Heidinger, R., Thies, W-G., Appel, H., Then, G.M., Hyperfine Interactions 35, 1007 (1987).

[9]    Stern, E.A., Newville, M., Ravel, B., Yacoby, Y., Haskel, D., Physica B 208 & 209, 117-120 (1995).

[10]   T. Ressler, J. Physique IV 7, C2-269 (1997).

[11]   Rehr, J.J., Albers, R.C., Zabinsky, S.I., Phys. Rev. Lett. 69, 3397 (1992).

[12]   Krstanovic, I.R., Acta Cryst. 11, 896 (1958).

[13]   Pohlmann, C., Degering, D., Geckeis, H., Thies, W-G, Hyperfine Interactions 120 & 121, 313-318 (1999).

[14]   Reich, T., Moll, H., Arnold, T., Denecke, M.A., Hennig, C., Geipel, G., Bernhard, G., Nitsche, H., Allen, P.G., Bucher, J.J., Edelstein, N.M., Shuh, D.K., J. Electron Spectr. and Related Phenomena 96, 237-43 (1998).

[15]   Östhols, E., Manceau, A., Farges, F., Charlet, L., J. Colloid Interface Sci. 194, 10-21 (1997).

[16]   Cheah, S-F., Brown, G.E., Parks, G.A., J. Colloid Interface Sci. 208, 110-28 (1998).

[17]   Fendorff, S.E., Lamble, G.M., Stapleton, M.F., Kelly, M.J., Sparks, D.L., Environ. Sci. Technol. 28. 284-89 (1994).

[18]   Linke, W.F., *Solubilities of Inorganic and Metal-organic Compounds*, D. Van Nostrand Company, Princeton 1958, p. 1178.

[19]   Dzombak, D.A., Morel, F.M., *Surface Complexation Modelling: Hydrous Ferric Oxide*, Wiley-Intersience, New York 1990.

[20]   Farquhar, M.L., Charnock, J.M., England, K.E.R., Vaughan, D.J., J. Colloid Interf.ace Sci. 177, 561-7 (1996).

[21]   Antonio, M.R., Soderholm, L., Ellison, A.J.G., J. Alloys and Compounds 250, 536-540 (1997).

[22]   Denecke, M.A. in *Proceedings of the Workshop on Speciation, Techniques, and Facilities for Radioactive Materials at Synchrotron Light Sources*, OECD Publications 1999, pp. 135-141.

**Table 1. Metrical parameters obtained from theoretical fits to the EXAFS spectra shown in Figure 6: co-ordination numbers (N), Hf-O bond lengths ($R_{Hf-O}$) and EXAFS Debye-Waller factors ($\sigma^2$). The scaling factor, $S_o^2$, was held constant at 0.84 and the shift in threshold energy, $\Delta E_0$, at 0.13 eV.**

| pH | N | $R_{Hf-O}$ [Å] | $\sigma^2$ [Å$^2$] |
|----|-----|------|-------|
| 1 | 6.9 | 2.01 | 0.020 |
| 2 | 7.2 | 2.01 | 0.160 |
| 3 | 8.6 | 2.01 | 0.022 |
| 4 | 5.3 | 2.04 | 0.005 |
| 5 | 5.7 | 2.08 | 0.003 |
| 6 | 8.8 | 2.02 | 0.012 |
| 7 | 8.2 | 2.04 | 0.007 |

**Figure 1. Uranium L3 edge X-ray absorption spectrum for $UO_2$ showing the division into XANES and EXAFS regions. The $UO_2$ XANES spectrum (…) is compared to those for $U_3O_8$ (---) and $UO_3$ (—) to display the absorption edge energy increase with increasing cation valence**

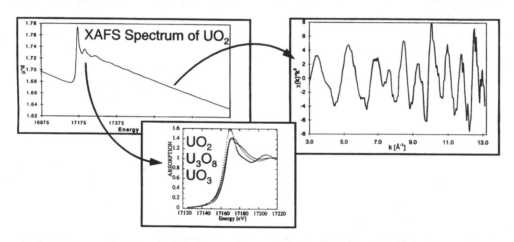

**Figure 2. Schematic representation of the physical processes responsible for EXAFS oscillations. EXAFS spectra minima correspond to where the photo electron wave function (and the phase shift it experiences) are such that the outgoing wave interferes destructively with the backscattered wave at the absorbing atom; maxima reflect the opposing situation where the interference is constructive.**

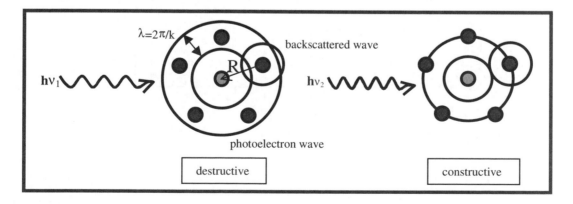

**Figure 3. Grazing incidence XAFS beamline BN2 at Bonn University ELSA**

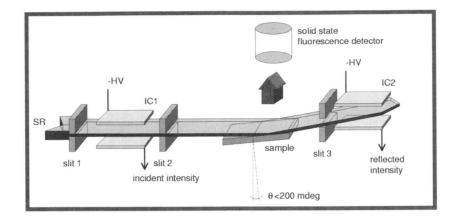

**Figure 4. TDPAC spectrum of Hf(IV) species sorbed a pH = 5 onto amorphous silica (left) and its corresponding Fourier transform (right). The peaks associated with the different Hf(IV) binding sites are indicated in the Fourier transform.**

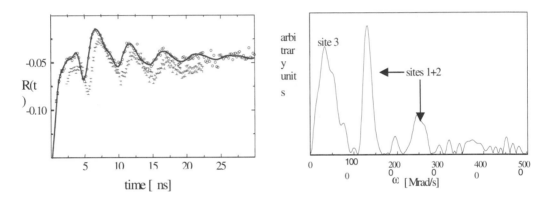

**Figure 5. Relative fractions of the three Hf(IV) sorption sites on an amorphous silica surface as a function of sample preparation pH observed in TDPAC spectra**

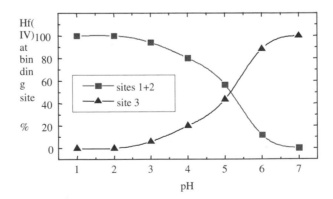

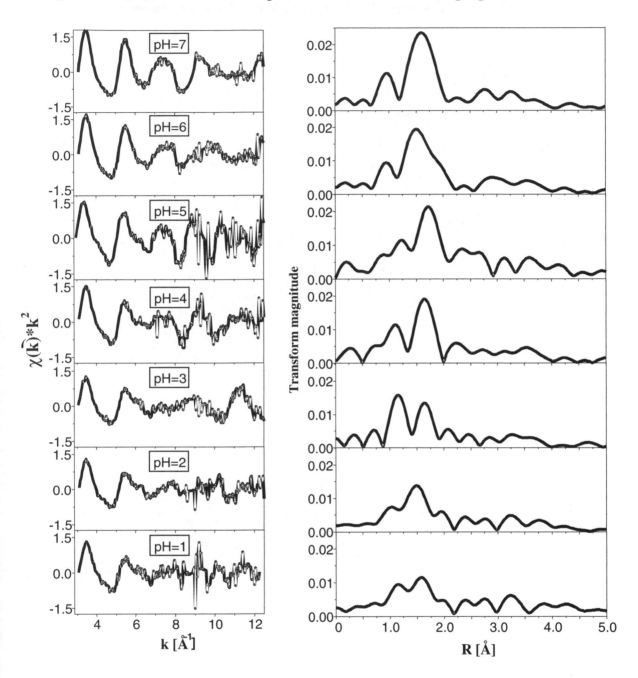

Figure 6. Hf L3 edge $k^2$-weighted EXAFS (left) and their corresponding Fourier transforms (right) for Hf(IV) sorbed onto amorphous silica at the nearest integer pH values indicated

**Figure 7. Hf L3 edge GIXAFS spectra of Hf(IV)-species sorbed onto the (0001) plane of mica at the pH values given. Left: k-weighted EXAFS. Middle: Fourier transforms. Right: Fourier filtered Hf-O co-ordination shell.**

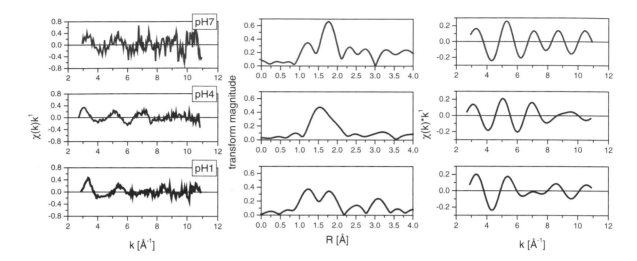

# MAGNETIC RESONANCE TECHNIQUES IN SPECIATION OF ACTINIDES AND FISSION PRODUCTS

**Jean F. Desreux, Bernard Lambert and Vincent Jacques**
Co-ordination and Radiochemistry, University of Liège,
Sart Tilman (B16), B-4000 Liège, Belgium

## Abstract

This paper presents an overview of the applications of magnetic resonance methods in the speciation of $f$ elements. The emphasis is on how to obtain information on the structure and the solution behaviour of complexes with extractants and humic acids by nuclear magnetic resonance. The applicability of each approach is discussed with a few examples taken from recent results obtained in the laboratory.

## Introduction

Nuclear magnetic resonance (NMR) is an extraordinarily powerful technique for unravelling the structure and the conformation of organic molecules and metallic complexes in solution. Successful applications have been reported in every branch of chemistry and biochemistry and the last few years have witnessed an exponential growth of new and more effective pulse techniques in two- and three-dimensional spectroscopy. It is now possible to solve problems that were unthinkable only a few years ago. Despite these remarkable progresses, NMR has rarely been applied to the speciation of radioactive nuclides. Two difficulties are at the origin of the lack of interest for NMR among radiochemists. Firstly, NMR is considered as a poorly sensitive technique especially when it is compared with radiotracers or fluorescence methods. Secondly, few spectrometers dedicated only to the study of radionuclides are available and a radioactive solution is rarely welcomed by spectroscopists. However, the ever increasing sensitivity of NMR should help making this technique more popular as the handling of very small samples in safe conditions becomes easier. High field superconducting magnets and better pulse techniques have considerably improved the sensitivity of NMR. One impetus behind theses improvements is the use of NMR as a detection technique in chromatography, whether HPLC or capillary electrophoresis [1]. Recording a two-dimensional spectrum of 5 µg (about 10 nmol) of a medium size organic molecule is becoming customary.

In the present overview of the NMR applications in nuclear chemistry, it is convenient to divide metallic complexes into two groups depending on whether the metal ions are dia- or paramagnetic. It is also simpler to consider separately metal ions in an S state, namely $Gd^{3+}$ and $Cm^{3+}$. Recent results will be discussed here according to that sequence and a few comments on solid state NMR, NMR microscopy and electron paramagnetic resonance (EPR) will be presented at the end of this overview.

## NMR spectroscopy of diamagnetic metal complexes

Spectroscopic studies of derivatives of diamagnetic ions such as $La^{3+}$, $Lu^{3+}$, $UO_2^{2+}$ and $Th^{4+}$ pose no special problems and the whole arsenal of NMR techniques can be fruitfully used (two-dimensional procedures such as COSY, NOESY, EXSY...). Models of the solution structures of metal complexes have been inferred from relative peak areas, J coupling constants and $^1H$-$^{13}C$ correlation 2-D spectra. A prerequisite for these studies is that the solution species are sufficiently stable and rigid. Moreover, some form of symmetry greatly simplifies the analysis. Dynamic properties such as equilibria between topomers, ligand exchange, hydration state and the related thermodynamic parameters are also easily analysed by the classical NMR methods. Finally multinuclear NMR spectra are yielding interesting information for instance on the co-ordination with water or with inorganic ions even if measurements are sometimes hampered by peak broadening due to quadrupolar relaxation, for instance in the case of $UO_2^{2+}$.

As an example of a detailed structural analysis, one could mention an NMR study of the complexation of $Th^{4+}$ by the macrocyclic ligand HEHA [2].

Contrary to all expectations, a kinetically stable topomer of ThHEHA [2] is formed upon mixing the ligand and $Th^{4+}$. On heating, a conversion to a thermodynamically stable form takes place with an exceedingly high activation energy ($\Delta G^I = 199$ kJ mol$^{-1}$). Using two-dimensional techniques and simulations, all $^1H$-$^1H$ coupling patterns can be interpreted and the geometry of the two topomers can then be established with the help of the Karplus equations. These equations yield torsion angles from coupling constants. The kinetically stable topomer adopts a $C_2$ symmetry while the symmetry of the thermodynamically stable form is remarkably high ($S_6$). NMR is an extremely powerful technique for such studies and yield information on structures, dynamics and kinetics that could not be easily obtained by other spectroscopic methods.

Speciation rather than conformation analyses are feasible by NMR provided the exchanges between the different forms in solution are slow enough so that each of them gives rise to a separate peak. Multinuclear spectra are often very useful in these studies, for instance for the identification of hydrolysed species or for the determination of hydration states. For instance, two hydrolysed forms of the $UO_2^{2+}$ ion have been clearly identified by $^{17}O$ NMR in addition to the free ion [3]. Stability constants are readily obtained from such spectra simply by peak integration. The same approach can be used for the speciation of complexes with simple anions, for instance using the nucleus $^{15}N$ as a tool for an investigation of the complexation of $Th^{4+}$ by $NO_3^-$ ions [4].

## NMR spectroscopy of paramagnetic metal complexes

The NMR spectra of solutions of paramagnetic ions are often more informative than the spectra of diamagnetic species because the unpaired electronic spins can induce large chemical shifts. Separate NMR peaks due to different exchanging species are thus more easily observed. Moreover, the induced paramagnetic shifts are directly related to the solution conformations by simple geometrical factors provided the interaction between the unpaired electronic spins and the nuclei takes place through space rather than through chemical bonds (dipolar $vs.$ contact). The electronic relaxation times also need to be sufficiently short. These conditions are met with $Yb^{3+}$ which is the ion of choice for studies on trivalent $f$ ions [5] and they could also be met in the case of $NpO_2^{2+}$ [6].

The chemical shifts $\delta_i$ due to the through-space dipolar interactions between nuclei i and the unpaired electronic spins on a metal ion are accounted for by the dipolar equation:

$$\delta_i = -D\left\langle \frac{3\cos^2\theta_i - 1}{r_i^3} \right\rangle - D'\left\langle \frac{\sin^2\theta_i \cos 2\psi_i}{r_i^3} \right\rangle \tag{1}$$

where D and D = are magnetic susceptibility parameters and where $r_i$, $\theta$ and $\nu_i$ are the polar co-ordinates of nucleus i in the set of axes of the magnetic susceptibility tensor. Applying this equation requires that the metal complexes in solution are stable and rigid and preferably display at least a two-fold symmetry. If numerous metal chelates are exchanging rapidly and yield resonances averaged over all species, the induced paramagnetic shifts are often small and cannot be related to a conformation in solution. On the contrary, if symmetric highly rigid metal derivatives are formed in solution, the relative magnitudes of the induced paramagnetic shifts directly lead to a structural model. The situation is even simpler if the investigated complex is axially symmetric (at least one $C_3$ axis) as equation 1 simplifies to:

$$\delta_i = D\left\langle \frac{3\cos^2\theta_i - 1}{r_i^3} \right\rangle \tag{2}$$

because there is no longer an anisotropy in the xy plane ($D' = 0$). Equations 1 and 2 have been extensively used in the past to interpret the NMR spectra of various organic substrates associated to lanthanide β-diketonates called shift reagents [5]. Because of the paramagnetism of the metal ions, the spectra of these substrates become first order and are readily interpreted. However, it rapidly appeared that the lanthanide β-diketonates are labile species for which equations 1 and 2 cannot be reliably used and no structural inferences can be made. Shift reagents thus fell out of favour with most spectroscopists. Despite these difficulties, attempts at using the dipolar equations to unravel the solution structure of lanthanides or $Pu^{4+}$ complexes with extractants such as monoamides have been reported [7] but structural inferences based only on the dipolar equations should be accepted with reservation. For non-symmetric chelates, the most reliable use of paramagnetism in NMR remains the analysis of exchange processes and of hydration states. Recent examples are the investigation of the hydration of $NpO_2^{2+}$ and $PuO_2^{2+}$ [8]or of the association of the $CO_3^{2-}$ with $NpO_2^+$ [9].

Confidence in the use of the dipolar equations 1 and 2 has recently been restored by studies of symmetric non-labile lanthanide complexes. Not surprisingly, most of those compounds belong to the family of macrocyclic ligands because these compounds usually feature some form of symmetry and are able to impose their steric requirements to their metal complexes [5]. Some macrocyclic ligands recently proved to be very selective extracting agents. NMR seems to be an appropriate method for finding the origin of this selectivity as equations 1 and 2 can be reliably used [10]. The $^1H$ spectrum and the COSY spectrum of the anhydrous $Yb^{3+}$ complex of calix[4]arene $(CMPO)_4$ in dry acetonitrile is shown in Figure 1. Despite the short electronic relaxation times of the $Yb^{3+}$ ion, well-resolved 2-D spectra are readily obtained with short pulse sequences and a full assignment of the peaks can be made. Without two-dimensional NMR spectra, assigning resonances is often difficult as coupling patterns are no longer visible because of the peak broadenings due to the paramagnetism of the metal ions. A solution structure of the $Yb^{3+}$ calixarene complex as deduced from equation 1 is depicted schematically in Figure 2. In addition, NMR allows a step by step approach to an investigation of the role of each parameters in an extraction system. Water or nitrate ions can be added to an anhydrous solution of the $Yb^{3+}$ calixarene complex shown above and the effect of this addition is clearly seen in the NMR spectra: the metal chelate becomes labile and hence the NMR spectra are poorly resolved because water and/or $NO_3^-$ ions penetrate into the first co-ordination sphere of the metal ions. However, the addition of $H_2O$ or $NO_3^-$ has no effects on the NMR spectra of other ligands of the calix[4]arene type that form more stable complexes [10].

## Relaxation in NMR

Measuring the spin-spin relaxation time $T_2$ of diamagnetic chelates at different temperatures yields details on the exchange processes between a complex and a ligand, for instance activation parameters, participation of solvent molecules in the exchange and other mechanistic information. Several multinuclear studies on actinide complexes have been reported including a recent study on the $UO_2^{2+}$ chelates of phosphonic acids [11].

Paramagnetic ions are again especially interesting as they modify the relaxation times of their ligands more profoundly than their diamagnetic analogues. With a half-filled *4f* shell, $Gd^{3+}$ is the ion of choice ($^8S_{7/2}$). Because of the spherical distribution of its electrons, $Gd^{3+}$ has a relatively long electronic relaxation time $T_{1e}$ (or $\tau_s$) that is more in tune with the proton = s frequency so that it drastically reduces the relaxation times of nuclei in its immediate surroundings. This property has been extensively used for improving the contrast of magnetic resonance images [12] but it is also of interest in speciation even if so far no studies have yet been published.

The frequency dependence of the $1/T_1$ relaxation rates of the *solvent protons* of a solution of a $Gd^{3+}$ ion or complex (also called relaxivity, in $s^{-1}$ $mmol^{-1}$) is well accounted for by the Solomon-Bloembergen-Morgan equations [5;12]:

$$\left[ \frac{1}{T_{1is}} = \frac{P_M q_{solvent}}{[solvent]} \frac{1}{T_{1M} + \tau_m} \right] \tag{3}$$

$$\frac{1}{T_{1M}} = \frac{2}{15} \frac{\gamma_H^2 g^2 S(S+1)\mu_B^2}{r^6} \left[ \frac{7\tau_c}{1+\omega_S^2\tau_c^2} + \frac{3\tau_c}{1+\omega_H^2\tau_c^2} \right] \tag{4}$$

According to these equations, the relaxation rate of solvent molecules in the inner co-ordination sphere of the $Gd^{3+}$ ion, $I/T_{1is}$, depends on $q_{solvent}$, the number of solvent molecules directly co-ordinated to the metal ion, on $P_M$, the molar fraction of the metal ion, on r, the distance between $Gd^{3+}$ and the solvent protons and on a correlation time $\tau_c$. In these equations, $\omega_S$ and $\omega_H$ are the Larmor frequencies of the electron and of the proton respectively and the other factors have their usual meaning. The correlation time $\tau_c$ is given by:

$$\frac{1}{\tau_c} = \frac{1}{\tau_s} + \frac{1}{\tau_m} + \frac{1}{\tau_r} \tag{5}$$

where $\tau_c$ depends on the smallest of three correlation times, $\tau_r$, the mean rotational correlation time of the $Gd^{3+}$ species, $\tau_s$, the metal electronic relaxation time or $\tau_m$, the residence time of the solvent molecules in the first co-ordination sphere of the metal ion. The frequency dependence of $\tau_s$ is accounted for by the equation:

$$\frac{1}{\tau_S} = \frac{1}{5\tau_{s0}} \left[ \frac{\tau_v}{1+\omega_S^2\tau_v^2} + \frac{4\tau_v}{1+4\omega_S^2\tau_v^2} \right] \tag{6}$$

where $\tau_{s0}$ and $\tau_v$ are the electronic relaxation time at zero field and the correlation time for the modulation of the zero field splitting respectively. Relaxation rates *vs.* Larmor frequency (or magnetic field strength) curves, also called nuclear magnetic relaxation dispersion curves (NMRD), give insight into the formation and the dynamic behaviour of the $Gd^{3+}$ complexes. For instance, the complexation of $Gd^{3+}$ by a ligand reduces the solvation of the metal ion and leads to a decreased relaxivity because solvent molecules relax more slowly in the bulk of the solution than in the immediate surrounding of the metal ion. Moreover, rapidly rotating small $Gd^{3+}$ chelates with small $\tau_r$ values display a S-shaped curve with an inflection point between 1 and 10 MHz. By contrast, a relaxivity maximum at about

20-30 MHz is observed for slowly tumbling species because the rotational correlation time is no longer the major contributor to relaxivity. It is the frequency dependence of the electronic relaxation time, $\tau_S$, that brings about relaxivity maxima as reported for $Gd^{3+}$ complexes of high molecular weight ligands.

Interesting information on the speciation of organic solutions of complexes formed between extractants and $Gd^{3+}$ can be gained by measuring the relaxivity either in function of the applied field or in titration experiments at fixed frequencies. Relaxivity titration curves in anhydrous acetonitrile are collected in Figure 3 for the calix[4]arene(CMPO)$_4$ macrocycle shown above that is substituted by donor groups on the wide (upper) rim and for a tetraamide calix[4]arene that bears its co-ordinating functions on the narrow (bottom) rim. For the latter ligand, one observes the expected behaviour: the relaxivity of the $Gd^{3+}$ solution linearly decreases until a plateau is reached for a 1:1 concentration ratio. Solvent molecules are thus progressively removed from the first co-ordination sphere of the metal ion and a highly stable $Gd^{3+}$ complex is fully formed in a stoichiometric metal-ligand mixture. On the contrary, the calix[4]arene(CMPO)$_4$ ligand features a totally unexpected titration curve. The relaxivity first increases until a maximum is reached for a ligand:$Gd^{3+}$ ratio of about 1. It then slowly decreases until a plateau appears for ratios higher than 2.5. The NMR studies mentioned above were not conducted in the experimental conditions used here but they clearly indicate that the CMPO ligand is able to encapsulate lanthanide ions. It can thus be assumed that a complexation takes place and that solvent molecules are progressively replaced by co-ordinating groups. The relaxivity maximum is assigned to the formation of oligomers that are tumbling sufficiently slowly so that the longer relaxation times of solvent molecules released to the bulk of the solution are more than compensated by the shorter relaxation times of solvent molecules co-ordinated to aggregates. This behaviour was systematically observed for calix[4]arene substituted on the wide rim and one is thus led to assume that an oligomerisation takes place because of the greater flexibility of the ligating functions on the wide rim [10]. The NMRD curves presented in Figure 4 are in keeping with this interpretation: a pronounced maximum at about 30 MHz is observed for the calix[4] arene that forms oligomers as indicated by the Solomon-Bloembergen equations. A best fit of the NMRD curves yields tumbling time $\tau_r$ values that are 51, 193 and 649 ps for free $Gd^{3+}$, the tetraamide and the CMPO calixarene respectively. The complex with the latter must form large aggregates as $\tau_r$ values of 700-900 ps have been reported for dendrimeric $Gd^{3+}$-containing polymers in water [12]. It thus seems that the selectivity of the CMPO-calixarene ligands cannot be interpreted without taking into account the formation of large oligomers.

The speciation of $Gd^{3+}$ complexes with humic acid in water can also be investigated by relaxivity methods. As shown in Figure 5, the relaxivity of a $Gd^{3+}$ solution markedly increases when humic acid is added at pH 5.5. A plateau is reached once the complex is fully formed and a discontinuity is observed at about the first half of the titration curve. The NMRD curves recorded for these two situations are reproduced in Figure 6 together with the corresponding curve obtained for free $Gd^{3+}$. Obviously, high molecular species are formed and tumbling times $\tau_r$ of 924 and 1240 ps were obtained by a best fit treatment of the data. Competition experiments with ligands such as EDTA are easily performed and stability constants can be extracted from the relaxivity data.

Relaxivity in speciation is a hardly broached but pregnant area of research. The sensitivity of the method is noticeably higher than in the case of NMR spectroscopy since one is looking at the solvent molecules rather than at the complexes themselves (the limit should be close to $10^{-6}$ M solutions of high molecular weight $Gd^{3+}$ complexes in water). However, the correlation between tumbling times and molecular sizes is still unclear: the $\tau_r$ parameters are mean values of different movements around a metal ion in a macromolecule. Moreover, the exact values of several parameters in the Solomon-Bloembergen equations are not known with a sufficient accuracy. However, some of them such as the exchange times $\tau_m$ can be determined independently for example from $^{17}O$ spectra of aqueous solutions at variable temperature and pressure [12]. Finally, it should be noted that $Cm^{3+}$ is also an S state ion but unfortunately the splitting of the ground level is much higher than for $Gd^{3+}$ and recent results show that the relaxivity effect is much smaller.

## Other magnetic resonance methods

Electron paramagnetic resonance (EPR) is a much more sensitive spectroscopic technique than NMR and has been applied nearly exclusively to the study of solid materials. These studies are not included in the present overview. Furthermore, it should be mentioned here that solid state $^{13}C$ NMR brings valuable information on the chemical composition of humic and fulvic acids [13]. NMR imaging is another magnetic resonance method that is of potential interest in speciation [14]. Barely used until now for that purpose, NMR microscopy should allow studies of the diffusion of $f$ elements in a variety of materials.

## REFERENCES

[1]    P. Gfrörer, J. Schewitz, K. Pusecker, E. Bayer, *Anal.Chem.*, 315A-320A (1999).

[2]    V. Jacques, J.F. Desreux, *Inorg.Chem.*, *35*, 7205-7210 (1996).

[3]    W-S. Jung, H. Tomiyasu, H. Fukutomi, *J. Chem. Soc., Chem. Commun.*, 372-373 (1987).

[4]    D.K. Veirs, C.A. Smith, J.M. Berg, B.D. Zwick, S.F. Marsh, P. Allen, Conradsen, S. D. *J .Alloy. Compd.*, *213/214*, 328-332 (1994).

[5]    J.A. Peters, J. Huskens, D.J. Raber, *Prog. Nucl. Magn. Reson. Spectrosc.*, *28*, 283-350 (1996).

[6]    V.A., Shcherbakov, E.V Iorga, *Radiokhimiya*, *17*, 875-880 (1975).

[7]    C. Berthon, C. Cachaty, *Solv. Extr. Ion Exch.*, *13*, 781-812 (1995).

[8]    N. Bardin, P. Rubini, C. Madic, *Radiochim. Acta*, *83*, 189-194 (1998).

[9]    D.L. Clark, S.D. Conradson, S.A. Ekberg, N.J. Hess, D.R. Janecky, D. M.P. Neu, P.D. Palmer, C.D Tait, *New J. Chem.*, *20*, 211-220 (1996).

[10]   B. Lambert, V. Jacques, A. Shivanyuk, S.E. Matthews, A. Tunayar, M. Baaden, G. Wipff, V. Böhmer, J.F. Desreux, *submitted*.

[11]   J. Muntean, K.L. Nash, P.G. Rickert, J.C. Sullivan, *J. Phys. Chem.*, *103*, 3383-3387 (1999).

[12]  P. Caravan, J.J. Ellison, T.J. McMurry, R.B. Lauffer, *Chem. Rev.*, *99*, 2293-2352 (1999).

[13]  G.R. Choppin, *Chemical Separation Technologies and Related Methods of Nuclear Waste Management*, Eds Choppin, G.R., Khankhasayev, M.K., NATO Science Series. Environmental Security, vol. 53, Kluwer (Dordrecht), 247 (1999).

[14]  N. Nestle, R. Kimmich, *Coll. Surf.*, *115*, 141-147 (1996).

**Figure 1. COSY spectrum of the anhydrous Yb$^{3+}$ perchlorate complex with calix[4]arene(CMPO)$_4$ in dry acetonitrile at 273 K (ligand metal ratio 0.59). Peak assignments: calix phenyl H: 1, 15; calix bridging C$\underline{H}_2$: 2, 3, 12, 14; C(O)-C$\underline{H}_2$-P(O): 18, 19; P-phenyl H: 5, 10 (para), 6, 13, (ortho), 10, 16 (meta).**

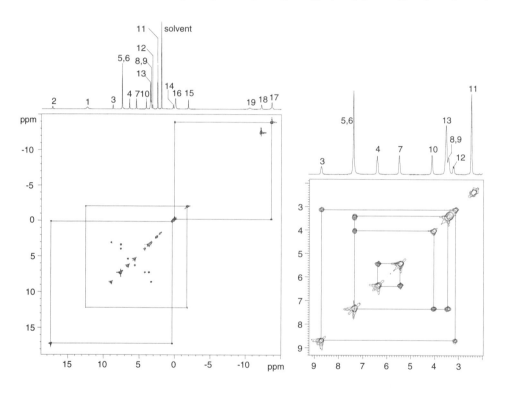

**Figure 2.** Schematic representation of the structure suggested for the $Yb^{3+}$ complex of calix[4]arene(CMPO)₄. Only one fourth of the complex is shown and lines are joining atoms in magnetically equivalent positions.

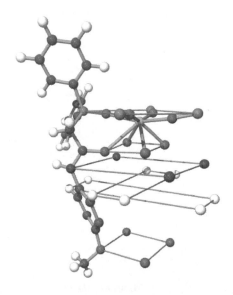

**Figure 3.** Relaxivity titration curves of the $Gd^{3+}$ complexes with calix[4]arene(CMPO)₄ (•) and calix[4]arene(amide)₄ (▽) in anhydrous acetonitrile. The donor groups in the former are located on the wide (upper) rim and in the latter on the narrow (lower) rim

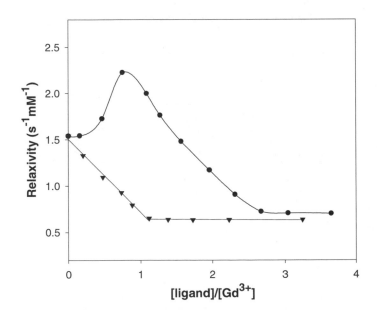

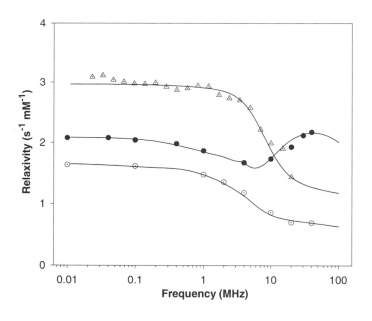

**Figure 4. NMRD curves of free Gd$^{3+}$ ($\triangle$) and of the Gd$^{3+}$ complex with calix[4]arene(CMPO)$_4$ ($\bigcirc$) and calix[4]arene(amide)$_4$.($\bullet$) in anhydrous acetonitrile**

**Figure 5. Relaxivity titration curve of Gd$^{3+}$**
**(solution $3 \times 10^{-5}$ M) by purified Aldrich humic acid at pH 5.5**

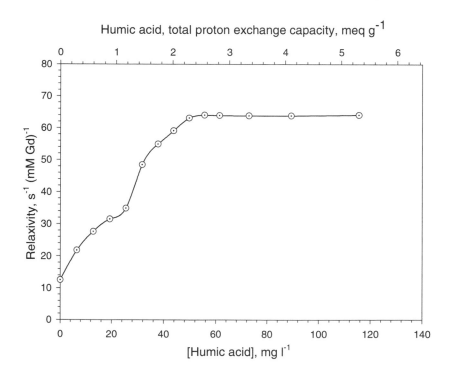

**Figure 6. NMRD curves of aqueous solutions of free Gd$^{3+}$ ($\triangle$) and of Gd$^{3+}$-humic acid mixtures (purified Aldrich, 30 ($\bigcirc$) and 50 mg l$^{-1}$ ($\square$) for a $3 \times 10^{-5}$ M solution of metal) at pH 5.5**

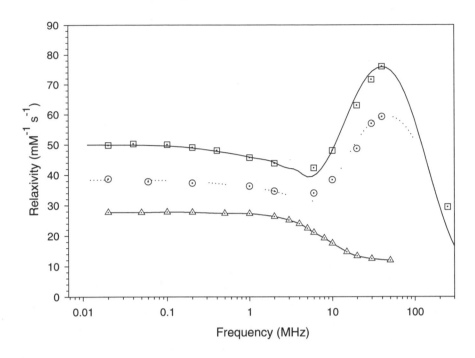

# SESSION D

## *Methods for Redox Speciation*

# AN OVERVIEW OF ACTINIDE REDOX SPECIATION METHODS, OTHER THAN CHEMICAL SEPARATION TECHNIQUES: THEIR APPLICABILITY, ADVANTAGES AND LIMITATIONS

**Dhanpat Rai and Nancy J. Hess**
Pacific Northwest National Laboratory
Richland, Washington, USA
E-mail: dhan.rai@pnl.gov

## Abstract

The development of reliable thermodynamic models for chemical systems requires knowledge of the structural nature and stoichiometry of the aqueous and solid phase species. However, because of the complexity of many systems, the limited applicability of different speciation techniques, or the inherent uncertainties of the methods, it is often difficult to confidently obtain such information using a single technique. Characterisation of aqueous and solid phase species in complex systems, therefore, typically requires the use of multiple techniques.

Determination of oxidation states is a first step towards characterising aqueous species. In addition to chemical separation methods, there are a large number of techniques that can be used to determine redox speciation. The most commonly used techniques are the absorption spectroscopies (UV-Vis-NIR, X-ray absorption spectroscopy, laser-induced photoacoustic spectroscopy), fluorescence spectroscopy, nuclear magnetic resonance (NMR) spectroscopy, and vibrational spectroscopy (Raman). This article is a review of the applicability of these various oxidation state determination methods in terms of specific oxidation states, elements, and detection limits, as well as the advantages and limitations of these methods.

## What is meant by speciation

In the strictest sense, speciation should involve identifying the stoichiometry and the structural nature of the aqueous complexes and chemical compounds. Such information is needed to develop reliable thermodynamic models for predicting the chemical behaviour of elements in different systems. However, in practice, because of the complexity of the chemical system or the limitations of a given technique, it is not always possible to obtain information on the stoichiometry and structure of the species. In these cases, one may have to be satisfied with incomplete knowledge of the system. This knowledge may be limited to identification of oxidation states, differentiation between soluble and colloidal fractions, identification of ligands, and differentiation among adsorbed, precipitated, or soluble complexes.

In this paper, we discuss methods of oxidation state analysis of actinide elements other than by chemical separation techniques, which are being addressed separately. Determination of oxidation state is important because the oxidation state controls the precipitation/dissolution and adsorption/desorption reactions, aqueous complexation, and reactivity and environmental behaviour of multivalent actinides. Among the actinides that may be of environmental concern (Figure 1), only U, Np, and Pu show multiple oxidation states that are stable under ordinary environmental conditions. Therefore, the applicability, advantages, and limitations of some of the most commonly used or available methods for oxidation state analyses of U, Np, and Pu are briefly discussed in this paper.

## Oxidation state analyses

Several different methods are available for quantifying oxidation state distributions in aqueous solutions and solids, but only a few of these methods can be applied at low concentrations, such as sorbed species or dilute solutions and solids. The available methods can be divided into two different categories: methods capable of simultaneous determination of multiple oxidation states and selective methods for single oxidation state determination.

### *Methods capable of simultaneous determination of multiple oxidation states*

Absorption spectroscopy (UV-Vis-NIR, laser photoacoustic [LPAS], X-ray absorption, and complexation followed by absorption spectroscopy) is the main method for simultaneous determination of U, Np, and Pu oxidation states. In general, absorption spectroscopy has two advantages:

- The absorption spectra is dependent on the oxidation state and the nature of complexes, thus allowing direct determination of species.

- It is non-invasive and non-perturbing.

The absorption spectra generated by UV-Vis-NIR and LPAS are essentially identical. Oxidation states of a given actinide element show a distinct absorption spectrum for each of the oxidation states (see Table 1; Refs. [1-3]), enabling identification and quantification of different oxidation states of the actinide. The limitations of UV-Vis-NIR include:

- The method is suitable for analyses of relatively concentrated samples (>~0.0001 M).

- The method is not appropriate for direct study of environmental samples.

- Information about the dominant species and their molar extinction coefficients is required for reliable quantification.

The limitations of UV-Vis-NIR apply equally to LPAS, with the exception that LPAS is generally 100 to 1 000 times more sensitive than the UV-Vis-NIR [4]. Similarly, X-ray absorption near edge spectra (XANES) of the different oxidation states of the actinide elements occurs at specific energies (Figure 2), allowing differentiation between the actinide elements and their oxidation states. A major advantage of X-ray absorption over UV-Vis-NIR spectroscopy is that X-ray absorption edge energies for a given actinide fall in a narrow energy range that is well separated from the other actinides (Figure 3), allowing discrimination of contributions from the other actinide elements. In addition, XANES is about a factor of 100 more sensitive than the UV-Vis-NIR. The major limitations of this method are:

- The shift in absorption edge energy is non-linear with oxidation state.

- It requires the transportation of samples off site to a synchrotron facility, thereby prohibiting analysis of time-sensitive complexes.

There is always a need to develop more sensitive methods for oxidation state analyses. If a ligand selectively forms a strong complex with a single oxidation state and the resulting species has a very high molar absorptivity, the concentration of the complex can be quantified by absorption spectroscopy. Arsenazo(III) forms a strong complex with Pu(IV) with a molar absorptivity of $6.8 \times 10^4$ Swanson and Rai [5] used arsenazo III to specifically complex Pu(IV) and analysed it at 660 nm wave length to quantify Pu(IV) when present at concentrations $>10^{-7}$ M. They also used specific oxidants and reductants to quantify Pu(III) and Pu(V)+Pu(VI). The advantages of this method are:

- It does not require exotic equipment.

- It provides much better detection limits for Pu(IV) than any of the other methods with the exception of chemical separation techniques.

The major limitations of this method are that it is invasive and perturbing.


## *Selective methods for determination of a single oxidation state*

Oxidation state analysis methods in this category include Nuclear Magnetic Resonance (NMR), fluorescence, and Raman spectroscopy. The vast majority of NMR studies of actinide complexes are conducted with the ligand atoms as the NMR active nuclide. However, when the actinide metal is selected as the NMR active nuclide, this method is isotope and oxidation-state specific (see Table 2, Ref. [6]). Only a few of the isotopes are NMR sensitive and, in general, only the actinyl species can be identified by this method. The major advantage of this method is that the chemical environment of the metal can be determined. The major limitations of this method are:

- Very high concentrations are required ($> \sim 0.01$ M).

- Only actinyl species can be detected.

Although this method can potentially be applied to both solid and solution studies, thus far no studies utilising the actinide metal as the NMR active nuclide have been conducted in solution. The vibrational spectrum of the actinyl species of U, Np, and Pu also show specific Raman shift (see Table 3; Refs. [7-9]) making it possible to identify the presence of the actinyl species of these elements. The advantages of this method are:

- It is non-invasive and non-perturbing.

- The magnitude of the Raman shift is sensitive to the chemical environment.

The limitations of this method include:

- Only actinyl species can be detected.

- Generally high concentrations ($>\sim10^{-4}$ M) are required.

Of all the methods, fluorescence is the most sensitive ($10^{-6}$ to $10^{-8}$ M or even lower) allowing speciation of sorbed species [10]. In addition, the fluorescence spectrum and lifetime are sensitive to the chemical environment, and by using time-resolved techniques individual species can be discriminated [11]. However, the major limitation of this method is that only a few elements and oxidation states fluoresce. Among U, Np, and Pu, only U(VI) fluoresces. Although these selective methods can only be used to quantify actinyl species, in the case of U and Np, where only two oxidation states are expected to be important under most environmental conditions (Figure 1), it should in principle be possible to estimate the tetravalent states by subtracting the actinyl concentrations from the total.

## Conclusions

There are many spectroscopic-based methods of oxidation state determination of the actinide elements, each with its advantages and limitations. Oxidation state determination of chemically complex systems often requires the use of multiple techniques, including chemical separations, to provide information over a range of actinide concentrations.

*Acknowledgements*

This research was conducted at the Pacific Northwest National Laboratory, operated by Battelle for the US Department of Energy (DOE) under Contract DE-AC06-76RLO 1830. The authors thank Drs. Herman Cho of Pacific Northwest National Laboratory and David Wruck of Lawrence Livermore National Laboratory for their input and useful discussion.

# REFERENCES

[1]     J.M. Cleveland (1979), *The Chemistry of Plutonium*, American Nuclear Society, p. 653.

[2]     J.J. Katz, G.T. Seaborg, L.R. Morss (1986), *The Chemistry of the Actinides*, Chapman Hall, New York, p. 1164.

[3]     J.J. Katz, E. Rabinowitch (1958), *The Chemistry of Uranium*, Atomic Energy Commission, p. 769.

[4]     M.K. Richmann, D.T. Reed (1996), *Materials Research Society Proceedings* 412, 623-630.

[5]     J.L. Swanson, D. Rai (1981), *Radiochemistry and Radioanalytical Letters* 50, 89-98.

[6]     R.D. Fischer (1973) in *NMR of Paramagnetic Molecules: Principles and Applications*, G.N. La Mar, W. DeW. Horrocks, Jr., R.H. Holm, eds., Academic, New York, pp. 521-553.

[7]     L.J. Basile, J.C. Sullivan, J.R. Ferraro, P. LaBonville, (1974), *Applied Spectroscopy* 28, 144-148.

[8]     L.J. Basile, J.R. Ferraro, M.L. Mitchell, J.C. Sullivan (1978), *Applied Spectroscopy* 32, 535-537.

[9]     C. Madic, D.E. Hobart, G.M. Begun (1983), *Inorganic Chemistry* 22, 1494-1503.

[10]    D.E. Morris, C.J. Chisholm-Brause, M.E. Barr (1994), *Geochimica et Cosmochimica Acta* 58, 3613-3619.

[11]    C. Moulin, I. Laszak, V. Moulin, C. Tondre (1998), *Applied Spectroscopy* 52, 528-535.

**Table 1. Selected absorption bands for the oxidation states for U, Np, and Pu**

| Species | Wavelength (nm) | Reference |
|---------|-----------------|-----------|
| $U^{4+}$ | 650 | [3] |
| $UO_2^{2+}$ | 415 | [3] |
| $Np^{4+}$ | 723 | [2] |
| $NpO_2^{+}$ | 980 | [2] |
| $NpO_2^{2+}$ | 1 223 | [2] |
| $Pu^{3+}$ | 600 | [1] |
| $Pu^{4+}$ | 470 | [1] |
| $PuO_2^{+}$ | 569 | [1] |
| $PuO_2^{2+}$ | 831 | [1] |

**Table 2. NMR observable oxidation states of the isotopes of actinide elements**

| Isotope | Observable oxidation state |
|---------|----------------------------|
| $^{229}Th$ | 4 |
| $^{231}Pa$ | 5 |
| $^{233}U$ | 6 |
| $^{235}U$ | 6 |
| $^{237}Np$ | 7 |

**Table 3. Actinyl vibrational frequencies in aqueous solution**

| Actinide | Solution conditions | Raman shift ($cm^{-1}$) | Reference |
|----------|---------------------|-------------------------|-----------|
| $UO_2^{2+}$ | 0.01 M $HClO_4$ | 872 | [8] |
| $NpO_2^{+}$ | 1.3 M $HClO_4$ | 767 | [7] |
| $NpO_2^{2+}$ | 0.01 M $HClO_4$ | 863 | [8] |
| $PuO_2^{+}$ | $Na_2CO_3$ | 755 | [9] |
| $PuO_2^{2+}$ | 0.01 M $HClO_4$ | 835 | [8] |

# Figure 1. Environmentally important oxidation states of different actinide elements

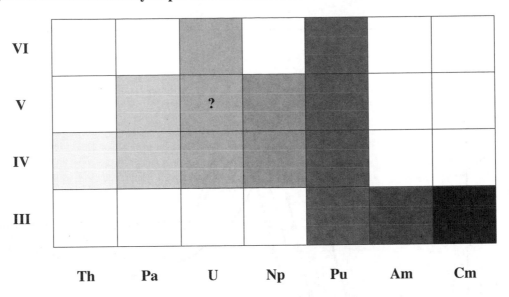

# Figure 2. X-ray absorption edge energies for U, Np, and Pu

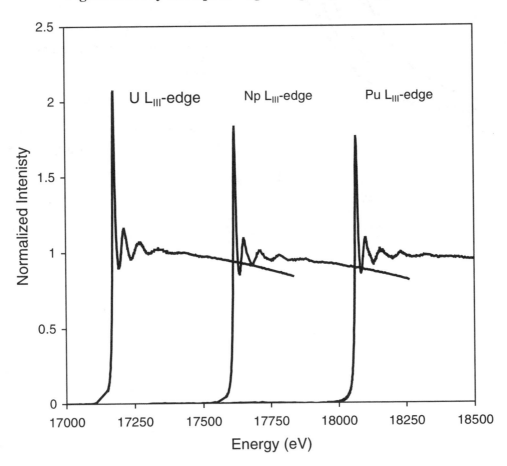

**Figure 3. Shift in absorption edge energy for the different oxidation states of Pu**

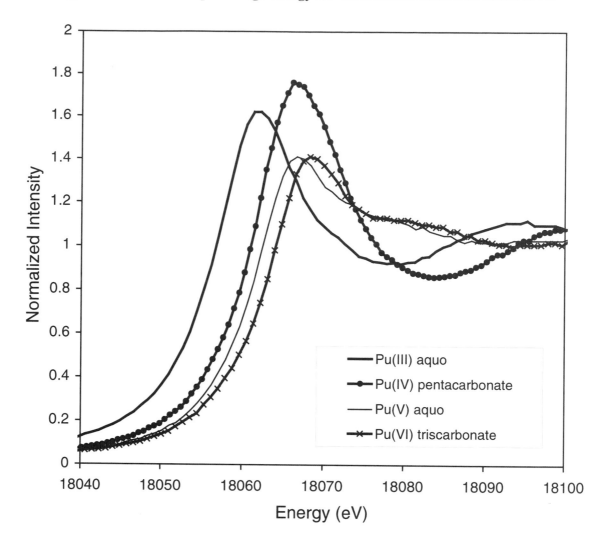

# CHEMICAL SEPARATION METHODS FOR REDOX SPECIATION OF ACTINIDES

**Akira Saito**
Chemistry Department, Tokyo Gakugei University
Koganei, Tokyo 184-8501, Japan

## Abstract

Chemical separation methods for redox speciation of actinides, particularly of plutonium, in tracer concentration were reviewed. The currently existing techniques are basically constructed on the principles of chemical behaviours of each species in precipitation, solvent extraction, and solid surface sorption systems. Based on the results of the applied investigations of the environmental behaviour of actinides, the methods were evaluated in terms of their reliability, reproducibility, advantages, or limitations. Furthermore, probable improvements, or directions of developing new techniques are briefly stated.

# Introduction

Energy production relies increasingly on nuclear power reactors. Accordingly the issues of spent nuclear fuel reprocessing, radioactive waste management is becoming a serious problem to be solved promptly. One of the major factors, which form a negative influence, is the potential hazard of the transuranium elements that is inevitably produced. A particular attention has been paid on plutonium because of its highest toxicity and of complexity of chemical behaviour. To decide the final form of actinide-bearing waste and the repository site, thorough understanding of the behaviour of transuranium nuclides in the environment must be accomplished. Particularly complicated behaviour of the transuranium elements comes from the fact that they can exist in several oxidation states at the same time under ambient conditions of the media. Therefore, knowledge of not only the total concentration, but also the distribution among various oxidation states, is of primary importance in the assessment of the environmental safety. In the present review, redox speciation means the quantitative determination of the distribution of transuranium nuclides among thermodynamically probable oxidation states, e.g. for plutonium, they are Pu(III), Pu(IV), Pu(V), and Pu(VI).

A few practically important procedures, which have been developed by different research groups, are reviewed and evaluated from various point of views to provide a useful guide to their applications and to provide suggestions for further researches to improve or to develop new techniques.

# General aspects

## *Relevant nuclides and the sources*

Transuranium elements, neptunium, plutonium, and americium have been produced and accumulated in enormous amounts. Because of their long half-lives and the high energy of $\alpha$-particles they emit, we have to keep them under strict control or in complete isolation from the public.

Among these transuranics, plutonium has aroused special interest in its environmental behaviour since it can exist in four oxidation states, III, IV, V, and VI simultaneously in natural waters and the past injection into the global ecosystems as fallouts or as effluent from spent fuel reprocessing plants. The plutonium nuclides of environmental concern are $^{238}$Pu, $^{239}$Pu, and $^{240}$Pu, as they are being produced in power reactors or used as nuclear weapons. The latter two nuclides emit $\alpha$-particles of too close energies to be detected separately by $\alpha$-spectrometry, and thus they are usually detected as combined activities, and are often designated as $^{239,240}$Pu. Among various sources of the nuclides, spent fuel reprocessing plants, high-level radioactive waste repository sites or burial sites will become major ones. Although the world inventory of plutonium has reached an enormous amount, concentrations as the pollutant of the ecosystems are extremely low even in the near field.

## *Systems of interests*

Environmental systems we have to pay attention to as probable polluted area consist of natural waters and solid phases. The former includes seawater, river water, lake water, groundwater, and aquifers, whereas the latter includes suspended particulates, sediments, soils, and rocks. Colloids are also important components of the aquatic systems, and are regarded to be in between solutes and particulates. In discussing the behaviour of pollutants, it seems a common strategy to separately deal with biological systems from those mentioned above.

The knowledge about the oxidation states of plutonium in solution is primarily needed for evaluation of its potential hazard on humans. The scope of this review is the redox speciation of plutonium in natural waters and at the interfaces. Many parameters characterise the natural water systems. Among them, Eh, pH, and concentrations of inorganic anions and of organic carbon, both as ligands are important in terms of the redox behaviour of plutonium.

## Redox speciation

Since redox speciation has to be executed in tracer level of concentrations, chemical behaviour in a specifically designated system is the only property that gives an identifying evidence of its oxidation state. Such systems include solvent extraction, precipitation, adsorption or ion exchange, where the nuclides are distributed between two phases with the concentration ratios according to the oxidation state. The distribution ratio must be large enough, or small enough, to be regarded as quantitative separation. Thus, the addition of any reagents required to the target system and processing of the system should not disturb the steady state of the redox transformation. Moreover, the speciation methods are required to be reproducible, applicable to diverse ranges of samples, and less time consuming.

## Review of currently available procedures

### Basic studies

Basic chemical properties of plutonium in III, IV, V and VI oxidation states were extensively studied since its discovery in 1940. As the use of nuclear power reactors has increased and the inventory of plutonium in global environment has been recognised, research interests in its behaviour in aquatic systems has increased [1,2].

Three research groups made rudimentary studies. In 1964, Foti and Freiling reported the procedures for the separation of U, Np, and Pu into different oxidation state fractions by TTA (4,4,4-trifluoro-1-(2-thienyl)-1,3-butane-dione ) extraction and $LaF_3$ precipitation [3]. This study laid the base for later development of more practical and extensive speciation schemes. More than a decade later, Bondieetti and Reynolds [4] reported several precipitation (co-precipitation) and solvent extraction methods for the separation of Pu(IV) and Pu(VI), which were tested for some real systems with satisfactory results. An important information that Pu(V) is the dominant species in the soluble Pu(V+VI) fraction in the water sampled from radwaste final settling basin was obtained by its application. The techniques, proposed in these studies, suggested that precipitation and solvent extraction methods must work satisfactorily in the redox speciation of tracer actinides. Inoue and Tochiyama [5] demonstrated the capability of sorption in distinguishing the oxidation states of actinides by the study of silica gel and $BaSO_4$ as the sorbents for Np.

### The method of Nelson and Lovett

The method developed by Nelson and Lovett [6,7] is an adaptation of the classical separation procedure [8] of Pu(III) and Pu(IV) from Pu(VI) by the co-precipitation of the former pair with a rare earth fluoride. They intended to distinguish Pu(III+IV) from Pu(V+VI), since the environmental chemical behaviour differs significantly according to these oxidation state groups. The whole scheme consists of two consecutive co-precipitation steps after the acidity of sample solution is adjusted by nitric and sulfuric acids: at first Pu(III+ IV) is separated with use of a holding oxidant, $K_2Cr_2O_7$, and then Pu in higher oxidation states in the filtrate is reduced to the lower states by $Fe^{2+}$, and followed by second co-precipitation. It is assumed that the addition of the acids and $K_2Cr_2O_7$ does not disturb the initial oxidation state distribution among two groups because the oxidation of Pu(IV) to Pu(V) or Pu(VI) is slow enough to be neglected when its concentration is low. Two nuclides $^{236}$Pu(VI) and $^{242}$Pu(IV), in different oxidation states as designated in parentheses, are used as internal standards to monitor the yields. Common overall yields exceeded 80%, and the separation of the two tracers typically was over 97%. The use of the internal standards also enables to estimate cross-contamination between lower and higher oxidation state species in the separated fractions. These authors claimed that Pu(III+IV) was contaminated by Pu(V+VI) of 0.4-1.2%, whereas Pu(V+VI) contained Pu(III+IV) of 1.0-2.2%.

## The speciation techniques of Choppin's group

Solvent extraction has been proved to be a powerful technique to investigate complexation equilibria of metal cations with wide variety of inorganic and organic ligands in aqueous media. The application of TTA extraction to actinide redox speciation was elaborated by seeking the most appropriate conditions such as the extractant concentration, the hydrogen ion concentration, and the equilibration time [9]. As fairly common in such basic studies, they used the " actinide analogues", i.e. Am(III), Th(IV), Np(V), and U(VI), in order to evaluate the method's performance. A scheme consisting of two parallel extractions at pH 0.6 and 4.5 can determine the distribution of neptunium and plutonium in oxidation states IV, V, and VI.

Since An(V) as $AnO_2^+$ cation is not extracted in the organic phase in these schemes, the activitiy remained in the aqueous phase in the above scheme is regarded as the pentavalent fraction. Also the $LnF_3$ precipitation method cannot distinguish An(V) and An(VI). Since two species differ significantly in their complexation ability, solubility, distribution ratio between solid surfaces and solution phase, quantitative separation from each other is desirable. Sorption behaviour of Np(V) and Np(VI) on silica gel(5), that of Pu(V) and Pu(VI) on $CaCO_3$ [10] were found to be useful for speciation of two oxidation states. Kobashi, Choppin, and Morse [11] reinvestigated these sorption systems by conducting thorough experimentation under systematically varied parameters. A practical scheme for Pu constructed according to the optimal conditions found that silica gel adsorbs more than 90% of An(VI), whereas $CaCO_3$ carries 80% of An(V) with small percentage of An(VI). They also showed an analytical procedure, which consists of a $LaF_3$ co-precipitation from acidified solution, a TTA extraction at pH 4-5, and two adsorption processes can determine total redox speciation.

## The speciation technique of Nitshe's group

Another scheme for Pu speciation in both high-level and trace-level concentrations was proposed by Nitsche, Lee, and Gatti [12]. The principle is not different from the ones already reviewed since it is a combination of solvent extraction and co-precipitation with $LaF_3$. In their scheme, use of three different types of compounds, TTA, HDEHP, and MIBK( Hexone ) as extractants conforms five extraction steps which provide five quantitative information about the distribution of Pu among III, IV, V, and VI oxidation states and polymeric species. The behaviour of Pu in each step of the scheme was made sure by using simulated oxidation state mixtures prepared by mixing Pu solutions of pre-adjusted oxidation states. It agreed satisfactorily with that expected from the oxidation state data by spectrophotometry. Also, the whole scheme was tested on samples prepared by mixing $^{242}$Pu in various oxidation states and was confirmed as an applicable technique with high reproducibility.

## Systems studied with application of the method and the results

The techniques briefly introduced in the preceding section have been applied to various systems of environmental significance. The method developed by Nelson and Lovett was used by themselves and other researchers for the field studies of Pu in aquatic environment, particularly in the Irish Sea where considerable amount of plutonium had been discharged from the nuclear fuel reprocessing plants [7,13-18]. These studies found how the discharged plutonium disperses, which of the oxidation states is the steady state in sea water, and what the depth profiles of concentration and of oxidation state distribution are. The separation is made only by $LnF_3$ precipitation, this technique is most suitable for large volume of solution samples. Very efficient use of two nuclides $^{236}Pu$ and $^{242}Pu$, oxidation state of which being adjusted to IV and VI, enables to determine absolute Pu concentrations of the reduced and the oxidised forms, and also to estimate the degree of cross-contamination of the separation of Pu in two fractions. The important findings are summarised in the last column of Table 1.

The TTA extraction technique was applied to laboratory studies of plutonium and neptunium behaviours in marine environments. Np and Pu in tracer concentrations were studied extensively in the systems consisted of natural and artificial seawaters and various mineral particles such as goethite, $\delta$-$MnO_2$, and calcite [19]. In addition to the measurements of adsorption isotherms, redox speciation of the tracer nuclides in the solution phases was conducted. Particularly for Pu, time course of its adsorption was followed to investigate the effect of mineral surfaces on the redox reactions of Pu. The important conclusions are that $PuO_2^+$ is unexpectedly susceptible to surface adsorption, and that mineral surfaces play significant roles in controlling the redox species distribution. The stability of Pu(V) and the effect of humic material on it's stability was also studied by the group as the application of their own techniques [20]. Pu tracer adjusted to pentavalent state was injected in artificial or real seawater and its re-distribution among other oxidation states was followed for a time span of a few tens of days. Conclusions of significance are given in Table 1. These studies on the behaviour of Pu(V) were made possible by the availability of a reliable preparation procedure of Pu(V) in tracer concentrations. The method was developed by Saito and Choppin [21] based on the experimental finding that Pu(VI) extracted into TTA is reduced to Pu(V) very quickly by the photochemical effect on irradiation of visible light [21].

Nitsche's group has undertaken the projects to study solubilities and redox speciation of transuranics in groundwaters and brines [12, 26-30]. Groundwater from a potential nuclear waste disposal site, Yucca Mountain, Nevada, USA, and synthetic brines were prepared as the simulants for the Waste Isolation Pilot Plant facility of US DOE, located deep underground in a bedded salt formation in south-eastern New Mexico, USA. Extensive solubility determination project provided the data of numerical solubility of Np, Pu, and Am, the composition of the solubility controlling solids, and the oxidation state distributions of species in the solution phases. The conclusions on the redox speciation are summarised in Table 1. The group also applied LPAS to the same sample of Pu concentration in the order of $10^{-5}$ M to compare the results with those obtained by their chemical separation technique. Discrepancy between two sets of data is within a permissible range [28].

## Evaluation of the techniques

Speciation techniques have to satisfy the requirements:

- The methods can distinguish Pu(III), Pu(IV), Pu(V), and Pu(VI) at tracer concentrations.

- The analytical processing of the sample should not disturb the steady state of the redox equilibrium.

- Reliability and reproducibility should be satisfying (satisfactory).

- The method should be applicable to wide variety of real and laboratory systems.

In the developmental stage, the actinide analogues, Am(III), Th(IV), Np(V), and U(VI), are usually used because of less possibility of oxidation state variation of these ions under relevant solution conditions. Reliability of the technique thus developed is evaluated according to the results of the experiments with the analogues. This is an easily accessible way for the evaluation of the accuracy, or methodical errors, of the speciation technique for tracer amount of plutonium. It should be noted that there are no direct experimental ways to evaluate the oxidation state variation during the analytical processing of the actual samples. Therefore, it is highly desirable that a scheme consists of less steps, uses smaller number of reagents as well as less amount of each reagent. Even the most sensitive instrumental method to date, LPAS, can only be used for samples of Pu concentration larger than $10^{-6}$ M.

In radiochemical separation techniques, detection limits are decided by the available instruments for $\alpha$-particle detection. Therefore, in field studies, sample volumes must be appropriately chosen to recover enough amount of activity. Volume sizes from a few tens to a hundred litres are common in the field studies of natural waters. This factor imposes severe restriction on the selection of the separation methods. On the other hand, in laboratory, Pu nuclides of wide range of half-lives, e.g. 45.4 d for $^{237}$Pu and $3.75 \times x \ 10^5$ year for $^{242}$Pu, have become available, which enables to run experiments at broad range of concentrations.

The speciation techniques described in the previous section have demonstrated their applicability to wide variety of systems. Particularly the method by Nelson and Lovett is useful for field studies, whereas that by Choppin's group is suitable for laboratory experiments and that by Nitsche's group has applicability to both types of studies because it can be regarded as the combination of the former two techniques.

All the techniques described are consisted of either parallel separation steps or sequential separation steps. In either case, each separation step is one of the following methods: precipitation (co-precipitation), sorption, and solvent extraction. No doubt these separation methods have advantages over the other analytical methods. A research group, H. Gehmecker, N. Trautmann, M. Mann, have reported two speciation techniques, one by ion exchange chromatography and the other by electrophoretic ion focusing [31,32]. These techniques seem to be rather time-consuming and the latter needs to use organic complexing agents, which potentially cause unfavourable redox reactions with probable disturbance of the original steady state.

In Table 2, evaluations of precipitants, sorbents, and extractants are summarised.

## Further researches needed

### *Improvements*

Precipitation, co-precipitation, adsorption, and solvent extraction are all well-established techniques. Their usefulness, capability in speciation have been well demonstrated. Improvements, if any, should be sought in new compounds as precipitants, sorbents or extractants. Particularly, selectivity of sorption should be improved. Even for $CaCO_3$ and silica gel, mechanisms and basic molecular level understanding of the adsorption phenomena do not seem to be fully understood.

Although TTA is a very useful extractant, it's photosensitive nature asks to avoid light exposure in the extraction of reducible species as observed for Pu(VI) [22]. Moreover, the solubility in aqueous solution tends to increase as the pH of the aqueous phase becomes higher. A more stable and weaker acid DBM (1,3-diphenyl-1,3-propanedione) was tested by Saito and Choppin [33] to provide a procedure for solutions of relatively high pHs. Two successive DBM extractions, first at pH 8 and then at pH 4-5 enable the separation of the actinides into III, IV, V, and VI oxidation state fractions. Satisfactory separation was fulfilled when the actinide analogs were used. The scheme performance was then tested for Pu with use of two nuclides, $^{239}$Pu and $^{242}$Pu. The oxidation state of each nuclide was initially adjusted to either IV or VI. Distribution of the nuclides among the four fractions was determined in four cases to each nuclide of IV or VI oxidation states, i.e. $^{239}$Pu(IV), $^{239}$Pu(VI), $^{242}$Pu(IV), or $^{242}$Pu(VI), and in two mixtures, i.e. $^{239}$Pu(IV) + $^{242}$Pu(VI) and $^{239}$Pu(VI) + $^{242}$Pu(IV). Natural water samples are mostly neutral, therefore, this method is applied without any adjustments of solution conditions. In the test experiments with uses of $^{239}$Pu(IV) and $^{242}$Pu(IV) in pH 8, 0.7 M NaCl, about 30% of the total activity was found in An(VI) fraction, and nearly 10% in the An(V) fractions. Although we tentatively explained the appearance of Pu(VI) as due to the disproportionation of Pu(IV), more experimental data should be accumulated. It is also found that the extraction kinetics of DBM at high pH, are rather slow and should be improved by the use of more efficient vial mixer or use of large organic solvent to aqueous phase volume ratios. On the recognition that Pu(VI) reduction in the TTA extraction cannot be avoided, Schramke, Rai, Fulton, and Choppin [34] constructed another extraction scheme with use of another extractant di-n-butyl-n-butylphosphonate( DBBP ) in addition to TTA to improve accuracy in Pu(V) and Pu(VI) speciation. As these examples demonstrate, modifications of currently existing methods will improve accuracy and reliability. Also they will be often required to design a practical scheme specifically suitable for systems of unique characteristics.

### Development of new techniques

In developing new techniques, the followings should be recognised: (1) relevant systems in which Pu redox speciation should be determined are very diverse, (2) the concentration of target nuclides is extremely low. Under these constraints principles of new speciation techniques must be basically similar to those given above: that is the speciation relies on the difference in chemical behaviour in a certain system produced by chemical processing of the sample.

Most plutonium discharged into the aquatic environment is adsorbed on particulates which tend to settle down to sediments layers, and that the adsorbed Pu is considered to be mainly in the reduced form. A study of Pu in the equatorial Equatorial Pacific surface sea water found that the species adsorbed on the solid surfaces control the solution phase concentration through remobilisation [35]. It has been also known that surfaces play important roles in the Pu oxidation state distribution, and therefore, solid-water interfaces should be regarded as one of the components of the aquatic systems. *In situ* techniques should be instrumental, such as photo acoustic absorption spectroscopy, fluorescence spectroscopy. But these techniques have applicability to samples of quite large nuclide contents. Total amount of actinide nuclides has been determined by digesting samples with acids. Therefore extraction of the species into solution phases is necessary for the speciation. Uses of acid extractants such as the aqueous solution of the same composition as used in LnF$_3$ precipitation or 1 M HClO$_4$ have been found effective [16]. However, it seems the studies in this respect are lacking compared to those in solution phase. Adsorption studies of actinides of specific oxidation states have been conducted to determine distribution coefficients, K$_d$, for various kinds of solids. Chemical extraction studies may complement the adsorption study and provide information about the oxidation states of the nuclides on the solid surfaces, although indirectly. Search of efficient reagents for chemical extraction and systematic study of the speciation of actinides on the surfaces of sediments, particulates, or colloids should be made.

## Conclusion

Currently existing techniques are able to provide satisfactory speciation of Pu in wide variety of aquatic systems. There are still needs of further modification and improvement for more accurate and reliable methods. Particularly, further researches are necessary for the speciation of Pu-polymers and of Pu adsorbed on solid material.

## REFERENCES

[1]     INTERNATIONAL ATOMIC ENERGY AGENCY (1976), "Transuranium Nuclides in the Environment", Proceeding of a Symposium, San Francisco, 17-21 November, 1975.

[2]     W.C. Hanson, ed., (1980), "Transuranic Elements in the Environment", DOE/TIC-22800, Technical Information Centre, US DOE.

[3]     S.C. Foti, E.C. Freiling (1964), *Talanta*, 11, 385- 392.

[4]     E.A. Bondieeti, S.A. Reynolds (1976) BNWL-2117, Proc. of the Actinide-Sediment Reactions Working Meeting, Seattle, Washington, USA, 1976, p. 505-530.

[5]     Y. Inoue, O. Tochiyama, (1977), *J. Inorg. Nucl. Chem.*, 39, 1443-1447.

[6]     D.M. Nelson, M.B. Lovett (1981), in *Techniques for Identifying Transuranic Speciation in Aquatic Environments*, Proc. Joint CEC/IAEA Technical Meeting, Ispra, 1980.

[7]     D.M. Nelson, M.B. Lovett (1978), *Nature*, 276, 599-601.

[8]     L.B. Magnusson, T.J. La Chappelle, (1948), *J. Am. Chem. Soc.*, 70, 3534-3538.

[9]     P.A. Bertrand, G.R. Choppin (1982), *Radiochim. Acta*, 31, 135-137.

[10]    A.L. Sanchez (1983), Ph. D. Dissertation, Univ. Washington, Seattle, Washington.

[11]    A. Kobashi, G. R. Choppin, J.W. Morse (1988), *Radiochim. Acta*, 43, 211-215.

[12]    H. Nitsche, S. C. Lee, R.C. Gatti (1988), *J. Radioanal. Nucl. Chem.*, Articles, 124, 171-185.

[13]    S.J. Malcolm, P.J. Kershaw, M.B. Lovett and B.R. Harvey (1990), *Geochim. Cosmochim. Acta*, 54, 29-35.

[14]    J. Hamilton-Taylor, M. Kelly, J.G. Titley, D.R. Turner (1993), *Geochim. Cosmochim. Acta* 57, 3367-3381.

[15]    D.M. Nelson, A. E. Carey, V.T. Bowen (1984), Earth Planet. Sci. Lett., *68*, 422-430.

[16] D.M. Nelson, M.B. Lovett (1981), in *Impact of Radionuclide Releases in to the Marine Environment*, Proc. of the Symposium held in Otaniemi, 30 June-4 July 1980.

[17] R.J. Pentreath, B.R. Harvey, and M.B. Lovett (1986), in *Speciation of Fission and Activation Products in the Environment*, eds. R. A. Bolman and J. R. Cooper, Elsevier, 312-325.

[18] D. Boust, P.I. Mitchell, K. Garcia, O. Condren, L. Leon Vintro, G. Leclerc (1996), *Radiochim. Acta*, 74, 203-210.

[19] W.L. Keeney-Kennicutt, J. W. Morse (1985), *Geochim. Cosmochim. Acta*, 49, 2577-2588.

[20] G.R. Choppin, R.A. Roberts, J.W. Morse (1986), in *Organic Marine Geochemistry*, M.L. Sohn, ed., ACS Symposium Series, Vol. 305, American Chemical Society, Washington, pp. 382-387.

[21] A. Saito, G.R. Choppin (1985), *Anal. Chem.* 57, 390-391.

[22] G.R. Choppin, A. Saito (1984), *Radiochim. Acta*, 35, 149-154.

[23] D.E. Meece, L.K. Benninger (1993), *Geochim. Cosmochim. Acta*, 57, 1447-1458.

[24] D.M. Nelson, A.E. Carey, V.T. Bowen (1984), *Earth Planet. Sci. Lett.*, 68, 422-430.

[25] A.L. Sanchez, J.W. Murray, T.H. Sibley (1985), *Geochim. Cosmochim. Acta*, 49, 2297-2307.

[26] H. Nitsche, A. Muller, E.M. Standifer, R.S. Deinhammer, K. Becraft, T. Prussin, R.C. Gatti (1992), *Radiochim. Acta*, 58/59, 27-32.

[27] C.F. Novak, H. Nitsche, H.B. Silber, K. Roberts, Ph.C. Torretto, T. Prussin, K. Becraft, S.A. Carpenter, D.E. Hobart, I. AlMahamid (1996), *Radiochim. Acta*, 74, 31-36.

[28] M.P. Neu, D.C. Hoffman, Kevin Roberts, H. Nitsche, Robert J. Silva (1994), *Radiochim. Acta*, 66/67, 251-258.

[29] H. Nitsche and N.M. Edelstein (1985), *Radiochim. Acta*, 39, 23-33.

[30] H. Nitsche, K. Roberts, R. Xi, T. Prussin, K. Becraft, I. Al Mahamid, H.B. Silber, S.A. Carpenter, R.C. Gatti (1994), *Radiochim. Acta*, 66/67, 3-8.

[31] H. Gehmecker, N. Trautmann, G. Herrmann (1986), *Radiochim. Acta*, 40, 81-88.

[32] M. Mang, H. Gehmecker, N. Trautmann and G. Herrmann (1993), *Radiochim. Acta*, 62, 49-54.

[33] A. Saito, G.R. Choppin (1983), *Anal. Chem.*, 55, 2454-2457.

[34] J.A. Schramke, D. Rai, R.W. Fulton, G.R. Choppin (1989), *J. Radioanal. Nucl. Chem.*, Articles, 130, 333-346.

[35] E. Sholkovitz (1983) *Earth-Science Rev.* 19, 95-161.

**Table 1. Systems studies; the methods have been applied**

| Method [Ref.] | Systems studied | Results |
|---|---|---|
| Nelson and Lovett [7, 13-18] | Field studies of $^{239,240}$Pu discharged in the Irish Sea and the North Pacific Ocean. | 1) Pu in the Irish Sea: 75% Pu(V+VI) in solution phase, 10% Pu(V+VI) on particulate matter.<br>2) Pu(IV) injected in filtered sea water is oxidised up to 60-80%.<br>3) Pu on sediments, in interstitial water is in the reduced state. |
| Choppin, *et al.* [19, 23-25] | Laboratory studies of actinides behaviour in marine environment; sea water/mineral surfaces and sea water/humic materials. | 1) Pu(V) is stable in sea water, but slowly reduced to IV.<br>2) Mineral surfaces play important roles to control Pu oxidation states.<br>3) Humic materials reduce Pu(VI) to Pu(IV).<br>4) Sunlight causes oxidation of Pu(IV) to Pu(V), or Pu(V) to Pu(VI). |
| Nitsche, *et al.* [12, 26-30] | Groundwaters from prospective waste repository sites, and natural and synthetic brines. | 1) Pu(IV) solubility in ground waters; 3 $10^{-7}$-1 $10^{-6}$ M.<br>2) Solubility controlling solid is amorphous Pu(IV) polymer.<br>3) Pu in brines; Pu(V) is stable, Pu(VI) is reduced to Pu(V). |

**Table 2. Evaluation of precipitants, adsorbents and extractants**

| | Precipitants | | | Adsorbents | | | Extractants | | |
| | | | | | | | β-diketones | | HDEHP |
| | LnF$_3$ | BiPO$_4$ | Zr(IO$_3$)$_4$ | Silica gel | CaCO$_3$ | TiO$_2$ | TTA | Others | |
|---|---|---|---|---|---|---|---|---|---|
| Advantages | Suitable for large volume samples. | Suitable for large volume samples. Dense, crystalline precipitate readily separable. Soluble in concentrated nitric acid | | Distinguish Pu(V) and Pu(VI). Used for large volume samples. Better separation efficiency. | Distinguish Pu(V) and Pu(VI). Used for large volume samples. Short contact time (5-10 min). | Distinguish Pu(V) and Pu(VI). Used for large volume samples. | Distribution ratio heavily depends on [H]. Quantitative determination. Quick extraction kinetics. | PMBP: more resistant, small pKa (4.1). DBM: large pKa (9.2). | Very stable |
| Disadvantages/ limitations | Necessary to make sample solutions acidic. Requires drastic treatment to effect re-dissolution of precipitate. | | Iodate oxidises Pu(III) to Pu(IV). Cause depolymerisation of Pu polymer | Careful pre-treatment of silica gel is necessary. Long contact time (2-4 h). | Careful pre-treatment of CaCO$_3$ is necessary. Poor separation efficiency. | Poor separation of Pu(V) and Pu(VI). | Photo-sensitive. Larger solubility and instability in neutral solution. | DBM: slow extraction kinetics. | Less selective for oxidation states. |

# SESSION E

## *Predictive Approach to Speciation*

# ESTIMATION OF THERMODYNAMIC DATA IN ASSESSING THE GEOLOGICAL DISPOSAL OF RADIOACTIVE WASTE

**David Read**
Enterpris Limited
Postgraduate Research Institute for Sedimentology
University of Reading
Whiteknights, Reading, United Kingdom, RG6 6AB

## Abstract

The paucity of data for safety critical species is commonly cited as the reason why relatively few thermodynamic parameters are employed directly in conventional safety calculations. This paper provides an overview of the information required and describes a number of data estimation methods that have been employed with considerable success in the geological sciences. Each is based upon a thermodynamic representation of the processes known to take place during rock-water interaction. It is hoped that in the future such techniques will allow a shift of emphasis from the collection of empirical "sorption constants" to a more fundamental appraisal of trace element immobilisation in waste disposal systems.

## Introduction

Thermodynamic data provide much of the scientific basis for assessing the performance of a radioactive waste disposal facility, even though relatively few thermodynamic parameters are employed directly in conventional safety calculations. Representation of chemical processes tends to reduce to a combination of leach rates from the waste form, solubility constraints and empirical "sorption coefficients" to account for retardation in the near field and host geology, respectively. A key factor, the "source term" is often treated in terms of "instant release fractions", concentrations that are invariant with time or those that vary as a simple function of source depletion. Thereafter, levels are constrained solely by dilution or equilibrium partitioning onto components of the engineered barriers and the surrounding rock. Such models are proving to be increasingly difficult to justify, as not only do they fail to adequately represent the processes occurring, it can often be shown that the consequences are not necessarily conservative [1,2].

Improving the rigour of safety analyses is the aim of all waste management agencies, which raises the question as to why thermodynamic methods are not used more extensively. In addition to resource constraints [3], three main reasons are commonly cited:

- Lack of detailed mechanistic understanding, particularly with respect to water-rock interaction processes.

- The inherent limitations of equilibrium thermodynamics, for example when applied to mineral dissolution.

- The paucity of data for safety critical aqueous species and solids.

The first two items are inter-related and, consequently, are addressed together in the next section. Thereafter, emphasis is placed on methods for estimating appropriate parameter values in the absence of direct experimental evidence. The present state of knowledge with respect to the actinide elements and the use of chemical analogy are discussed in more detail in an accompanying paper [4].

## Geochemical processes and their representation in performance assessment

Over the past twenty years, a consensus has emerged regarding the processes that are critical in a performance assessment and the necessary prerequisites for ensuring safe disposal. Positive attributes include a stable geological environment, low permeability to groundwater flow and reducing chemical conditions. However, there is considerable debate on how best to represent these processes and a large degree of uncertainty in choosing appropriate parameter values. With respect to chemical aspects the key factors are:

- Corrosion of the primary containment.

- Dissolution of the waste form.

- Evolution of the backfill, typically bentonite clay.

- Interaction with the host rock.

The biosphere is normally regarded as a simple dilution medium although many near surface processes have the potential to concentrate as well as to disperse radionuclides [5,6].

Corrosion of metal containers, be they steel, copper, iron, titanium, alloys or composites is amenable to thermodynamic treatment and considerable research has been carried out on this subject in recent years [7]. The formation of a wide variety of mixed corrosion products such as Fe-U-Cu constitutes a challenging area for further research; however, detailed discussion of this topic is beyond the scope of the current paper.

Models for predicting source terms require an adequate description of the inventory characteristics and depend, therefore, on the complexity of the waste form. Though thermodynamic models have been developed for highly variable, multi-mineralic encapsulants such as blended cement [8], leach rates for assessment purposes are more commonly derived from accelerated leach tests or wash out experiments. In contrast, the simulation of spent fuel dissolution relies far more on fundamental principles. The kinetics of $UO_2$ leaching both in the form of fuel pellets and natural uraninite have been reported as a function of pH. Eh, complexant concentrations and particle size [9,10]. These experiments indicate very slow breakdown under reducing conditions controlled by the low solubility of tetravalent U. Depending on the extent of radiolysis, relatively high oxidation potentials, exceeding 200 mV are required to promote oxidative dissolution. This accompanies a change in the stoichiometry of the oxide beyond $U_3O_7$ towards $U_3O_8$. Published dissolution rates show a decrease with time and an increase with decreasing particle size, values in the 200 μm region being typically ~ 1 mg m$^{-2}$ d$^{-1}$. In natural waters a complex assemblage of U(VI) phases may result depending on the composition of the groundwater, the most common being the autunite series of phosphates ($M^{2+}[UO_2]_2[PO_4]_2.10H_2O$) and silicates such as uranophane ($Ca[UO_2SiO_3OH]_2.5H_2O$).

It is apparent from these studies that the formation of surface alteration products is an essential factor in the oxidation-dissolution mechanism. These phases tend to be less stable than $UO_2$ and play an important role in controlling aqueous concentrations. The availability of thermodynamic data for secondary uranium minerals is discussed briefly in the following section.

Despite the progress made to date, calculation of source terms for released nuclides is not straightforward even with spent uranium fuel, the most stoichiometric of waste forms. Whereas the major controls on dissolution of the $UO_2$ matrix itself are reasonably well established, this is certainly not the case for trace transuranic and fission product impurities. Johnson and Tait [10] provide an outline of the issues with respect to performance assessment studies in Sweden.

Once released from the waste form, equilibrium partitioning onto components of the engineered barrier (buffer, backfill) and the host rock is normally assumed. "Sorption" is the means via which almost all performance assessment models constrain contaminant migration. There is, of course, ample evidence of sorptive uptake of trace elements on a variety of mineral surfaces. Equally, however, it has been demonstrated that the process is not entirely reversible and surface sites can be saturated relatively easily [11]. Sorption is increasingly being regarded as a short-term phenomenon more relevant to laboratory investigations than migration in natural systems on geological time scales. Indeed, many workers consider the processes of surface complexation, surface precipitation and co-precipitation as steps towards the formation of new minerals [12]. For example, detailed re-evaluation of U "sorption" at natural analogue sites (Table 1; Figure 1) has revealed the presence of previously unidentified, microcrystalline uranium-bearing phases [1,2].

Thus, though quasi-thermodynamic, surface complexation models provide a means of quantifying sorption behaviour they strictly apply only to reversible uptake of trace ions on amphoteric (usually iron) oxide surfaces [13]. This idealised situation is unlikely to obtain for rocks and soils particularly in the case of actinide ions (e.g. U(IV), Th(IV), Pu(IV)). It is essential to establish whether the transuranic elements are firmly bound at lattice sites within these secondary phases or more loosely attached to the surface. Evidence from alteration of natural ore deposits highlights the importance of secondary mineral formation in immobilising uranium, thorium and related elements such that solubility rather than "sorption-desorption" constants are required. These data are relatively scarce in stark contrast to the plethora of empirical batch "Kd" values listed in the literature. Methods of estimating the appropriate thermodynamic parameters are described below.

## Estimation of thermodynamic data for geochemical systems

### *Rationale*

The limitations of equilibrium thermodynamics when applied to geochemical systems are well known. Whether a given system will attain equilibrium is dependent not only upon reaction kinetics but also upon the physical properties of the media. These determine the rate at which constituents can be transported to and from the reactive zone such that where reactions are fast relative to transport processes (e.g. diffusion) local equilibrium can often be assumed. However, heterogeneous reactions, particularly those involving dissolution of crystalline mineral solids are generally much slower than those in solution and the local equilibrium assumption is less valid. Such reactions can only be studied experimentally at elevated temperatures where phase transformations may occur and, further, the degree of order will have a significant effect on the Gibbs free energy [14]. For most mineral phases the state of order is not known as a function of temperature or, in some cases, even at reference conditions.

Notwithstanding the difficulties outlined above, thermodynamic methods are inherently more defensible in an assessment context than a purely empirical approach. They can be used to at least assess those chemical reactions that are theoretically possible should equilibrium ultimately be attained. Even where the release of a given radioelement from a matrix is kinetically controlled, for example the oxidative dissolution of uranium dioxide, its concentration in solution may be determined by the solubility of secondary phases that are amenable to thermodynamic treatment. Examples of recent studies using a combination of controlled laboratory and field observations in conjunction with thermodynamic models include those of Ollila [15] and Bruno and co-workers [9].

### *Scope of the task*

Enormous resources have been devoted to the compilation of thermodynamic data by the nuclear industry. Despite this, a critical review of available listings carried out in the early 1990s [16] concluded that none were adequate for quantitative analysis of natural series radionuclide behaviour, though several were purportedly constructed for this purpose. Among the widely-used databases evaluated, one contained no entries for U(VI) silicates [17], another no data for U(VI) silicates, carbonates or phosphates [18] and several [19] no data for natural uranium mineral phases at all. Indeed, the OECD/NEA review of uranium thermodynamics [20], which is generally considered to be the most authoritative yet undertaken, did not originally include mineral phases owing to the perceived lack of published experimental determinations of thermodynamic properties. This has been rectified in a subsequent review [21] from which the strengths and shortcomings in the available data are clear. The most complete set is that for oxide-hydroxide phases, reflecting the interest in spent fuel dissolution, and is followed by that for the phosphates, the most abundant group of secondary U(VI) phases. Very few attributable values exist for the corresponding silicates, arsenates, carbonates or sulphates and the situation for the neighbouring elements, which do not occur naturally, is far worse. New experimental data are required urgently, but given the extent of the task [22] will not make a substantial difference in the short term. The need for reliable methods of estimating thermodynamic values is readily apparent.

## Data estimation methods

### Aqueous species

The remainder of the paper is devoted to data estimation methods on the premise that no thermodynamic compilation will ever be complete. Published listings differ substantially in the range of parameters considered and so a decision is required at the outset regarding the quantities to be estimated. In a recent survey of extant databases [23], the OECD/NEA distinguished three main types: (i) often extensive, pure substance databases at standard state, (ii) very specific compilations for metal-oxide systems and (iii) more generally applicable databases designed to model a variety of aqueous systems and incorporating data for solid and gaseous species. The last category includes the databases commonly employed in radioactive waste disposal studies. The range of thermodynamic quantities listed and also the stringency of acceptance criteria decreases in the order (i) to (iii), reflecting the availability of parameter values. Thus, the majority of "geochemical" compilations described contain only formation constants (log $\beta$) and solubility products ($K_{sp}$) at 25°C for aqueous species and mineral solids, respectively.

Numerous approaches have been developed for estimating the above constants over the past 15 years, predominantly for aqueous species. These methods are based on correlation between ions of similar charge-size characteristics and the same or similar configuration of valence electrons. Empirical rules have been devised to describe the relative affinities between various donor ligands and metal ions, providing a guide to the likely stability constants for complexes where no direct experimental determinations are available [24]. The approaches are distinguished by the empirical relationships employed and the relative weighting attached to complexation affinities with various ligands. Additionally, structural information on the preferred geometry of a metal ion with different ligands in binary or ternary configurations provides a means of estimating the stoichiometry of postulated complexes in multi-component systems. Finally, a reasonable assessment of likely behaviour may be obtained by simple chemical analogy without recourse to complex algorithms. For example, the behaviour of actinide (III) ions such as $Am^{3+}$ can be inferred from the properties of the corresponding lanthanides, Nd or Sm. The use of chemical analogy as applied to the actinide elements is discussed in more detail in an accompanying paper [4].

### Mineral solids

In the case of mineral solids, data estimation relies heavily on structural information as it has been shown that approximately linear relationships exist between molar volumes and entropy for many isostructural compounds [25]. Departures from this idealised behaviour have been attributed to crystal field effects for which corrections can be applied [26]. More recent hybrid approaches combine the above with statistical theory. Thus, many of the thermodynamic properties and functions required for estimating petrological phase equilibria can be derived from basic principles and several "internally consistent" data sets exist [25,27-29]. The two most commonly used are those of Berman [28] and Holland and Powell [29].

The different philosophies behind these compilations have been reviewed by Gottschalk [14]. The starting point, for each, is a subset of thermodynamic properties and functions for each phase considered, *i.e.* the standard enthalpy of formation ($\Delta_f H°$), the standard entropy (S°), heat capacities over the relevant range of temperatures ($C_p$) and the molar volumes as a function of temperature and pressure (V°, $\alpha$, $\beta$), where $\alpha$ and $\beta$ are the coefficients of thermal expansion and compressibility, respectively. If phase transitions occur (e.g. aragonite to calcite) or composition varies in response to temperature or pressure gradients then the corresponding changes in these properties must be known.

Additionally, if a phase is impure, then activity-composition relations are required. Thereafter, the data sets differ in terms of the techniques used in their derivation and in the extent of their derived properties. Holland and Powell [29], for example, employ least squares regression to refine $\Delta_f H°$, assuming all other parameters are known and can be fixed. This approach has both theoretical and procedural shortcomings. First, as only $\Delta_f H°$ values are optimised, uncertainties in other properties such as $S°$, which may be considerable, are ignored. Second, the algorithm employed always assigns the highest probability of equilibrium to cases where the enthalpy values are at the centre of the experimentally measured range, thereby precluding any weighting for high quality or highly uncertain results.

In contrast to the above, Berman [28] considered all thermodynamic parameters and functions to contain some experimental error and, therefore, to be targets for optimisation. For every experimental datum an inequality of the form:

$$0 \overset{<}{_{>}} H°-T\Delta S° + RT\ln K \tag{1}$$

is written and the best solution derived using an objective function. Since the objective function is quadratic rather than linear, mathematical programming is used to obtain the optimised solution.

The main disadvantage of this approach arises from the linear programming algorithm used, which requires a fully consistent set of measurements and the fact that it attempts to optimise too many parameters. Thus, potentially contradictory experimental observations must be eliminated by *a priori* decisions that inevitably bias the final outcome (Figure 2).

An iterative method developed recently [14] builds on these methods and overcomes their shortcomings to a considerable extent. By rearranging Eq. (1) and plotting $1/T$ versus $\ln K$, the equilibrium of a chemical reaction reduces to a straight line (Figure 3), which can be used to check the consistency of both experimental data and estimates. It follows that if the reaction is formulated in such a way that $\Delta H°$ is always positive, then results where the products are stable should plot above the line. Conversely, a plot below the line denotes where the reactants are stable. All available data can be included and outliers are readily apparent.

*Estimation of equilibrium constants at elevated temperatures*

The temperature of a high level waste repository is likely to exceed 100°C, for which few direct measurements are available. As a consequence, extrapolation of standard state data for the major groundwater ions has been studied extensively by geochemists since the 1950s. Established methods include the correspondence principle [30], electrostatic theory [31], the isocoulombic principle [32] and the use of density functions [33]. Of these, the isocoulombic principle is the most flexible, easiest to use and requires the least number of experimental parameters. For these reasons, it is described briefly below.

For isocoulombic reactions (those where there is no change in the charges on the reactants and products) the temperature dependence of free energy change ($\Delta G$) reduces to:

$$\Delta G_T° = \Delta G_{T0}° - (T - T_0)\Delta S_{T0}° + A\Delta C_p° \tag{2}$$

where $A = T - T_0 - T\ln T/T_0$

As noted by Gu, *et al.* [32], The heat capacity term in such reactions is nearly constant and close to zero. Further, since the change in entropy is related to ion-solvent interaction and, hence, the charge on the ions, a balanced isocoulombic reaction should also minimise the preceding entropy term in Eq. 2, giving:

$$\Delta G^{\circ}_{T} = \Delta G^{\circ}_{T0} = -RT \ln K_{T} \tag{3}$$

Thus, only the free energy change at one reference temperature is required to estimate equilibrium constants at any temperature of interest. This simple approach has clear advantages, as heat capacity data, in particular, are difficult to obtain. Obviously, where enthalpy, entropy, molar volume and heat capacity data are available then more elaborate models would normally be favoured. However, the method has been tested against experimental data with excellent results [32] and, in a number of cases, appears to perform better than more complex models requiring a larger number of adjustable parameters (Figure 4).

## Concluding remarks

Performance assessment of geological repositories depends to a very large extent on estimates of radio element solubility in the near field and, subsequently, quantifying their interaction with the host rocks in the far field. The highly empirical models that have been applied over the past twenty years are becoming increasingly difficult to justify, as they do not adequately represent the processes known to occur. Further, evidence from natural ore deposits suggests that assumptions previously thought to be "conservative" may not necessarily be so. A defensible thermodynamic basis is clearly required in future assessments.

Some of the reasons cited in the past for not adopting a more rigorous approach, for example limited computing resources, are no longer valid. However, the lack of high quality thermodynamic data is a real concern. Even in the case of spent nuclear fuel, essentially $UO_2$, the potential range of alteration products is very large and few reliable data exist for these phases. In critical areas, such as plutonium redox chemistry for example, there is little alternative to commissioning new experimental measurements. This is obviously not a feasible approach in all cases. It is accepted that all possible reactive species should be included in any given simulation. Thus, prioritisation of data requirements should distinguish those reactions for which new experimental data are essential and those for which estimates may suffice.

In this short paper, it has been possible to mention only a small number of the various techniques available to estimate thermodynamic quantities. These methods have gained widespread acceptance in the geochemical community and now need to be applied to radiologically critical elements in a waste disposal context. It is hoped that in the future emphasis will shift from the collection of empirical "sorption" data to a more fundamental appraisal of trace element immobilisation in rock-water systems coupled with a thermodynamic representation of the processes taking place.

# REFERENCES

[1]     T. Murakami, T. Ohnuki, H. Isobe, and T. Sato, *Am. Min.*, 82, 888-899 (1997).

[2]     D. Read, K. Rasilainen, C. Ayora and T. Ruskeeniemi, Proc. 8th CEC NAWG Workshop, Strasbourg, France (1999).

[3]     U. Vuorinen, S. Kulmala, M. Hakanen, L. Ahonen and T. Carlsson, POSIVA Report 98-14 (1998).

[4]     H. Moriyama, these proceedings (1999).

[5]     D. Read, D.G. Bennett, P.J. Hooker, M. Ivanovich, G. Longworth, A.E. Milodowski and D.J. Noy, *J. Contam. Hydrol.*, 13, 291-308 (1993).

[6]     A.E. Milodowski, *Nature*, 347, 465-467 (1990).

[7]     E. Smailos, I. Azkarate, J.A. Gago, P. van Iseghem, B. Kursten and T. McMenamin, Proc. 4th European Conf. "Management and Disposal of Radioactive Waste", CEC Report EUR 17543 (1997).

[8]     D.G. Bennett, D. Read, M. Atkins and F.P. Glasser, *J. Nucl. Mat.*, 190, 315-325 (1992).

[9]     J. Bruno, I. Casas, I. Puigdomènech, *Geochim. Cosmochim. Acta*, 55, 647-658 (1991).

[10]    L.H. Johnson, J.C. Tait, SKB Report 97-18 (1997).

[11]    D. Read, D. Ross and R.J. Sims, *J. Contam. Hydrol.* 35, 235-248 (1998).

[12]    J. Bruno, J. De Pablo, L. Duro and E. Figuerola, *Geochim. Cosmochim. Acta*, 59, 4113-4123 (1995).

[13]    D.A. Dzombak and F.M.M. Morel, Surface Complexation Modelling, John Wiley, New York (1990).

[14]    M. Gottschalk, M. *Europ. J. Min.*, 9, 175-223 (1997).

[15]    K. Ollila, POSIVA Report 98-06 (1998).

[16]    W.E. Falck, UK DoE Report HMIP/RR/92.064 (1992).

[17]    M. Chandratillake, W.E. Falck and D. Read, UK DoE Report HMIP/RR/92.094 (1992).

[18]    J.E. Cross and F.T. Ewart, UK Nirex Report NSS/R212 (1990).

[19]    J. Bruno and I. Puigdomènech, *Mat. Res. Soc. Symp. Proc.*, 127, 887-896 (1989).

[20]  I. Grenthe, J. Fuger, R.J.M. Konings, R.J. Lemire, A.B. Muller, C. Nguyen-Trung and H. Wanner, "Chemical Thermodynamics of Uranium", North Holland (1992).

[21]  R.J. Silva, G. Bidoglio, M.H. Rand, P.B. Robouch, H. Wanner and I. Puigdomènech, "Chemical Thermodynamics of Americium", North Holland (1995).

[22]  Draft final report, CEC contract FI4W-CT96-0029, T. Fanghänel and D. Read, eds. (1999).

[23]  OECD/NEA, Nuclear Science Committee Report NEA/NSC/DOC (96)27 (1996).

[24]  I. Grenthe, *et al.*, in "Modelling in Aquatic Chemistry", Edition 1, OECD/NEA (1997).

[25]  H.C. Helgeson, J.M., Delany, H.W., Nesbitt and D.K. Bird, *Amer. J. Sci.*, 278A, p. 229 (1978).

[26]  B.J. Wood, in "Advances in Physical Geochemistry", Vol. 1, R.C. Newton, A. Navrotsky, B.J. Wood, eds., Springer, Berlin (1981).

[27]  B.S. Hemingway, J.L. Haas and G.R. Robinson, *Geol. Surv. Bull.*, 1544, p. 70 (1982).

[28]  R.G. Berman, *J. Petrol.*, 29, 445-522 (1988).

[29]  T.J.B. Holland and R. Powell, *J. Metamorphic Geol.*, 8, 89-124 (1990).

[30]  C.M. Criss and J.W. Cobble, *J. Amer. Chem. Soc.*, 86, 5385-5390 (1964).

[31]  H.C. Helgeson, D.H. Kirkham and G.C. Flowers, *Amer. J. Sci.*, 281, 1249-1516 (1981).

[32]  Y. Gu, C.H. Gammons and S. Bloom, *Geochim. Cosmochim. Acta*, 58, 3545-3560 (1994).

[33]  G.M. Anderson, S. Castet, J. Schott, and R.E. Mesmer, *Geochim. Cosmochim. Acta*, 55, 1769-1779 (1991).

**Table 1. Uranium fixation at natural analogue sites**

| Location | Dominant processes |
| --- | --- |
| Poços de Caldas | Reduction |
| Koongarra | Secondary mineralisation |
| Broubster | Organic complexation |
| Needle's Eye | Organic complexation + reduction |
| South Terras | Secondary mineralisation |
| Steenkampskraal | Secondary mineralisation |
| El Berrocal | Co-precipitation |
| Palmottu | Secondary mineralisation |

**Figure 1.** (a) Backscattered electron image of saleeite (bright areas) in quartz-chlorite schist from Koongarra, (b) partly and (c) almost completely replacing apatite (from [1])

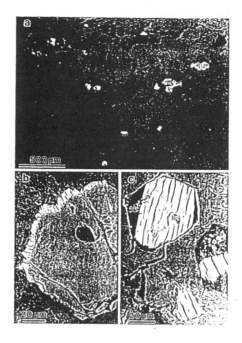

**Figure 2.** Illustration of bias introduced by omitting contradictory results. (a) four consistent constraints (grey area is region of feasibility), (b) two additional but contradictory constraints (I and II) are introduced. No region for a feasible solution exists. In (c) constraint (II) and in (d) constraint (I) is cancelled. Note the regions of feasibility in (c) and (d) do not overlap (from [14]).

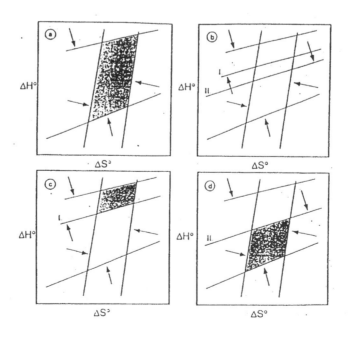

**Figure 3.** 1/T vs. lnK plots for the reactions (a) 1 calcite + 1 muscovite + 2 quartz ⇔ 1 anorthite + 1 sanidine + 1$CO_2$ + 1$H_2O$, (b) 1 diopside + 3 dolomite ⇔ 2 forsterite + 4 calcite + 2 $CO_2$. Filled and empty symbols represent stable products and reactants, respectively (from [14]).

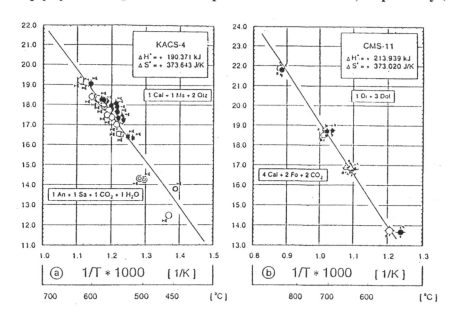

**Figure 4.** Equilibrium constants for zinc chloride and bisulphide ligand exchange. Experimental data denoted by solid circles, one-term and two-term isocoulombic extrapolations by solid and dashed lines, respectively (from [32]).

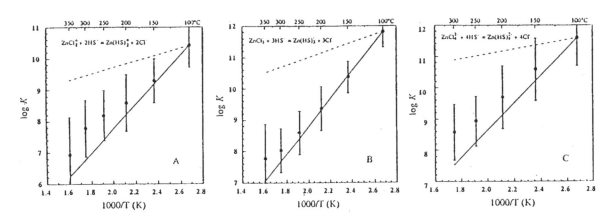

# CHEMICAL ANALOGY IN THE CASE OF HYDROLYSIS SPECIES OF F-ELEMENTS

**Hirotake Moriyama, Hajimu Yamana**
Research Reactor Institute, Kyoto University
Kumatori-cho, Sennan-gun, Osaka, 590-0494, Japan

**Kenso Fujiwara**
Department of Nuclear Engineering, Kyoto University
Yoshida, Sakyo-ku, Kyoto 606-8501, Japan

**Akira Kitamura**
Waste Management and Fuel Cycle Research Centre
Japan Nuclear Cycle Development Institute
Tokai-mura, Naka-gun, Ibaraki 319-1194, Japan

## Abstract

In spite of much importance, some of the thermodynamic data of actinide elements are still lacking, and the chemical behaviour of these elements is often predicted by considering chemical analogy, that is by taking the known data of the same and similar group elements. It is thus quite important to establish the basis and conditions for applying the chemical analogy. In the present study, some topics are discussed which are related with the chemical analogy for the hydrolysis behaviour of actinide ions. Systematic trends of the thermodynamic data of actinide ions, namely the solubility products and hydrolysis constants, are discussed by considering the results of recent measurements and by considering possible contributions of non-electrostatic interactions of actinide ions in addition to those of ordinary electrostatic ones.

**Introduction**

Because of the characteristics of 5f-electrons, actinides are known to exhibit complicated chemical behaviours in aqueous solutions. For example:

- Actinides, especially lighter ones are much sensitive to redox conditions and are dissolved in various valence states in aqueous solutions. Very different chemical behaviours are observed for different valence states.

- Actinide ions are usually of multi-valence and are easily hydrolysed in aqueous solutions. Very low solubility and colloidal behaviour are often observed.

- Actinide ions of multi-valence are likely to interact with various complexing anions present in solutions. Because of strong tendencies for complexation, carbonate and some organics are recognised to be important in natural waters. The solubility is enhanced in some cases and pseudo-colloids are formed in other cases.

- Actinide ions are also likely to interact with mineral components in geological formations. Very high distribution ratios are known for rock specimen of many kinds.

Much effort has been dedicated to understand such complicated behaviours of actinide ions, but it is still needed for a complete understanding. In fact, some of the thermodynamic data of actinide ions are still lacking which are required for the safety assessment of geological disposal of radioactive wastes. Thus, the chemical behaviour of these elements is often predicted by considering chemical analogy, that is by taking the known data of the same and similar group elements, in the safety assessment programs. So, not only from a scientific but also from a practical point of view, it is quite important to establish the basis for applying the chemical analogy.

In the present study, some topics are discussed which are related with the chemical analogy for the hydrolysis behaviour of actinide ions. Systematic trends of the thermodynamic data of actinide ions, namely the solubility products and hydrolysis constants, are discussed by considering the results of recent measurements and by considering possible contributions of non-electrostatic interactions of actinide ions in addition to those of ordinary electrostatic ones. It is suggested to take into account the contributions of 5f-electrons and orbitals for such non-electrostatic interactions.

**Solubility products of actinide oxides**

The solubility product of actinide oxides is one of the most important factors to be considered for the safety assessment of geological disposal since it directly controls the actinide solubility. However, the reported values are often scattering and, especially for tetravalent actinide oxides, there are very large differences of several orders of the magnitude from the lowest to the highest values. Since there are a number of experimental difficulties, it is likely that the differences in the reported values are mostly due to experimental problems in their measurements. In the case of tetravalent actinide ions, for example, the concentrations of the solution species are very low and are not so easily determined by direct means. This is one of the main reasons for such large differences, in addition to the effect of ageing of the solid phase.

In their extensive studies on the solubility products of tetravalent actinide hydrous oxides, Rai, *et al.* [1] have found that the experimental log $K_{sp}$ values of tetravalent actinide hydrous oxides show a linear relationship with the inverse square of the $M^{4+}$ ionic radii as shown in Figure 1. This means that the 5f-electrons of the actinides are well shielded from the environment similarly to the case of the

4f-electrons of the lanthanides, and that the degree of covalency in these elements is relatively small compared to the covalency in d-group elements. Such a relationship is much useful to reliably estimate the log $K_{sp}$ values for other tetravalent actinide hydrous oxides as well, and the values are estimated to be -50.6 for Pa(IV), -52.6 for U(IV), -58.0 for Am(IV), -59.1 for Cm(IV), -60.7 for Bk(IV), and -61.4 for Cf(IV).

It is interesting and important to compare the estimated log $K_{sp}$ values with the experimental ones to establish this relationship. In Table 1 are shown some of the experimental values for comparison. For U(IV) hydrous oxides, Rai, *et al.* have reported two different log $K_{sp}$ values of -52.0 ± 0.8 [2] and -53.45 [3]. The latter value is about 1.5 orders of magnitude lower than the former one which has been calculated from the short-term data, and it is believed that this difference is due to the effect of ageing of U(IV) hydrous oxides. In Figure 1, one can see that the value of -53.45 is not so different from the above estimated value of -52.6. However, a much lower value of -55.66 has been reported by Yajima, *et al.* [4] recently. In their experiment, the filtrate has been subjected to solvent extraction by 10% TBP-CCl$_4$ in order to eliminate the effect of U(VI) which may coexist with U(IV). From this fact, it is expected that the higher log $K_{sp}$ value of -53.45 is due to possible coexistence of U(VI) or U(V) with U(IV) in the case of Rai, *et al.* [3]. A direct measurement of the solution species of U(IV) is needed for confirmation.

For Pu(IV), Rai [5] reported the log $K_{sp}$ value of -56.85 ± 0.36 which was used to obtain the linear relationship in Figure 1. However, his experiment was carried out in nitrate medium and rather large uncertainty would result from nitrate complexation. In fact, a lower log $K_{sp}$ value of -58.3 ± 0.3 was obtained in perchloric medium by Capdevila and Vitorge [6]. In order to determine a very low $Pu^{4+}$ concentration in their experiments, the speciation was carefully controlled by choosing the chemical conditions to reach a steady state between a Pu(IV) hydrous oxide and $PuO_2^{2+}$, $PuO_2^{+}$, $Pu^{3+}$ aqueous species in a concentration range high enough to be directly and independently measured by spectrophotometry. The redox potential of the solution was calculated from the measured $[PuO_2^{2+}]/[PuO_2^{+}]$ ratio and the known potential of this couple, and then the $Pu^{4+}$ concentration was deduced from the calculated redox potential, the measured $[Pu^{3+}]$ and the known $Pu^{4+}/Pu^{3+}$ potential. Also, Kim and Kanellakopulos [7] reported a much lower value in their direct measurement of the $Pu^{4+}$ concentration by spectroscopy. Although their calculations were not sufficiently documented for evaluating the $Pu^{4+}$ activity coefficient, a very low value of -60.73±0.05 would be obtained from their results by taking the specific ion interaction theory (SIT) for ionic strength corrections [8,9]. Further studies are thus suggested for this difference.

## Hydrolysis constants of actinide ions

### *Correlations for hydrolysis constants*

The stability of hydrolysis species has been discussed in many literatures in terms of the charge and ionic radius of the central metal ion [10-15]. For example, Davies discussed the first hydrolysis constants of the 1A, 2A and some other group elements in terms of the ratio $z^2/r$, where z is the formal cationic charge and r the ionic radius [10]. Following this, Irving and Williams [11] and Williams [12] suggested the use of the ratio of z/r, the ionic potential, or the above function, $z^2/r$. On the other hand, Baes and Mesmer [13], considered the function z/d, where d is the interatomic distance, M-O, rather than the ionic radius. Although there are also some other functions proposed considering the effects of cations upon their anionic neighbours, it may be generalised, according to Huheey, that any function $z^n/r^m$ can be used with similar results [14].

Although a number of models has been proposed so far, the unified theory of metal ion complex formation constants which has been proposed by Brown and Sylva is one of the most successful models considering the effects of cations upon their anionic neighbours [15]. Following the development of the Sylva-Davidson equation [16], it has been shown that, for hydroxo complexes, the formation constant $\beta_{pq}$ could be predicted from chemical and physical properties of the reacting metal ion as given by:

$$\log \beta_{pq} = q\text{Int}_1 + (p-1)\text{Int}_2 + [q\text{Slp}_1 + (p-1)\text{Slp}_2] \, g_1 \, (z_M/r_M^2 + g_2) - \log U_{pq} \tag{1}$$

where $\text{Int}_1$ and $\text{Slp}_1$, and $\text{Int}_2$ and $\text{Slp}_2$ are, respectively, the least squares slope and intercept values of linear equations for the (1,1) species and the polynuclear (p,q) species. The $\log U_{pq}$ term accounts for the increasingly poorer estimates provided by the earlier Sylva-Davidson equation [16] as the molecularity increases and is given by:

$$\log U_{pq} = \log [(q-p+1)!] - 0.5(q-p)(q-p+1) \log k \tag{2}$$

where k is a constant. Also, in Eq. (1), $z_M$ is the charge of the metal ion and $r_M$ its ionic radius, and $g_1$ and $g_2$ are defined by:

$$g_1 = (1 + 2S + D)(z_M + 2) \tag{3}$$

$$g_2 = g(n)(z_M-1) + 0.1d(n-3)^2(1-S) \tag{4}$$

where S depends on the absence (S = 0) or presence (S = 1) of s-electrons in the outermost shell of the ion (those ions exhibiting the inert pair effect), D = 1 if d-orbitals are available for bonding (otherwise D = 0), and g(n) is a Slater function [17] (g(n) = 1 if n, the principal quantum number, is greater than unity, otherwise g(n) = 0), and d is the number of d-electrons in the outermost shell of the ion.

Thus, the model considers the electronic structure of the metal ion as well as the charge and ionic radius and is used to describe quantitatively the hydrolysis and complexation behaviours of most cations in the periodic table. Using this model, one can predict stability constants for a considerable number of aqueous species composed of metal ions and ligands which are important for the safety assessment [18]. However, this model still tends to give poorer estimates for higher co-ordination species of actinide ions [19], and needs to be improved especially for the $\log U_{pq}$ term.

Recently a simple hard sphere model has been proposed to describe the systematic trends for such higher co-ordination species [19,20]. In the model, an octahedral structure is assumed for all the hydrolysis species of actinide ions, and a central actinide ion, its ligands of water molecules and hydroxide ions are all treated as hard spheres, including oxide ions of the oxo-complex, i.e. $AnO_2^+$ and $AnO_2^{2+}$. The assumption is only for simplicity and, in principle, some other structures may also be considered.

By considering the Coulomb interactions between the hard spheres, the electrostatic potential energy E of each species is given as:

$$E = \sum_{i \neq j} (Z_i Z_j / \varepsilon \, d_{ij}) \tag{5}$$

where $Z_i$ and $Z_j$ are the electric charges of hard spheres i and j, respectively, $\varepsilon$ the dielectric constant, and $d_{ij}$ the distance between hard spheres i and j. As a substitute for the dipole moment, water molecules are assumed to have an effective charge.

Since water molecules are substituted by hydroxide ions in hydrolysis reactions, the mononuclear hydrolysis constant $\beta_{1n}$ is related with the potential energy difference $\Delta E$ between the hydrolysis species and non-complexed actinide ion as expressed by:

$$\beta_{1n} = \exp(-\Delta E/RT) \tag{6}$$

where R and T denote the gas constant and absolute temperature, respectively.

As shown in Figure 2, the model has successfully been applied to the hydrolysis behaviour of higher co-ordination species. The $\beta_{1n}$ values increase with increasing co-ordination number, but the rate of increase is not simply porportional to the co-ordination number due to the increasing repulsive interactions between the negatively charged ligand ions.

## Non-electrostatic interactions of actinide ions

In Figure 3, the selected values for the first hydrolysis constants of actinide ions are plotted as a function of the atomic number [20]. In spite of some scatters, it can be found that the first hydrolysis constants of $An^{3+}$ ions increase with increasing atomic number. This trend can be explained by considering the effects of the actinide contraction [21]; the ionic radii of actinide ions decrease with increasing atomic number due to decreased shielding by 5f-electrons, and the electrostatic interactions for hydrolysis increase. For the other ions, however, such trends are hardly observed, but rather opposite trends are observed for $An^{4+}$ excepting $Th^{4+}$ and especially for $AnO_2^+$ in Figure 3. In the case of $AnO_2^{2+}$, no definite trend is found in Figure 3, but the opposite trend to the actinide contraction has already been pointed out [21].

From the above facts, it may be considered that actinide ions have not only the electrostatic interactions but also some additional non-electrostatic interactions. In order to evaluate a contribution of the non-electrostatic interaction, the simple hard sphere model has been extended to include not the formal but the effective charges of central actinide ions [20]. The values of the effective charges have been evaluated by applying the model to the analysis of hydrolysis constants of actinide ions. As shown in Table 3, the obtained effective charges of $An^{3+}$, $An^{4+}$, $AnO_2^+$ and $AnO_2^{2+}$ are found to be larger than the formal charges of +3, +4, +1 and +2, respectively. This result suggests that these actinide ions have not only the electrostatic interactions but also additional non-electrostatic interactions.

It is widely recognised that actinides represent a unique part in a periodic table of the elements and that the 5f electrons are not so localised and are likely to contribute to chemical bonding. Extensive efforts have thus been devoted to explain the bulk properties of actinides by considering the electronic structures [22]. For example, ab initio calculations of the electronic structures have been performed to understand bonding in metals and some compounds. However, it is still difficult to predict such properties as the stabilities of the solution species mainly because of the lack of structural information. In order to discuss the observed trends for the contributions of 5f orbitals, an empirical analysis has also been performed [20].

It may be considered that the radial distribution of 5f orbitals is one of the most important factors for the contributions of non-electrostatic interactions and, as a measure of the contributions, the ratios of the 5f orbital radial expectation values [23], $<r>$, to the ionic crystal radii [21], $r_c$, have been used. As shown in Table 2, the $<r>$ values are usually larger than the $r_c$ values, and the larger $<r>/r_c$ values are expected to result in the larger contribution of 5f orbitals to the additional interactions. With increasing atomic number, in general, the $<r>/r_c$ values decrease with atomic number and the contribution of 5f orbitals will decrease.

Figure 4 shows the effective electric charges as a function of the $<r>/r_c$ value [20]. The literature values of the effective charges, which have been given by Choppin and Rao for $AnO_2^+$ and $AnO_2^{2+}$ [24], are also shown in this figure, and are found to agree with the present values within the uncertainties. The effective charges for $An^{4+}$, $AnO_2^+$ and $AnO_2^{2+}$ increase with increasing $<r>/r_c$ values, suggesting that the contribution of non-electrostatic interactions is much important for $An^{4+}$, $AnO_2^+$ and $AnO_2^{2+}$. On the other hand, no clear dependence is expected for $An^{3+}$ because of the smaller $<r>/rc$ values, but rather opposite dependence may be found in Figure 4. Although this may indicate some uniqueness of $An^{3+}$, further confirmation will be suggested for $An^{3+}$, and especially for $Bk^{3+}$, $Cf^{3+}$, $Es^{3+}$ and $Fm^{3+}$ of which the experimental measurements are very limited.

It is interesting and important to compare the contributions of non-electrostatic interactions among $An^{3+}$, $An^{4+}$, $AnO_2^+$ and $AnO_2^{2+}$. In Figure 5, the additional charges which are the differences between the effective and formal charges are plotted as a function of the $<r>/r_c$ value [20]. The dependencies of the additional charges on the $<r>/r_c$ value are found to be almost the same irrespective of valence states, and the additional charges, $Z_{add}$, are expressed by the following empirical equation:

$$Z_{add} = 0.0217(<r>/r_c)^{8.23} \qquad (7)$$

This empirical equation is only temporal, and may be improved by further studies not only experimental but also theoretical. At the present stage, however, it will be useful for checking the experimental data and for predicting the unknown data from the systematic point of view.

## Conclusions

Although extensive studies have been performed so far, some of the thermodynamic data of actinide elements are still lacking, and the chemical behaviour of these elements is often predicted by applying the concept of chemical analogy for the safety assessment of geologic disposal of radioactive wastes. It is needed to consider the basis of this concept.

For ordinary electrostatic interactions, it is enough to consider the difference of the ionic radii, i.e. the difference of the Coulomb interactions. In addition to these interactions, however, non-electrostatic interactions, which are possibly due to the formation of hybrid orbitals, are also considered to contribute to the stability of the hydrolysis species of actinide ions in solution. For predicting the hydrolysis constants of actinide ions by applying chemical analogy, it should be remembered that the true values would be larger than those predicted only by taking into account the usual electrostatic interactions. However, this does not necessarily mean the higher solubility values since the non-electrostatic interactions are expected not only for the solution species but also for the solid phase species. It is thus quite important to study the contribution of the non-electrostatic interactions to the stability of the solid phase species.

A similar conclusion may be given for the complexation of actinide ions. There are various complexing anions present in substantial amounts in natural waters and, because of its strong tendency for complexation, carbonate is recognised to be one of the most important for the safety assessment. However, large discrepancies are still found among the literature data for the carbonate complexation, and sometimes the formation of different carbonate species have been suggested even under similar conditions. Similarly to the case of hydrolysis species, some correlation has been found by considering the chemical characteristics of metal ions and ligands, but any models have not been fully successful to comprehensively describe the complexation behaviour of actinide ions. Further studies are also suggested for the complexation of actinide ions.

# REFERENCES

[1]    D. Rai, J.L. Swanson, J.L. Ryan, *Radiochim. Acta*, 42, 35 (1987).

[2]    D. Rai, A.R. Felmy, J.L. Ryan, *Inorg. Chem.*, 29, 260 (1990).

[3]    D. Rai, A.R. Felmy, S.M. Sterner, D.A. Moore, M.J. Mason, *Radiochim. Acta*, 79, 239 (1997).

[4]    T. Yajima, Y. Kawamura, S. Ueta, *Mat. Res. Soc. Symp. Proc.*, 353, 1137 (1994).

[5]    D. Rai, *Radiochim. Acta*, 35, 97(1984).

[6]    H. Capdevila, P. Vitorge, *Radiochim. Acta*, 82, 35 (1998).

[7]    J. Kim, B. Kanellakopulos, *Radiochim. Acta*, 48, 145 (1989)

[8]    I. Grenthe, J. Fuger, R.J.M. Konings, R.J. Lemire, A.B., Muller, C. Nguyen-Trung, H. Wanner, "The Chemical Thermodynamics of Uranium", North Holland, Amsterdam (1992).

[9]    R.J. Silva, G. Bidoglio, M.H. Rand, P.B. Robouch, H. Wanner, I. Puigdomenech, "The Chemical Thermodynamics of Americium", North Holland, Amsterdam (1995).

[10]   C.W. Davies, *J. Chem. Soc.*, 1256 (1951).

[11]   H. Irving, R.J.P. Williams, *J. Chem. Soc.*, 3192 (1953).

[12]   R.J.P. Williams, *J. Chem. Soc.*, 3770 (1952).

[13]   C.F. Baes, Jr., R.E. Mesmer, "The Hydrolysis of Cations", John Wiley and Sons, New York, (1976).

[14]   J.E. Huheey, "Inorganic Chemistry", 2nd ed., Harper and Row, New York (1976).

[15]   P.L. Brown, R.N. Sylva, *J. Chem. Res.*, (S) 4-5, (M) 0110 (1987).

[16]   P.L. Brown, H. Wanner, "Predicted Formation Constants Using the Unified Theory of Metal Ion Complexation", OECD/NEA, Paris (1987).

[17]   R.N. Sylva, P.L. Brown, *J. Chem. Soc.*, Dalton Trans., 31 (1983).

[18]   J.C. Slater, *Phys. Rev.*, 36, 57 (1930).

[19]   H. Moriyama, M.I. Pratopo, K. Higashi, *Radiochim. Acta*, 66/67, 73 (1994).

[20]   H. Moriyama, A. Kitamura, K. Fujiwara, H. Yamana, *Radiochim. Acta*, in press.

[21] J.J. Katz, G.T. Seaborg, L.R. Morss, "The Chemistry of the Actinide Elements", 2nd ed., Chapman and Hall, London (1986).

[22] M.S.S. Brooks, B. Johansson, H.L. Skriver, "Handbook on the Physics and Chemistry of the Actinides", A.J. Freeman and G.H. Lander, eds., North Holland, Amsterdam (1984), Vol. 1, p. 153.

[23] J.P. Desclaux, A.J. Freeman, "Handbook on the Physics and Chemistry of the Actinides", A.J. Freeman and G.H. Lander, eds., North Holland, Amsterdam (1984), Vol. 1, p. 1.

[24] G.R. Choppin, L.F. Rao, *Radiochim. Acta*, 37, 143 (1984).

### Table 1. Solubility products of An(IV) hydrous oxides

| Solid phase | Media | log $K_{sp}$ at I = 0 | References |
|---|---|---|---|
| $UO_2 \cdot xH_2O$ | I = 0.05 ± 0.01, NaCl | -52.0 ± 0.8 | Rai, *et al.*[2] |
| | upto 6 M NaCl, upto 3 M MgCl$_2$ | -53.45 | Rai, *et al.*[3] |
| | 0.1 M NaClO$_4$ | -55.66 ± 0.4 | Yajima, *et al.*[4] |
| $PuO_2 \cdot xH_2O$ | Nitrate medium | -56.85 ± 0.36 | Rai [5] |
| | 0.1 to 3 M NaClO$_4$ | -58.3 ± 0.3 | Capdevila & Vitorge [6] |
| | 1 M HClO$_4$ | -60.73 ± 0.05* | Kim & Kanellakopulos[7] |

* Extrapolated from the experimental data by taking the specific ion interaction theory (SIT) [8,9] for ionic strength corrections.

### Table 2. Ionic radii $r_c$ of actinide ions for co-ordination number 6 [21]
### and 5f orbital radial expectation values <r> [23] ($10^{-10}$ m)

| Valency | <r>/$r_c$ | | | |
|---|---|---|---|---|
| | III | IV | V | VI |
| Ac | –/1.12 | | | |
| Th | –/– | –/0.93 | | |
| Pa | –/1.05 | 1.36/0.91 | –/0.78 | |
| U | 1.38/1.03 | 1.30/0.89 | 1.24/0.76 | (1.19)[a]/0.73 |
| Np | 1.31/1.01 | 1.25/0.87 | 1.20/0.75 | 1.15/0.72 |
| Pu | 1.26/1.00 | 1.20/0.86 | 1.15/0.74 | 1.12/0.71 |
| Am | 1.21/0.98 | 1.16/0.85 | 1.12/(0.73)* | 1.08/– |
| Cm | 1.17/0.97 | –/0.84 | | |
| Bk | 1.12/0.96 | 1.09/0.83 | | |
| Cf | 1.09/0.95 | | | |
| Es | 1.06/0.93 | | | |
| Fm | 1.03/0.92 | | | |

* Estimated by extrapolation.

## Table 3. Effective charges of actinide ions obtained from the analysis by using the hard sphere model [20]*

| Valency | III | IV | V | VI |
|---------|-----|-----|-----|-----|
| Th | | (4) | | |
| Pa | | 4.52 ± 0.17 | 3.38 ± 0.35 | |
| U | | 4.37 ± 0.15 | | 3.23 ± 0.20 |
| Np | 3.13 ± 0.41 | 4.34 ± 0.14 | 2.40 ± 0.46 | 3.19 ± 0.23 |
| Pu | 3.36 ± 0.37 | 4.27 ± 0.12 | 2.55 ± 0.50 | 3.18 ± 0.24 |
| Am | 3.13 ± 0.25 | | 2.08 ± 0.61 | |
| Cm | 3.35 ± 0.24 | | | |
| Bk | 3.37 ± 0.37 | | | |
| Cf | 3.45 ± 0.36 | | | |
| Es | 3.49 ± 0.35 | | | |
| Fm | 3.70 ± 0.32 | | | |

\* The dielectric constant and effective electric charge of water molecules were obtained to be 7.60 ± 5.17 and -0.57 ± 0.18, respectively.

## Figure 1. Variation of solubility products of actinide dioxides.
## Open marks are of hydrous oxides and full marks of crystalline oxides.

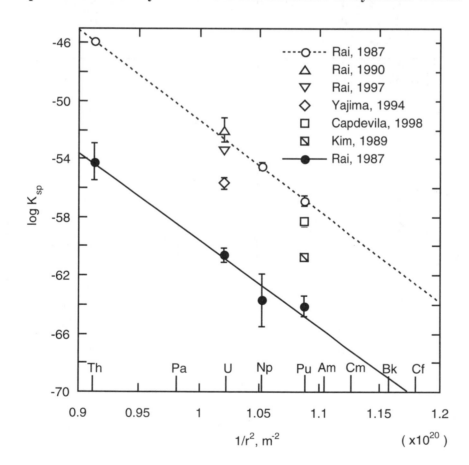

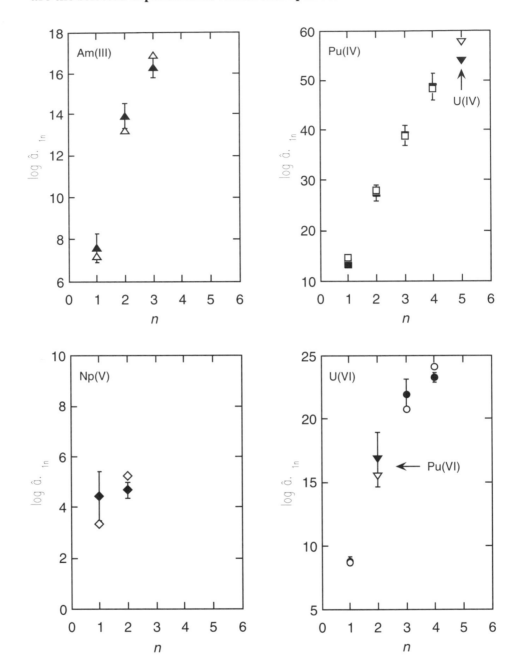

Figure 2. Step-wise hydrolysis constants of some actinide ions [20]. Full marks are the selected experimental values and open marks the calculated ones.

**Figure 3. Atomic number dependence of the first hydrolysis constants of actinide ions [20]. Open marks are the averaged literature values and full marks are the values recommended in the NEA TDB project [8,9].**

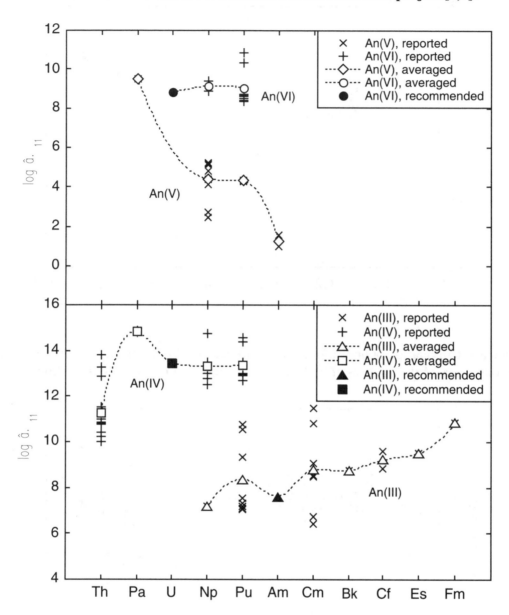

**Figure 4. Effective charges of actinide ions versus the $<r>/r_c$ value [20]. Open marks are of the analysis of hydrolysis constants and full marks of the literature [24].**

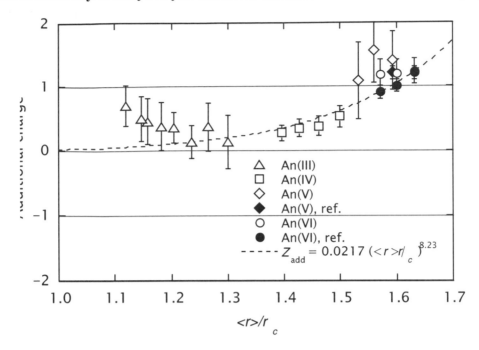

**Figure 5. Correlation of additional charges of actinide ions with the $<r>/r_c$ value [20]. Open marks are of the analysis of hydrolysis constants and full marks of the literature [24].**

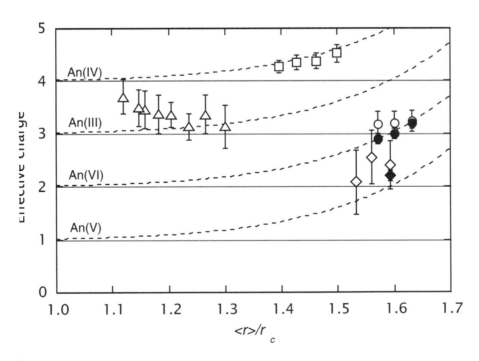

# POSTER SESSION

## Part A: Methods for Trace Concentration Speciation ($<10^{-6}$ M)

# USE OF CAPILLARY ELECTROPHORESIS FOR SPECIATION PURPOSES INVOLVING F-ELEMENTS; APPLICATION TO THE SYSTEMS $Nd^{3+}/H_nPO_4^{(3-n)-}$ AND $UO_2^{2+}/IO_3^-$

**M. Illemassène, S. Hubert**
Institut de Physique Nucléaire
F-91406 Orsay Cedex, France

**P. Tarapcik**
Department of Analytical Chemistry Slovak Technical University
Radlinského 9, 81237 Bratislava, Slovaquia

## Abstract

Two different systems of interest for the problematic of a direct uranium-based fuel storage or disposal have been investigated by capillary electrophoresis (CE) and, in the case of uranyl, by time resolved laser induced fluorescence. For the $UO_2^{2+}/IO_3^-$ system, the mobility variations can be predicted from the speciation diagrams. CE appears thus as an interesting method to confirm and even to precise the involved formation constants. The mobility results concerning the Nd/phosphate system point out a disagreement at pH 3 between calculated and experimental values. They also point out the limits of the method, at pH 5, when sorption and precipitation processes occur.

## Introduction

Several complementary methods are often necessary to obtain a realistic speciation diagram of an actinide in complexing media and new analytical methods have to be developed in this direction. Moreover, in near neutral aqueous solutions, the concentration of the element can be very low, due to sorption or precipitation processes. Methods relevant to radiochemistry, partition and transport methods are thus particularly interesting in this case.

Among the former methods, solvent extraction is generally used but leads very often to erroneous results (high ionic strength, ...) and the solubility method developed in the case of thorium in phosphate containing solutions [1] is limited to high ligand concentrations. Among the latter methods, capillary electrophoresis (CE) appears as a promising one for three reasons:

- The experiments are simple to carry out and each run lasts only a few minutes.

- The volume of sample needed for an experiment is very small (around a few nanolitres). This point is of interest concerning actinides, because micro-chemistry leads to a minimum of wastes, in accordance with a strict management of radioactive solutions.

- The usual UV-vis. detector can be replaced by a $\gamma$ or $\beta$ counting system [2,3,4].

CE has been recently applied to the speciation of uranium in biological medium (complexation between U and human transferring) [5] and the purpose of the present work is to test the possibilities and the limits of the method when applied to two different systems for which only few thermo-dynamical data can be found in the literature:

- Neodymium, a homologous element of americium, in aqueous solutions containing varying concentrations of phosphates.

- Uranyl in presence of iodate anions.

## Experimental

*Solutions*

All solutions were prepared with deionised water. Creatinine, $C_4H_7N_3O$ (> 99% FLUKA-Biochemika) was used either as a reference (mobility equal to $3.7 \ 10^{-4} \ cm^2 \ V^{-1} \ s^{-1}$) or as the absorbing buffer (at 220 nm; pKa = 4.85) at a concentration ranging between $10^{-2}$ to $10^{-3} \ mol \ l^{-1}$. In the latter case, the pH of the buffer solution (3 and 5) was adjusted to the desired value by adding a few drops of 0.1 N HCl (Merck).

The concentration of the $NdCl_3$ (99.9%) stock solution was checked to be close to 0.1 mol $l^{-1}$ using back titration with EDTA. This value is also in agreement with the measured UV-vis. spectrum (absorbance of 0.339 at 353.7 nm).

$UO_2(ClO_4)_2$ has been prepared by dissolving 1g of $UO_2(NO_3)_2$ (FLUKA) in 25 ml of concentrated perchloric acid and by evaporating the resulting solution at 150°C in a sand bath. For a complete elimination of the nitrates, the procedure was repeated.

On one hand, $KH_2PO_4$ (FLUKA Microselect) diluted to 0.1 mol $l^{-1}$ was used to make vary the phosphate concentration of the buffer solution. On the other hand, $NaIO_3$ (FLUKA puriss p.a.) was mixed in varying proportions to $(Na,H)ClO_4$ solutions such that the resulting ionic strength was equal to 0.005 or 0.01 mol $l^{-1}$.

## Devices

The CE apparatus is a modular system consisting of a Spectraphoresis 100 injector (hydrodynamic mode) coupled with a high voltage (0-30 kV) power supply (Prime Vision VIII from Europhor) and a scanning UV-visible detector (Prime Vision IV from Europhor). In the present study, a voltage of 25 000 V resulting in an electric field of 333 V/cm was applied and a fixed wavelength of 220 nm (creatinine) or 238 nm (uranyl) was chosen for the absorbance measurements. The external temperature was maintained equal to 25°C (ventilated capillary in an air-conditioned room). The acquisition of the electropherograms and the data treatment were performed using the chromatography software BORWIN (JMBS Développements).

For TRLIF experiments, samples ($10^{-4}$ mol $l^{-1}$ $UO_2^{2+}$ in 0.01 mol $l^{-1}$ $(Na,H)ClO_4$ solutions) were introduced in a quartz cell of 4 ml. A Nd:YAG laser (continuum mini-light) operating at 266 or 355 nm was used as the excitation source. The fluorescence was analysed using a monochromator (Jobin Yvon Spex 270 M) equipped with a CCD camera of 1 024 pixels (Princeton Instrument Inc.).

## Procedure

Capillary (SUPELCO) is in silica and is 75 cm (= L) long (40 cm (= d) from injector to detector and 35 cm from detector to outlet end). It was first conditioned by successive washes with 0.1 N NaOH (5 min), deionised water (10 min) and buffer solution (15 min). It was rinsed three minutes with the buffer between two runs and kept filled with deionised water during the night. Only one nanolitre of sample was injected for each run.

In order to avoid the uncertainties due to temperature fluctuations and other experimental variations (current, electro osmotic flow due to the silanol groups present on the capillary walls), one (creatinine for $UO_2^{2+}$) or two elements (potassium and sodium for neodymium in the following proportions: 2 K for 1 Nd and 1 Na) of known mobility were added to the samples. In the former case, the electrophoretic mobility of the cation $u_i$ is given by:

$$u_i = u_{ref} + ((d\ L)/V\ (1/t_i - 1/t_{ref})$$

V being the applied voltage and t the time of migration of the ion, i, or of the reference, ref. When two standards are used, the value of $u_i$ becomes completely independent of the experimental parameters:

$$u_i = u_{ref} + (u_{ref2}-u_{ref1})\ (1/t_i-1/t_{ref1})/(1/t_{ref2}-1/t_{ref1})$$

# Results

## $Nd^{3+}/H_nPO_4^{(3-n)-}$ system

Each charged species present in solution has a specific mobility value. Different complex forms can coexist and the kinetics of the equilibria involved are generally rapid compared to the time of migration. Consequently, only one peak is detected which corresponds to the sum of the weighted contributions due to the different species in equilibrium. The variation of this peak (position + intensity) versus pH or pC (ligand) should give the expected information on the equilibria really involved. This variation can be predicted, at least qualitatively, when the thermodynamic constants are known, assuming that the ratio of the mobility of the aquo and complex species is directly proportional to the ratio of the charges of these species (the size effect can be neglected in this first approach).

In the present case, hydrolysis and phosphate complexation are competing. The corresponding formation constants used for the speciation calculations are reported in Table 1. From these data, hydrolysis is expected to occur around pH 6.8, but the mobility versus pH variation derived from diffusion experiments on europium indicates a shift of one pH unit towards lower pH values [6].

CE experiments have been carried out on $10^{-5}$ and $5\ 10^{-5}$ mol $l^{-1}$ Nd in 0.01 mol $l^{-1}$ creatinine solutions containing varying amounts of phosphate and at pH 3 and 5. Two standards have been added to these solutions, $K^+$ ($u = 7.503\ 10^{-4}$ cm$^2$ V$^{-1}$ s$^{-1}$) and Na ($u = 5.114\ 10^{-4}$ cm$^2$ V$^{-1}$ s$^{-1}$). As can be seen from Figure 1, the electropherograms obtained under these conditions show that the phosphate anions have an effect on the intensity of the peak corresponding to Nd (situated between the two standards): at pH 5, it decreases for $C_{PO4} \geq 2\ 10^{-4}$ mol $l^{-1}$. The position of the peak is also modified.

The experimental results can be compared in Figure 2 to the mobility values calculated as explained above. We notice that the experimental points are always situated under the calculated curves, the shift being more important when pH increases or when the surface charge of the capillary walls increases. The migration of Nd is thus reduced by an interaction between the trivalent cation and the silanol groups. But the interest of the Figure 2 is essentially in the comparison of the slope changes of the u versus $pC_{PO4}$ curves. These changes appear, as expected, around $C_{PO4} = 2.5\ 10^{-4}$ mol $l^{-1}$ at pH 5 (formation of $NdHPO_4^+$ species) but at $C_{PO4} = 1.6\ 10^{-4}$ mol $l^{-1}$ instead of $1.6\ 10^{-3}$ mol $l^{-1}$ at pH 3 (formation of $NdH_2PO_4^{2+}$ species). The value of $log\beta11$ given in Table 1 is thus probably under-estimated by one order of magnitude (equal to 3.3 instead of 2.31).

Moreover, the peak of Nd disappears ($C_{Nd} < 5\ 10^{-5}$ mol $l^{-1}$) at pH 5 for $C_{PO4} = 5\ 10^{-4}$ mol $l^{-1}$ (or $pC_{PO4} = 3.3$) due to the precipitation of $NdPO_4$. Under these conditions and on the basis of the data given in Table 1, the concentration of $PO_4^{3-}$ anions is estimated to be $6.8\ 10^{-13}$ mol $l^{-1}$ and 90% of $Nd^{3+}$ ions are not complexed (1% of $NdPO_4$ is formed). Consequently, a solubility product of $NdPO_4$ can be derived: $log\ Ks = -16.5$ at an ionic strength of 0.01 mol $l^{-1}$. Using the SIT method to correct the activity coefficients of $Nd^{3+}$ and $PO_4^{3-}$ (respectively 0.167 and 0.163), the Ks value can be extrapolated to zero ionic strength: $log\ Ks° = -18.1$, value in disagreement with the value reported in Table 1. An explanation can be found in the fact that the solubility of a crystallised compound is generally lower than the solubility of an amorphous or hydrated compound precipitated in situ.

## $UO_2^{2+}/IO_3^-$ system

### CE

The CE procedure used for this system is similar to the procedure previously described for the $Nd^{3+}/H_nPO_4^{(n-3)-}$ system. The concentration of uranyl was fixed to $10^{-4}$ mol $l^{-1}$ and the concentration of iodate was varying from $5\ 10^{-6}$ to $10^{-2}$ mol $l^{-1}$. The supporting electrolyte was a mixture of $NaClO_4$ and $HClO_4$ solutions in such proportions that the pH was equal to 3 and the ionic strength was equal to $5\ 10^{-3}$ mol $l^{-1}$. Creatinine ($10^{-4}$ mol $l^{-1}$) was added to the injected uranyl solution as a reference element.

The speciation diagram derived from the hydrolysis constants reported in Table 2 shows that the aquo $UO_2^{2+}$ ion prevails (around 99%) at pH 3. It is also not necessary to take into account the formation of polycationic species. Under these conditions, the mobility variation of uranyl is only assigned to the complexation by iodate anions. Two complex forms are expected to occur (see Table 2 for the corresponding formation constants): $UO_2IO_3^+$ and $UO_2(IO_3)_2$.

The electropherograms (examples given in Figure 3) show that the migration time of uranyl increases when the concentration of $IO_3^-$ anions increases. The peak of uranyl is even very close to the peak of creatinine at $C_{IO3} = 4\ 10^{-3}$ mol $l^{-1}$.

The calculated and experimental u versus $pC_{IO3}$ curves are plotted in Figure 4. A good agreement is obtained between the two curves, using a value of $4.5\ 10^{-4}$ $cm^2$ $V^{-1}$ $s^{-1}$ for the theoretical mobility of the aquo uranyl ion. Two edges can be pointed out on the experimental curve, for $C_{IO3} = 10^{-3}$ mol $l^{-1}$ and $C_{IO3} = 4\ 10^{-3}$ mol $l^{-1}$ (see Figure 4). They correspond respectively to the appearance of $UO_2IO_3^+$ and $UO_2(IO_3)_2$. As a matter of fact, we can roughly estimate (no correction of ionic strength), from the data given in Table 2, that 10% of uranyl is under the first complex form at $pC_{IO3}$ equal to 3 and that 6% of uranyl is complexed with two ligands at $pC_{IO3}$ equal to 2.4.

*TRLIF*

The fluorescence spectrum of a perchloric uranyl ion ($10^{-4}$ mol $l^{-1}$) has been recorded as a reference. This spectrum ranges from 470 nm to 590 nm and exhibits the well known features of uranyl ion with four intense characteristic bands at 488 nm, 510 nm, 534 nm and 560 nm. The lifetime of the fluorescence was measured to be $(1.7 \pm 0.2)$ μs at pH 3, in agreement with published data [7,8].

The solutions prepared for the CE experiments were then investigated by TRLIF. The spectra so obtained (Figure 5) show that the iodate ions have no influence up to $C_{IO3} = 10^{-3}$ mol $l^{-1}$. For higher iodate concentrations, a broadening of the bands was observed, as well as a shift in energy towards higher wavelengths and a decrease of the fluorescence intensity. The modifications in the spectra indicate a change in the environment of the uranyl ion, which corresponds to the complexation by the iodate anions, as expected from the CE results.

In order to avoid the possible formation of $I_3^-$ ions (wide bands centred on 287 and 351 nm) under UV laser irradiation, the fluorescence decay has been recorded after a laser excitation at 355 nm. But, under these conditions, the fluorescence intensity is very low and a rigorous interpretation of the fluorescence decays is difficult. Moreover, if the emission spectra are registered with a delay of 1 μs, in order to eliminate the contribution due to the complex species with a short lifetime, no more aquo uranyl ion seems to be present in the solutions containing more than $3\ 10^{-3}$ mol $l^{-1}$ of iodate. Quenching problems are probably involved since this result is not confirmed by the CE results, which show that the aquo $UO_2^{2+}$ species prevails for $pC_{IO3} > 2$.

**Conclusion**

Through the two systems investigated in the present work, CE appears as an interesting method to precise the conditions leading to the formation of soluble complex species. The limitations are essentially due to the sensitivity of the detection and to sorption or precipitation phenomena onto the capillary walls, which tend to increase with pH. For example, a rare earth concentration as high as $10^{-5}$-$10^{-6}$ mol $l^{-1}$ leads to the precipitation of the phosphate or hydroxide compound in neutral and basic media. Another difficulty is the choice of the absorbing buffer, which cannot cover a large pH range.

Some solutions are under test, such as the use of capillary made in not adsorbing material associated to the use of a γ or β-ray detector and the use of a tracer with a relatively high specific radioactivity. But, some results discussed in this paper are in disagreement with the literature data and already show the interest of the CE method as a complementary method for speciation purposes.

# REFERENCES

[1]   B. Fourest, N. Baglan, R. Guillaumont, G. Blain, Y. Legoux, *J. of Alloys and Compounds*, 213/214 (1994), 219-225.

[2]   S.L. Pentoney, Jr., R.N. Zare, *Anal. Chem.*, 61 (1989) 1642-1647.

[3]   K.D. Altria, C.F; Simpson, A.K. Bharij, A.E. Theobald, *Electrophoresis*, 11 (1990), 732-734.

[4]   G.L. Klunder, J.E. Andrews, Jr., P.M. Grant, B.D. Andresen, *Anal. Chem.*, 69 (1997), 2988-2993.

[5]   S. Scapolan, thesis, University Paris XI, Orsay, 1998.

[6]   F. Kepak, J. Kriva, J. Inorg, *Nucl. Chem.*, 33 (1971), 1741.

[7]   C. Moulin, P. Decambox, V. Moulin, J-G. Decaillon, *Anal. Chem.*, 67 N° 2 (1995), 15.

[8]   C. Moulin, I. Laszak, V. Moulin, C. Tondre, *Applied Spectroscopy*, 52 N° 4 (1998), 528-535.

[9]   Baes and Mesmer, "The Hydrolysis of Cations", Wiley-Interscience (1976).

[10]  J.H. Lee, R. H. Byrne, *Geochim. and Cosmochim. Acta*, 56 (1992), 1127.

[11]  R.H. Byrne, Ki-H. Kim, *Geochim. and Cosmochim. Acta*, 5 (1993), 519.

[12]  F.J. Millero, *Geochim. and Cosmochim. Acta*, 56 (1992) 3123.

[13]  I. Grenthe *et al.*, "Chemical Thermodynamics of Uranium", North Holland, NEA/OECD, 1992, p. 51.

[14]  F.I. Khalili, G.R. Choppin, E.N. Rizkalla, *Inorg. Chim. Acta*, 143 (1988), 131-135.

**Table 1. β1i constants used for the speciation of neodymium and corresponding to the equilibrium : $Nd^{3+} + iL^{j-} <=> Nd(L)_i^{(3-ij)+}$**

| ligand L | logβ11 | | logβ12 | | logβ13 | | logβ14 | | logKs | |
|---|---|---|---|---|---|---|---|---|---|---|
| OH⁻ | 6.0 | [9,10] | 11.10 | [10] | 15.50 | [10] | 18.90 | [10] | -23.89 | [9] |
| PO₄³⁻ | 11.78 | [10] | 19.50 | [10] | | | | | -25.42 | [11] |
| HPO₄²⁻ | 5.18 | [12] | 8.66 | [12] | | | | | | |
| H₂PO₄⁻ | 2.31 | [12] | | | | | | | | |
| CO₃²⁻ | 7.13 | [12] | 11.80 | [12] | | | | | | |
| HCO₃⁻ | 1.83 | [12] | | | | | | | | |

**Table 2. β1i constants used for the speciation of uranyl and corresponding to the equilibrium: $UO_2^{2+} + iL^{j-} <=> UO_2(L)_i^{(2-ij)+}$**

| ligand L | logβ11 | | logβ12 | | logβ13 | | logβ14 | | logKs | |
|---|---|---|---|---|---|---|---|---|---|---|
| OH⁻ | 8.8 | [13] | 17.7 | [13] | 22.8 | [13] | 23 | [13] | -23.89 | [13] |
| PO₄³⁻ | 2.00 | [14] | 3.59 | [14] | | | | | | |

# DEVELOPMENT PROGRAMME OF ANALYTICAL TECHNIQUES FOR ULTRA TRACE AMOUNTS OF NUCLEAR MATERIALS IN ENVIRONMENTAL SAMPLES FOR SAFEGUARDS – FROM A VIEWPOINT OF SPECIATION TECHNOLOGY

**Y. Hanzawa, M. Magara, F. Esaka, K. Watanabe, S. Usuda,**
**Y. Miyamoto, K. Gunji, K. Yasuda, H. Nishimura, T. Adachi**
Japan Atomic Energy Research Institute,
Tokai-mura, Naka-gun, Ibaraki, 319-1195 Japan

## Abstract

As a part of the strengthened safeguards system, the IAEA has adopted the environmental sample analysis method to verify the non-existence of undeclared nuclear facilities and activities by analysing samples taken at inside and outside of facilities. So the authors have started to develop analytical techniques for ultra trace amounts of nuclear materials in environmental samples. In this paper, the program and current situation for the development of the bulk and particle analysis techniques are reported and their possible application as speciation techniques is discussed.

## Introduction

As a part of the strengthened safeguards system based on "Programme 93+2" of the International Atomic Energy Agency (IAEA), the environmental sample analysis method was introduced [1]. Its aim is to verify the non-existence of undeclared nuclear facilities and activities by analysing environmental samples taken at inside and outside of facilities, based on a concept that any activities handling nuclear materials should leave their trace footprints in the environment around them. Amounts of nuclear materials contained in such samples are so minute and much attention should be paid to the analysis of them to obtain reliable results. Therefore, the authors have started to develop analytical techniques for ultra trace amounts of nuclear materials in environmental samples, as well as to prepare a clean chemistry laboratory, the purpose of which is to avoid cross contamination in the analysis from nuclear materials in the environment.

The techniques for safeguards environmental sample analysis are categorised into a "bulk analysis" and a "particle analysis". In the bulk analysis, soil, water and vegetation samples are chemically treated (decomposition, separation and purification) and average values for isotope ratios of nuclear materials are evaluated. It is effective for undeclared facilities. On the other hand, in the particle analysis, isotope ratios of uranium and/or plutonium in individual particles on "swipe samples" taken from facilities like smears are determined. This technique is powerful for undeclared nuclear activities, even in the declared facilities.

In order to detect undeclared nuclear facilities or activities by the bulk and/or particle analysis techniques, it is significant to consider the relation between such nuclear activities and their trace footprints left in the environment. It means the speciation of nuclear materials, as the footprints should be of importance. Besides, it is expected that the analytical techniques to be developed would be applied not only to safeguards purpose but also to various research fields, including speciation as one of the powerful analytical tools.

In this paper, the programme and current situation for the development of the analytical techniques is described and their possible application as speciation techniques is discussed.

## Trace footprints of nuclear activities

Isotope ratios of nuclear materials are the most important footprint of nuclear activities left in the environment. For example, the ratio of $^{235}U/^{238}U$ in natural uranium is 0.00725. In nuclear fuels for commercial power plants, $^{235}U$ is enriched to 3-5%. In uranium for a nuclear explosive device, content of $^{235}U$ is over 90%. Then, if a higher $^{235}U/^{238}U$ ratio than that of natural uranium is observed in an environmental sample, it is suggested that there are some nuclear activities. If the ratio exceeds that of commercial use, it suggests that an undeclared activity might be going on to produce highly enriched uranium. $^{234}U$ is enriched along with enrichment of $^{235}U$. It means that if content of $^{234}U$ detected is higher than that in natural uranium, there might be an enrichment plant. $^{236}U$ is not contained in natural uranium but is generated in nuclear reactors. So, if $^{236}U$ is detected, existence of a reprocessing plant is suggested.

On the other hand, plutonium is an artificial element generated in nuclear reactors. In plutonium recovered from a commercial nuclear reactor fuel whose burn up is 30-45 GWd/t, not only $^{239}Pu$ but also $^{238,240,241,242}Pu$ are generated, and the latter nuclides are not suitable for nuclear weapons. The $^{240}Pu/^{239}Pu$ ratios found in the global fallout and in the commercial spent nuclear fuels are $0.176 \pm 0.014$ [2] and near to 0.5, respectively. Then if $^{240}Pu/^{239}Pu$ ratio observed in an environmental sample is lower than that found in the global fallout, there is a possibility of undeclared activity.

# Evaluation of the footprints in the environment

In an evaluation of trace footprints released in the environment, chemical forms of nuclear materials could give information on their origins. This information would be a good help to speculate activities performed in a facility. For example, uranium released from an enrichment plant is considered to be in the chemical form of $UO_2F_2$ particle, which is generated through hydrolysis of $UF_6$ [3]. From a nuclear fuel fabrication plant, sintered $UO_2$ or $UO_2$-$PuO_2$ is supposed to be released. On the other hand, nuclear materials in metal form could be released from a military plant. Some information related to chemical forms could be obtained through the bulk and particle analysis.

In the particle analysis, it is possible to distinguish characteristics of individual particles on swipe samples. This is a great advantage of this technique, especially for the determination of isotope ratio of uranium, because uranium is a naturally occurring element in the environment. Trace of enriched uranium, if any, should be diluted when an average value of isotope ratio in the sample is estimated. Moreover, the particle analysis also gives information on morphology of each particle, which relates to its origin and history of released materials. Much knowledge should be accumulated on how morphology of particles is affected by various nuclear activities in facilities.

On the other hand, in the bulk analysis, it is necessary to take into account migration behaviours of nuclear materials in the environment after their release. For example, adsorption to soil would be affected by redox potential and/or pH. Organic substances and/or colloids in fresh water could increase solubility and/or mobility of nuclear materials, which would cause selective removal of such materials from soil. Plants would accumulate specific elements selectively. Behaviours of aerosol particles are also important. Therefore, when one wants to get information on nuclear activities from results of the bulk analysis, possible influences by such migration behaviours should be estimated.

# Development of analytical techniques

The main target of the authors' current development program of techniques for the bulk and particle analyses is to determine accurate and precise isotope ratios of nuclear materials. A flow diagram of the analysis is shown in Figure 1.

For the bulk analysis, thermal ionisation mass spectrometry (TIMS) and inductively coupled plasma mass spectrometry (ICP-MS) are adopted as analytical method. By TIMS analysis, precise isotope ratios of ultra trace amounts of nuclear materials will be obtained and its target value for the least amounts is $10^{-15}$ g/sample. Recently, ICP-MS is expected to give not only concentration values, but also good data for isotope ratios in simple procedures. The authors have introduced both high resolution type and multiple collector type ICP-MS, and the target value of the determination is $10^{-13}$ g/ml of nuclear materials.

Chemical treatment performed prior to the measurement of the bulk analysis should be carried out in suitable ways for ultra trace analysis, considering chemical forms of nuclear materials contained in samples, for example, the decomposition of sintered plutonium oxide. Conversely, during chemical treatment, information on chemical forms of nuclear materials concerned, such as existence of sintered plutonium oxide, could be obtained. It is also a main target of the bulk analysis to perform chemical treatment in simple and efficient ways using fewer amounts of reagents such as mineral acids.

In the current situation of the bulk analysis, instruments for ICP-MS (a high-resolution type and a multiple collector type) have been already installed and performance tests of them have been carried out. Subsequently, evaluations of the background level and influences of molecular ions such as $^{238}UH^+$, which interferes with the measurement of $^{239}Pu$, is being performed. As a preliminary work for chemical treatment, decomposition of samples by microwave digestion is examined using environmental reference materials produced by the US National Institute of Standards and Technology (NIST).

As tools of the particle analysis, electron probe microanalysis (EPMA) and secondary ion mass spectrometry (SIMS) have been adopted. EPMA is applied to the determination of elemental compositions of particles recovered from a swipe sample and to the identification of the particles containing nuclear materials among others. Then, the isotope ratios of nuclear materials, contained in the particles, are measured by SIMS. EPMA is also applied to the observation of morphology of particles. Our targets of this technique are to obtain isotope ratios for $<10^{-12}$ g of nuclear materials and particle size of less than 1 μm in diameter.

In the current situation of the particle analysis, the installation of instruments for SIMS and EPMA has been finished. After performance tests of them, determination of lead isotope ratio for particulate reference materials by NIST has been successfully performed by the combination of EPMA and SIMS techniques. These analytical results have provided us with some information on the origins of the particulate samples [4].

In the future analysis using a clean chemistry laboratory, screening of samples by preliminary radiometric measurements will be carried out to judge whether a sample is admitted to be handled in a clean room or not. The development target of the screening techniques is to obtain information on suitable treatment methods for samples, which contain ultra trace amounts of nuclear materials.

## Future perspective

There are a lot of common issues between safeguards environmental sample analysis and speciation of nuclear materials in the environment. Analytical techniques to be developed for safeguards purpose will provide us with much useful information for characterisation of ultra trace amounts of nuclear materials in the environment including information on isotope ratios and matters related to particles. For example, it was reported in a study that isotope ratio of plutonium was utilised as a tracer and that migration behaviour of plutonium in underground water was discussed [5]. Of course, these analytical techniques could be applied not only to uranium and plutonium but also to other actinides. Moreover, provided these techniques are connected to other attempts, which can clarify chemical states of elements concerned, they must be powerful tools for speciation of actinides in the environment.

*Acknowledgements*

A portion of this work is being performed by JAERI under the auspices of the Science and Technology Agency of Japan.

# REFERENCES

[1]   B. Pellaud, "IAEA Safeguards – Experience and Challenges", Proceedings of IAEA Symposium on International Safeguards, 13-17 Oct. 1997, Vienna, Austria, STI/DAT/3, An IAEA CD-ROM.

[2]   P.W. Krey, E.P. Hardy, C. Pachucki, F. Rourke, J. Coluzza and W.K. Benson, "Mass Isotopic Composition of Global Fall-out Plutonium in Soil", Proceedings of a Symposium on Transuranium Nuclides in the Environment, IAEA-SM-199-39, pp. 671-678 (1976).

[3]   J.A. Carter, D.M. Hembree, Jr., "Formation and Characterisation of $UO_2F_2$ Particles as a Result of $UF_6$ Hydrolysis", K/NSP-777 Task A.200.3 ISPO-406 (1998).

[4]   F. Esaka, K. Watanabe, M. Magara, Y. Hanzawa, S. Usuda, K. Gunji, H. Nishimura, T. Adachi, "Characterisation of Individual Particles Containing Lead by Combination of SIMS and EPMA", Proceedings of 12th International Conference on Secondary Ion Mass Spectrometry, 5-10 September 1999, Brussels, Belgium, submitted.

[5]   A.B. Kersting, D.W. Efurd, D.L. Finnegan, D.J. Rokop, D.K. Smith, J.L. Thompson, "Migration of Plutonium in Ground Water at the Nevada Test Site", *Nature*, 397, 56-59 (1999).

**Figure 1. Flow diagram of safeguards environmental sample analysis**

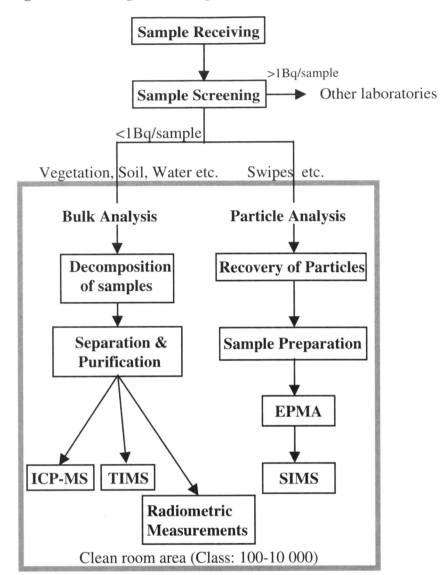

# MATRIX-ASSISTED LASER DESORPTION/IONISATION TIME-OF-FLIGHT MASS SPECTROMETRY AS A TOOL FOR SPECIATION ANALYSIS?

**Josef Havel, Dagmar Gajdošová and Julio Soto-Guerrero**
Department of Analytical Chemistry, Faculty of Science, Masaryk University
Kotlarska 2, 611 37 Brno, Czech Republic
E-mail: havel@chemi.muni.cz

## Abstract

The possibility of using MALDI-TOF MS for speciation analysis was investigated. The technique can be applied to determine molecular weight of species formed by an interaction of the reacting components (metal ions-ligand) and to get some information on complex formation or even partly on the speciation in solution. The main limitation is that aqueous samples are to be dried before analysis, which substantially changes the composition. Anyway, the information about solid phase or on formation of species in solution can be obtained and MS spectra reflect in some way the composition of the aqueous phase. In spite of this limitation, the method seems to be perspective because of high sensitivity. It has been applied for uranium determination in waters measuring intensity of $UO_2^+$ ion, while low sample volumes ($\approx 1\ \mu L$) are needed.

## Introduction

Mass spectrometry (MS) is a powerful tool for analysis of various kind of compounds and applicable also to speciation analysis. Newly developed ionisation methods include such that produce the conversion by the application of a high electric field to the sample (field desorption), or by the bombardment of the sample with energetic ions or atoms, or by the formation of ions directly from small, charged liquid droplets and/or by short-duration, intense pulses of laser light (laser desorption, LD).

Of these techniques, especially electrospray ionisation and matrix-assisted laser desorption/ ionisation with time-of-flight detector (MALDI-TOF) mass spectrometry, appear to hold a great promise for mass spectrometric analysis of biopolymers in the molecular mass range between a few thousands up to 1 000 000 Daltons. It is one of the latest laser ionisation methods developed, introduced only ten years ago, almost simultaneously in Germany and in Japan [1] and now belongs to the most prominent techniques to analyse biomolecules and DNA, but it can also be applied to analyse small ions and molecules, and in some cases to study speciation and complex equilibria in solution [2].

MALDI-TOF MS is a powerful analytical technique with detection limits in some cases down to attomoles and as such, excellent candidate as highly sensitive detector for hyphenated techniques, e.g. in separation methods including HPLC and capillary electrophoresis.

In MALDI-TOF MS the sample is mixed with a suitable matrix which isolates sample molecules, absorbs energy from the pulsed laser beam and provides photoexcited sites for ionisation of analyte in ion/molecule collisions. The matrix molecules prevent the formation of sample clusters, which reduce the abundance of molecular ions, thus decreasing the sensitivity of analysis.

Basic principals of MALDI-TOF analyser are schematically given in Figure 1.

MALDI-TOF MS has up to now scarcely been applied for inorganic analysis and because of being promising technique for speciation analysis [2,3] with very high sensitivity, in the present work this possibility was examined in details either in Laser Desorption Ionisation (LDI) or in matrix assisted modes.

## Instrumentation

The instrument used was a Kratos Kompact MALDI III (Manchester, UK) controlled by Kratos Kompact V5.2.0 software, equipped with a nitrogen laser VSL-337D-10-TTL of Laser Science Inc. (Franklin, MA, USA) operating at 337 nm with pulse energy 250 mJ, average power at 10 Hz is 2.5 mW and peak power 85 kW. The mass spectra presented were obtained as the average taken over one hundred spectra. The averaged spectrum was further smoothed by applying the five-channelled Savitzky-Golay algorithm. Peak maxima were determined via the gradient-centroid procedure.

## Results and discussion

We have studied the possibility to use MALDI-TOF MS for analysis of aqueous solutions or various samples or reaction mixtures, e.g. formed during complex equilibria studies.

## Preparation of the samples

There are two principal modes on how to use the instrumentation. In LDI mode, a sample is deposited to a metallic plate (aqueous sample is to be dried) and with a short pulse of the laser, the desorption and ionisation is done. In MALDI mode, the sample is mixed with a suitable matrix or deposited over a matrix, dried and analysed as mentioned above (Figure 1).

## Determination of uranium in waters

For determination of uranium in geological and environmental samples, mostly ICP or direct analysis of solids with Laser Ablation (LA) is used in literature, and to lower the detection limits, MS detection is advisable [4]. Also, other techniques are used, such as flow-injection analysis with fluorimetric detection [5], remote time-resolved laser/induced fluorescence [6], laser-induced thermal lensing spectroscopy [7], optical emission spectroscopy on laser-produced plasma [8] and other.

MALDI-TOF MS for the speciation and/or determination of uranium(VI) in waters was examined in this work. Several matrices were tried, like α-cyano-4-hydroxy-cinnamic acid; 3,5-dihydroxy-benzoic acid; 3,5-dimethyl-4-hydroxy-cinnamic acid; 5-methyl-salicylic acid; 2-(4-hydroxyphenylazo)-benzoic acid; 5-chloro-salicylic acid; dithranol, and others. However, it was found that no matrix is suitable as intensities of the peaks were low or not observed at all (e.g. for 2.5-dihydroxybenzoic acid).The best results were obtained with dithranol as a matrix or without the use of any matrix, i.e. in LDI mode. As optimal energy of the laser the value ~ 140 units (arbitrary relative units of laser energy, which can be changed on Kratos instrumentation between 0 to a maximum of 180). For lower or higher values, the decrease of intensity of the signal of uranium at $m/z = 270$, corresponding to $UO_2^+$ ion, was observed. It was found that for most of the organic matrices uranyl is reduced or transformed also to uranium carbide $UC^+$ ion and peaks around $UO_2^+$ are diminished. In LDI mode positively charged $UO_2^+$, $UO^+$ and $U^+$ ions were observed in spectra, similarly as in laser ablation ICP-MS technique [4,9]. This mode was applied to analysis of uranium salts and for uranium analysis in mineral waters.

The example of spectra obtained in positive linear mode is given in Figuress 2 and 3. Curve 1 in Figure 2 is an example for analysis of uranium acetate solution. Mass spectrum was obtained in LDI mode, while 1 μL of 10 mM uranium acetate was used. In agreement with literature [4,10], we have observed peaks with $m/z$ values corresponding to $UO_2^+$, $UO^+$, $U^+$ ions. The spectrum (Figure 2, Curve 2) is a demonstration of the sensitivity of uranium determination in mineral water (Podebradka, Czech Republic, the composition cf. Figure 4) spiked with $10^{-5}$ mol/L uranium. The detection limit for uranium determination in waters was estimated to be in the region $(1-5) \times 10^{-12}$ mole, while only 1 μL of the sample is necessary for analysis. However, with respect to the concentration value, the detection limit is only ~ $(1-5) \times 10^{-6}$ mol/L. The advantage of MALDI TOF MS analysis is that very low volume of sample is needed. With suitable pre-concentration technique, for example like that one described earlier [5], the detection limit can be decreased 2-3 orders of magnitude.

In Figure 3, it is demonstrated that in addition to the formation of simple positively charged $UO_2^+$, $UO^+$, and $U^+$ ions, uranium clusters with $m/z$ values equal to 576.8, 867.1, 1 156.6, 1 449.1, 1 731.6, and 1 998.5 were observed. Accompanying secondary clusters were: 560.2, 851.1, 1 140.6, 1 430.8, and 1 717.8. The $m/z$ values observed here are different from those reported for FAB technique [11]. The formation of the clusters with laser desorption ionisation was observed also for rhenium halides [12]. The presence of sample clusters is undesired when a high sensitivity is required.

## Speciation of uranium in waters

An example of the uranium speciation diagram in mineral water (Podebradka, Czech Republic) spiked with uranium and thorium, as calculated from composition of this water and database of equilibrium constants is for case of uranyl given in Figure 4. LDI mass spectrum, measured in negative mode (Figure 5), is demonstrating that speciation of uranium in mineral water is possible. In agreement with the distribution diagram, a peak with $m/z$ value ~ 197.3 was observed, corresponding probably to $UO_2CO_3)_2^{2-}$ species. In addition, negatively charged peak with $m/z$ ~ 328.3, which can correspond to $ThSO_4$ species, and many other unidentified peaks were observed.

## Other examples of MALDI TOF MS analysis

Other examples include e.g. analysis of platinum group metal anti-cancer compounds; complex formation of these and other metal ions (Pd(II), Cu(II), etc.) with reagents like phthalocyanine, cyclam and cyclen macrocycles, and other ligands [2,13]. In case of the formation of robust complexes (e.g. Pd(II) with phthalocyanins), the speciation determination is possible and MALDI-TOF MS can be even used to follow kinetics of the reactions in aqueous solutions. The technique has also been used to follow kinetics of the reactive dyes hydrolysis [3].

## Conclusions

It was found that mass spectra obtained by MALDI-TOF in either MALDI or LDI mode partly reflect the composition of aqueous phase and can therefore be, in some cases, used for the speciation analysis. The limitation of the method use for aqueous samples is that drying of the solution is necessary (as the sample must be introduced into a deep vacuum), and this dramatically changes the sample composition. In spite of this for robust and for some other inorganic complexes the speciation is possible. Concluding, MALDI is promising technique for speciation analysis with high sensitivity but further research in the field is needed. Such work continues in this laboratory with the aims to make direct analysis of aqueous solutions.

*Acknowledgement*

SHIMADZU Handelsges, MBH, Korneuburg, Austria and SHIMADZU Handels GmbH PRAGUE, Czech Republic, are greatly acknowledged for supporting this work via sponsoring Demonstration Laboratory of Shimadzu at Department of Analytical Chemistry, Faculty of Science, Masaryk University, Brno, Czech Republic.

J.S.G. would like to thank the Mexican National Council of Science and Technology (CONACYT) for the fellowship given for his PhD study at Masaryk University.

# REFERENCES

[1]  M. Karas, F. Hillenkamp, *Anal. Chem.* 1988, 60, 2299. K. Tanaka, H. Waki, Y. Ido, S. Akita, Y. Yoshida, T. Yoshida, *Rapid Commun. Mass Spectrom.* 1988, 8, 151.

[2]  J. Havel, "Matrix Assisted Laser Desorption Ionisation (MALDI) – Time of Flight (TOF) Spectrometry – A New Tool to Study Complex Equilibria?', SIMEC'98, C8, Girona, Spain, 2-5 June 1998.

[3]  H. Chromá, J. Havliš, J. Havel, *Rapid Commun. Mass Spectrom*, 13, 1 (1999).

[4]  M. Gastel, J.S. Becker, G. Kueppers, H.J. Dietze, *Spectrochim. Acta*, 1997, 52B, 2051.

[5]  J. Havel, M. Vrchlabsky, Z. Kohn, *Talanta* 1992, 39, 795.

[6]  C. Moulin, S. Rougeault, D. Hamon, P. Mauchien, *Appl. Spectrosc.*, 1993, 47, 2007.

[7]  Y. Enokida, M. Shiga, A. Suzuki, *Radiochim. Acta*, 1992, 57, 101.

[8]  W. Pietsch, A. Petit, A. Briand, *Spectrochim. Acta, B* 1998, 53, 751.

[9]  W. Tuszynski, R. Angermann, J.O. Metzger, *Nuclear Instruments & Methods in Physics Research*, 1994, 88, 184.

[10]  G.G. Managadze, I.Yu. Shutyaev, "Exotic Instruments and Applications of Laser Ionization Mass Spectrometry in Space Research", in A. Verter, R. Gijbels, F. Adams, "Laser Ionization Mass Analysis", John Wiley & Sons, New York (1993), p. 546.

[11]  T.J. Kemp, K.R. Jennings, P.A. Read,. *J. Chem. Soc. Dalton Trans.*, 1995, 885.

[12]  N. Carter Dopke, P.M. Treichel, M.M. Vestling, *Inorg. Chem.*, 1998, 37, 1272.

[13]  D. Kalny, J. Havel, to be published.

# Figure 1. Scheme of the principles of MALDI and of MALDI-TOF mass spectrometer

*G – ground, D – detector, VS – vacuum source*

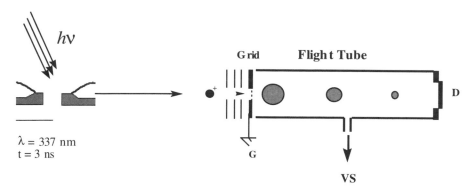

# Figure 2. Mass spectra of uranium(VI) from several samples

*1 – uranium acetate, 2 – mineral water spiked with uranium ($10^{-5}$ mol/L)*

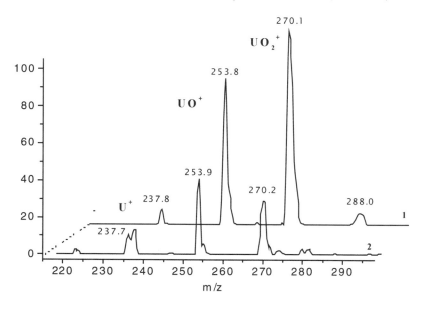

# Figure 3. Clusters of uranium(VI) observed during LDI

*1 µL of 10 mM uranyl acetate, laser energy 150*

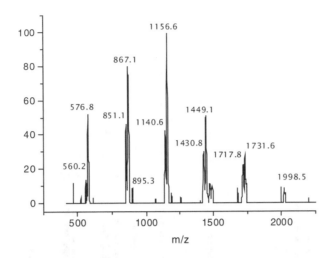

# Figure 4. Distribution diagram of uranium(VI) in Podebradka mineral water spiked with uranium

*Composition of water: $Na_{tot} = 22.17$, $K_{tot} = 61.2$, $Mg_{tot} = 2.90$, $Ca_{tot} = 4.30$,*
*$UO_{2\,tot} = 0.01$, $Cl_{tot} = 12.58$, $SO_{4\,tot} = 0.78$, $CO_{3\,tot} = 4.60$ mmol/L, and $F_{tot} = 47$ µmol/L*

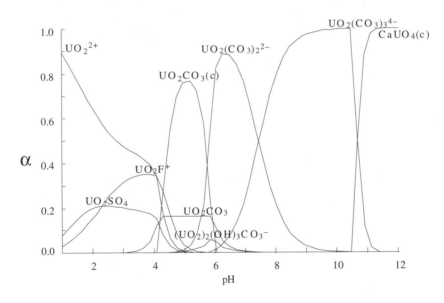

**Figure 5. LDI Mass spectrum obtained for mineral water and measured in linear negative mode laser energy 150, composition of water the same as given in Figure 4**

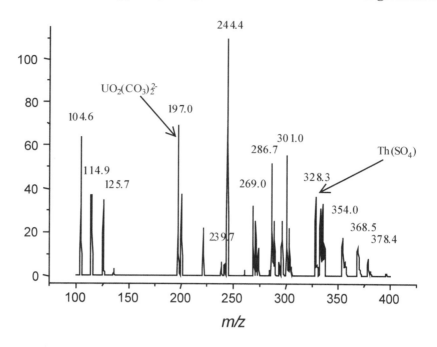

# ASSOCIATION OF ACTINIDES WITH DISSOLVED ORGANIC MATERIALS IN ORGANIC-RICH GROUNDWATERS

**S. Nagao, Y. Sakamoto, T. Tanaka, H. Ogawa**
Department of Fuel Cycle Safety Research, Japan Atomic Energy Research Institute
2-4 Shirakata Shirane, Tokai-mura, Naka-gun, Ibaraki 319-1195, Japan
E-mail: nagao@sparclt.tokai.jaeri.go.jp

## Abstract

Size fractionation method was used for the speciation of actinides in three types of groundwaters with DOC concentration of 2-81 mg C/L and p H8. Pu and Am were added to the groundwater samples, and were fractionated into various sizes by ultra filtration technique. The major parts of Pu and Am in the humus-poor groundwater were present as colloidal and particulate forms, respectively. In the humus-rich groundwater, Pu and Am were mainly associated with colloidal materials, but were present in different molecular size ranges (Pu: 30 000-10 000 Daltons, Am: more than 100 000 Daltons). Pu and Am added to the saline groundwater were present as soluble complexes with organic ligands as humic substances. Variations in the size distribution may be related to the size distribution and characteristics of dissolved organics.

## Introduction

Information on the speciation of actinides is essential for predicting the mobility and biological uptake in natural environment. Size fractionation methods utilising centrifugation, gel filtration, dialysis, filtration and ultra filtration have been used to investigate the particulate, colloidal and dissolved forms of actinides in natural waters, and especially to study the association with dissolved organic materials such as humic substances (HS) [1-4]. These speciation techniques have been proven to be simple and convenient methods, and normally free from contamination. Large variations are observed in the relative fraction of actinides present as colloidal forms in different water systems [1-4]. The purpose of this study is to investigate the association properties of actinides with naturally occurring organics in organic-rich groundwaters with different water chemistry. We have chosen three groundwater samples ($Ca^+$-$HCO_3^-$ and $Na^+$-$Cl^-$ types) with DOC concentration of 2-81 mg C/L for this study. Ultra filtration technique was applied to fractionate Pu and Am added to the groundwaters into various molecular sizes.

## Materials and methods

Groundwater samples were collected in boreholes at sedimentary rocks and peat bed. A double packer system was used to extract groundwater from tuffaceous rock at depth interval of 121-126 m in a bore hole from Imaichi in Tochigi Prefecture, Japan. The groundwater from peat bed was pumped up from 30 m depth below surface at Tokachigawa in Hokkaido Prefecture, Japan. The saline groundwater was collected from an artesian flowing well at Mobara in Chiba Prefecture, Japan. The perforated, screened intake section (PVC or stainless steel) collects water at 792-1 202 m depth below surface (alternation of strata, sandstone and mudstone). All of groundwaters were filtered with 0.45 μm filters. Chemical compositions of the samples are given in Table 1.

The groundwater organics were characterised by ultra filtration, UV-VIS spectroscopy and three-dimensional excitation emission matrix (3-D EEM) spectroscopy. The ultra-filtration was sequentially carried out with ultra-filters, Millipore Ultra Free CL, with molecular weight cut-off of 100 000 (100 k), 30 000 (30 k), 10 000 (10 k), and 5 000 (5 k) Daltons. The distribution of organics was estimated from absorbance at 280 nm in each size fraction. The UV-VIS spectra were measured by a Hitachi U-3300 spectrophotometer. The 3-D EEM spectra were measured by a Hitachi F-4500 fluorescence spectrophotometer. Relative fluorescence intensity (RFI) of the samples is expressed in terms of standard quinine unit (QSU). A QSU corresponds to fluorescence intensity of standard quinine sulfate (10 μg/L in 0.1 M $H_2SO_4$) at an excitation wavelength of 350 nm and an emission wavelength of 455 nm The maximum-intensity wavelengths are denoted as Ex. 350 nm/Em. 455 nm in this paper.

Stock solution of $^{238}Pu$, provided by CEA ($^{238}Pu$: plutonium nitrate in 3M $HNO_3$), was prepared by diluting 5 times in Milli-Q water. This stock solution was diluted 11 times in a Milli-Q water to obtain the solution used for the association experiments. The preparation was carried out just before the experiments. Americium-241 in 1 M $HNO_3$ was supplied from CEA. Stock solution of $^{241}Am$ was prepared by diluting with Milli-Q water and adjusted to pH 3 with NaOH solution.

The experimental procedure was as follows. 100 μL of $^{238}Pu$ tracer solution was added to 10 mL of the groundwater sample in a 30 mL Teflon centrifugation tube. 100 μL of $^{241}Am$ solution was added to 10 mL of the groundwater in a 10 mL glass centrifugation tube. The initial concentrations of $^{238}Pu$ and $^{241}Am$ were $6.7 \times 10^{-9}$ M and $7.6 \times 10^{-9}$ M, respectively. After seven days under shaking in an oven at 25°C, the solution was filtered with 0.45 μm Millipore filters and measured for pH. The final pHs of the solutions after the experiments were pH 8.0-8.7. The ultra filtration was sequentially carried out by the ultra-filters with 100 k, 30 k, 10 k and 5 k Daltons. The radioactive concentrations of $^{238}$Pu and $^{241}$Am in each size fraction were determined by liquid scintillation counting in a Packard Tri-Carb 2550 liquid scintillation counter. The percentage of these nuclides in each size range was estimated from the radioactivity concentration in each molecular size. Blank experiments for Pu and Am in NaClO$_4$ solution and artificial sea water were carried out with the same procedure. The association experiments were performed in duplicate or triple except for Pu and Am in the Imaichi and Pu in the Mobara.

## Results and discussion

### *Characteristics of groundwater organics*

The Imaichi groundwater shows yellowish colour, and the Tokachigawa and Mobara exhibit brownish colour. UV-VIS spectra of the groundwaters are shown in Figure 1. The absorption spectrum of the Mobara has two broad absorption peaks at 241 nm and 280 nm. The Tokachigawa sample exhibits featureless curve. The Imaichi groundwater has three broad peaks at 265 nm, 379 nm and 417 nm. The absorbance at three peaks consist mainly of organics (89-94%) with molecular size less than 5 k Daltons

The EEM contour plots for the groundwater samples are shown in Figure 2. The Imaichi groundwater has two distinct peaks at Ex. 220 nm/Em. 320 nm and Ex. 275 nm/Em. 320 nm, and a broad peak at Ex. 330 nm/Em. 415 nm. The Tokachigawa and Mobara samples have two broad peaks. These broad peaks are detected at similar wavelengths among these groundwaters, and are consistent with the HA and FA isolated from the groundwater samples [6, Nagao unpublished data]. The excitation and emission spectra of HS show broad peaks typically occurring around 260, 310-350 nm and 380-460 nm, respectively [7-9]. The protein-like fluorescence arises from the fluorescence of aromatic amino acids, either free or as protein constituents, and is observed at an emission of 300-305 nm (tyrosine-like) and 340-350 nm (tryptophan-like) from excitation at 220 and 275 nm [10]. Therefore, the fluorescent organics of Tokachigawa and Mobara groundwaters consist mainly of HS. The RFI is increased in the order of the Imaichi< the Tokachigawa< the Mobara. The Imaichi groundwater contains protein-like materials because the fluorescence peak disappears after the ultra filtration with ultrafilters having 100 k Daltons. The RFI is 7-30 times higher than that of humus-like materials.

Figure 3 shows molecular size distribution of UV-absorbing organics, humic acid (HA) and fulvic acid (FA) isolated from the groundwater by the XAD extraction procedure. These size patterns of organics are almost similar with the humic substances (HA and FA). Therefore, the groundwater UV-absorbing organics are mainly occupied by HS.

These results indicate that the groundwater samples can be classified into three groups:

- The Imaichi sample, humus-poor groundwater containing protein-like materials.

- The Tokachigawa sample, humus-rich groundwater.

- The Mobara sample, humus-rich saline groundwater.

## Molecular size distribution of Pu and Am in groundwaters

Figure 4 shows size distribution of Pu and Am in the absence of organics. Pu and Am added to $NaClO_4$ solutions and artificial sea water are mostly retained by the 0.45 μm filters, although about 20% of Pu is found in molecular size less than 5 k Daltons. There is no difference in the size distribution of Pu and Am in $NaClO_4$ solutions and artificial sea water. The both nuclides may undergo hydrolysis and form particles.

Figure 5 shows the size distribution of Pu and Am in organic-rich groundwaters together with that of dissolved organics. The 73% of the humus-poor groundwater from Imaichi is found in molecular size from 0.45 μm-100 k Daltons. On the other hand, the dominant fraction of Pu in the humus-rich groundwaters is 30k-10k Daltons for the Tokachigawa and <5 k Daltons for the Mobara. These size fractions correspond to the dominant size ranges of organics as shown in Figure 3.

The 90% of Am added to the Imaichi groundwater is retained by 0.45 μm filters, but particulate forms of Am in the Tokachigawa and Mobara are only 10%. The dominant fraction of Am is 0.45 μm-100 k Daltons for the Tokachigawa and <10 k Daltons for the Mobara groundwater.

## *Association of Pu and Am with groundwater organics*

Size distribution of Am in the presence of humic substances (HS) is compared with that in the groundwater. The similar variation occurs among Am-groundwater, Am-HA and Am-FA systems as shown in Figures 6a and b. The dominant size ranges are almost similar with those of the groundwater organics and HS (Figure 3). Am added to the humus-rich groundwaters, therefore, appears to be associated with humus-like materials. On the other hand, there are differences in the Am distribution between the saline groundwater and the HS (Figure 6c). The dominant size of Am in 0.01 M $NaClO_4$ was shifted to the larger size in comparison with those in the saline groundwater and artificial sea water. The size distribution of Am depends on water chemistry and size distribution of HS.

The percentage of Pu and Am in each size fraction is plotted as a function of the percentage of organics in humus-rich groundwater. Data plotted in Figure 7 is the results from the association experiments for the groundwaters (Figure 5). There is a positive correlation ($r^2 = 0.92$) between Pu and UV-absorbing organics, which are mainly occupied by humus-like materials. The relationship between Am and organics is not linear, but the amount of Am associated with organics increases with increasing the amount of organics in each size fraction. These results suggest that the association of Pu with humus-like materials may be controlled by the size distribution of humus-like materials.

Pu in the humus-poor groundwater is dissolved as colloidal forms, though major part of Pu (60-80%) in the absence of organics is retained by 0.45 μm filters (Figure 4 and 5). Most of Am in the groundwater and $NaClO_4$ solutions is present as particulate forms. To understand association of Pu with humus- and protein-like materials, the percentage of Pu and the organics in colloidal and solute phases is summarised in Table 2. The 70-90% of Pu and protein-like materials is found in size ranges from 0.45 μm-100 k Daltons. One possibility is considered that Pu added to the groundwater may be associated with protein-like materials rather than HS because of its low concentration, but the binding with inorganic colloids can not be eliminated.

# REFERENCES

[1]  N.A. Marley, J.S. Gaffney, K.A. Orlandini, M.M. Cunningham, *Environ. Sci. Technol.*, 27 (1993), 2456.

[2]  D.I. Kaplan, P.M. Bertsch, D.C. Adaiano, K.A. Orlandini, *Radiochim. Acta*, 66/67 (1994), 181.

[3]  J.F. Gaffney, N.A. Marley, K.A. Orlandini, "The Use of Hollow-Fibre Ultrafilters for the Isolation of Natural Humic and Fulvic Acids", *Humic and Fulvic Acids* (J.S. Gaffney, N.A. Marley, S.B. Clark, eds., ACS, Washington (1996), 26-40.

[4]  H. Amano, T. Matsunaga, S. Nagao, Y. Hanzawa, M. Watanabe, T. Ueno, Y. Onuma, *Org. Geochem.*, 30 (1999), 437.

[5]  G. Kamei, Y. Yusa, T. Arai, *Appl. Geochem.*, 15 (2000), 141.

[6]  S. Nagao, T. Tanaka, Y. Nakaguchi, Y. Suzuki, S. Muraoka, K. Hiraki, K., "Molecular Size Distribution of Np, Pu and Am in Organic Rich, Saline Groundwater", Proceedings of 9th Int. Conference of Int. Humic Substances Soc. (in press).

[7]  P.G. Coble, *Mar. Chem.*, 51 (1996), 325.

[8]  J.J. Mobed, S.L. Hemmingsen, J.L. Autry, L.B. McGown, *Environ. Sci. Technol.*, 30 (1996), 3061.

[9]  S. Nagao, Y. Suzuki, Y. Nakaguchi, M. Senoo, K. Hiraki, *Bunseki-Kagaku*, 46 (1996), 335.

[10] O.S. Wolfbeis, "The Fluorescence of Organic Natural Products", Molecular Luminescence Spectroscopy, Part I: Methods and Applications (S.G.Schulman, ed.), Wiley, New York (1985), 167-370.

# Table 1. Chemical compositions of the groundwater samples

| Sample | pH | Na$^+$ | K$^+$ | Ca$^{2+}$ | Mg$^{2+}$ (mg/L) | Cl$^-$ | SO$_4^{2-}$ | HCO$_3^-$ | DOC | Abs.$^a$ |
|--------|-----|--------|-------|-----------|------------------|--------|-------------|-----------|------|----------|
| Imaichi$^b$ | 8.0 | 16 | 0.7 | 21 | 2 | 7 | 15 | 80 | 2.2 | 0.009 |
| Tokachigawa$^c$ | 7.9 | 176 | 6.2 | 1.9 | 0.5 | 50 | 9 | 414 | 81.1 | 0.310 |
| Mobara$^d$ | 7.9 | 10 700 | 3 020 | 229 | 315 | 18 800 | 22 | 903 | 55.7 | 0.777 |

$^a$ Abs. = absorbance at 280 nm, $^b$ Shimada (unpublished data), $^c$ Shibata (unpublished data); $^d$ Kamei, *et al.* [5].
DOC concentration and absorbance at 280 nm were measured in this study.

# Table 2. Comparison of the percentage of Pu, protein-like and humus-like materials in two size fractions for the humus-poor groundwater sample from Imaichi

| | Sample | Molecular size fractions (%) | |
|---|--------|------------------------------|---|
| | | 0.45 µm-100 k Daltons | <5 k Daltons |
| Pu | Imaichi groundwater | 73 | 22 |
| | 0.01M NaClO$_4$ (absence of organics) | 2 ± 1 | 26 ± 5 |
| Protein-like materials | RFI at Ex. 275 nm/Em. 320 nm | 86 ± 1 | 14 ± 1 |
| | RFI at Ex. 225 nm/Em. 320 nm | 83 ± 2 | 17 ± 2 |
| Humus-like materials | Absorbance at 280 nm | 11 ± 1 | 89 ± 5 |
| | RFI at Ex. 330 nm/Em. 415 nm | 0 ± 3 | 104 ± 13 |

RFI – relative fluorescence intensity.

Mean values ± standard deviation of the percentages are presented in this table. The experiment for Pu in the groundwater was only carried out for one sample.

# Figure 1. UV-VIS spectra of the groundwaters

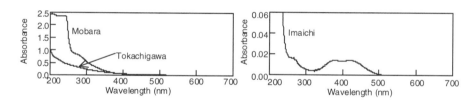

# Figure 2. EEM contour plots for the groundwaters. The intervals of contour are 0.3 QSU for the Imaichi, 3 QSU for the Tokachigawa and 16 QSU for the Mobara groundwater.

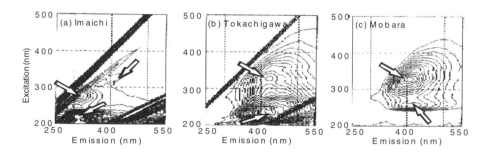

**Figure 3. Molecular size distribution of groundwater organics, HA and FA. The concentration of HA and FA was 10 mg/L in 0.01 M NaClO₄ (Tokachigawa) and artificial sea water and artificial sea water (Mobara) at pH 8.**

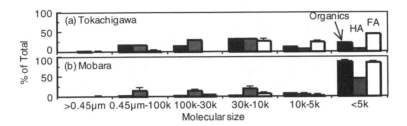

**Figure 4. Molecular size distribution of Pu and Am in NaClO₄ solutions and artificial sea water at pH 8**

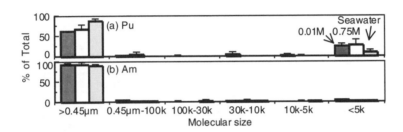

**Figure 5. Distribution of Pu, Am and UV-absorbing organics in various size fractions of the grounwaters**

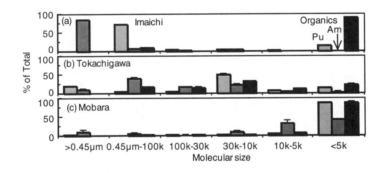

**Figure 6. Size distribution of Am in the presence of the groundwater HS and organics at pH 8. The concentration of HS is 10 mg/L (a) and 100 mg/L (b) and (c).**

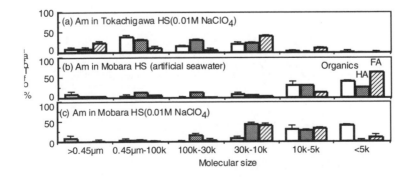

# Figure 7. Relation in size distribution between actinides and organics

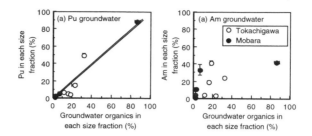

# SPECIATION OF "HOT PARTICLE" BY FISSION PRODUCT ELEMENT IDENTIFICATION

**Ihor Dryapachenko, Borys Rudenko**
Scientific Centre of National Academy
Institute for Nuclear Research
47 Nauka Avenue, Kyiv 03680, Ukraine
E-mail: dryapach@marion.iop.kiev.ua

## Abstract

The contribution of micro dispersed fractions of fuel and building materials ("hot particles") into Chernobyl radiation contamination is discussed. A method is proposed for measuring the hot particle distribution over the area. The micro- and macroscopic properties of the particles and the influence, of these properties on the efficiency of decontamination are discussed.

## Introduction

The general problems in studying of speciation of actinide and fission product elements in nuclear waste and nuclear reprocessing streams after Chernobyl disaster on the enormous territory of Ukraine have been not only technological but also more important social meaning. These include an individual and collective dose loading, a risk factor formation and so on. Firstly is the magnification of risk dying of cancers, which will be caused by a contact to an environment, contaminated by plutonium isotopes.

Because of absence of $\gamma$-decay in the plutonium isotopes, they are the most difficult to be measured. The $\alpha$-particle energies for $^{239,240}$Pu are very closely distributed. Much more complicated situation is for $^{241}$Pu, which is decayed by $\beta$-transition with 20 keV energy.

There are principal peculiarities in the environment soiling by transuranium elements and their decay products for example after Chernobyl disaster or nuclear weapon explosions. It is assembling into the melted glassy particles ("hot") which are weakly interacted with environment and save their threat potential on the very long time. There are the micron size fuel particles which are containing fuel and fission fragments ($^{134,137}$Cs, $^{144}$Ce, $^{90}$Sr, actinides, etc.), and their basic components-uranium oxides. The traditional radiochemical measurements are expensive and possible only in the first class laboratories. Moreover hot particles are irreversibly lost after radiochemical processes that is undesirables now for the unique hot particles after weapon bursts in 50-60. Study of physical and chemical properties of hot particles because of performances of their migration and density of contamination territory is carried out not only scientific but also by the practical interest. The greatest mass amount of soil is connected with $^{238}$U among all radionuclides that have dropped out in biosphere. The determination of the surface particle contamination by uranium in ground is hampered in connection with a rather high content of natural uranium and large non-uniformity of hot particles shedding.

Techniques of the determination of an isotope ratio in uranium were developed and the attempt for the identification of the contribution "Chernobyl" uranium by results of isotope ratio measurement in tests selected on different deleting from Chernobyl NPP was undertaken [1-3]. The detail tests and the methodological researches on model mixtures with the purpose of a choice of optimum conditions of extraction of uranium from ground and aerosol filters containing radioisotopes of U, Pu, Am and Cm were carried out Belarussia [1]. The technique of selection and concentration of uranium is based on ion exchange and extraction used in the given work. The radiochemical analysis of tests includes some stages. There are preparation for the analysis, decomposing of the ashen probes by mineral acids with destruction of a silicate structure or by leaching of the radionuclides with a nitric acid at boiling temperature, separating and selection of definite radionuclides or radionuclide groups by column chromatography and extraction. It is realised by $\alpha$-spectrometric or mass-spectrometric measurements with the purpose to determine radionuclides.

## Results and discussion

The Kyiv Institute for Nuclear Research has made two investigations on speciation of hot particles. These can use any methods for speciation in the trace concentration so as to apply a low-energy $\gamma$-spectroscopy or $\beta$-detection. The low-energy (10-100 keV) $\gamma$-quanta from the uranium isotope $^{234-236}$U excited states after $\alpha$-decay of the $^{238-240}$Pu are results of the internal conversion after decay of these excited states. It is possible to measure activities of the $^{238,240}$Pu and $^{241}$Pu by $^{241}$Am with high precision. These results represented the special interest for evaluation of the heat producing element in a campaign because half-life of $^{241}$Pu is 14 years and its accumulation is most sensitive during the campaign [2].

The clear results were obtained regarding the high internal stability of hot particles and preservation of their physical and chemical properties, which were determined in the environmental soil [3]. The β-irradiation intensities were measured on the same area unit near Kyiv for one year. The testing measurements of Chernobyl soil probes confirmed high reliability of the identification of the each hot particle as β-emitter. The field measurement results have confirmed their position fixation. There are fully retained spatial peculiarities of soiling according to the obvious possibility of changing of those spatial parameters because the natural climate changes in the environment.

In 1986, the task was practically important to establish, both the role of hot particles in contamination of an environment after Chernobyl accident and their physical and chemical properties. The first measurements were carried out with the emulsion track detectors, which were on display for a long time on the ground, and after appropriate handling time, the results were seen through a microscope. The hot particles were fixed as an accumulation of the α-particle tracks from decay of the transuranium elements contained in the particle. For identification of hot particles and detailing of the soil contamination, it was necessary to develop a technique based on devices to permit fast and effective measurements.

For this purpose, a simple device was developed which permitted to register a β-particle on a defined site of a surface for the beforehand-defined time interval. The device consists of a β-particle counter, a box made of lead with a slit collimator, a counter which accumulates number of events with digital indication in an assigned time interval. The device was compact and could work in field conditions. The measurement procedure was reduced to the following: for a rather clean ground site with a bar (length 1.5 m) having divisions through every 10 mm. Along the bar the box moved with the β-particle counter. The number of a β-particle registered from a surface $(1 \times 10)$ cm$^2$ in unit time was determined in each position of a box in one cm pitch. The box had the rectangular window with size $(1 \times 10)$ cm$^2$ through which a β-particle came into the counter.

If the hot particle was in the slit window boundaries, the device measured an increased number of counts in accordance with the activity of the hot particle. For a measurement of γ-rays background, the box slit was covered by a plate, which cut β-particles with energy up to 5 MeV. It was possible by such a scanning method to study distribution of β-activity on an objective site rather fast and effectively. It is necessary to underline the slit size of $(1 \times 10)$ cm$^2$ to determine a suitable position resolution of the device. It is necessary to reduce the slit size to obtain a higher resolution power. This naturally increases a measurement time and it is extremely undesirable (for example, in field conditions). We have applied the other mode for the definition of the true contribution of hot particles in averaged β-activity distribution. The top layer of ground in depth 5-10 mm was removed after a measurement of β-activity distribution in field conditions. The removed layer corresponded practically to full deleting of the β-activity. This removed ground was already stored in laboratory conditions and was diffused in a rather thin strip layer less than 10 cm in width and in length up to the full arrangement of the sample (up to 5-6 m). The prepared strip was measured by a method of scanning step-by-step with one-cm pitch. The arrangement of the ground sample on the large square was virtually forced in resolving power of the device approximately in 30-40 times and it enabled to reveal true distribution of hot particles. The measurement results show that the contribution of hot particles to the contamination of an environment was more than 90%.

The β-activity measurements required for a few distributions had spent in total about 50 h in the given work. It is necessary to develop a device permitting to receive data with a high positional solution in the reduced measurement time and deriving of a full information about distribution and activity of hot particles on the surface. It is possible to develop a device permitting to receive spatial distribution of radioactivity $(100 \times 100$ mm$^2)$ on the base of scintillation detectors after adjustment for the electron range in substance with a solution of the order 2 mm$^2$. The indicated solution is approximately in 500 times better than a device, used in the given work. The series of similar methods done by us were already described in the special literature [4].

It is important to investigate "migration" square of hot particles and to present a scenario of their destruction in natural conditions. The results of the β-activity distribution measurements in natural conditions, which had carried out by a scanning method by autumn 1986 and by summer 1988, have shown that from a view point of non-uniform attenuation of activity, the β-activity distribution was completely reproducing. The non-uniform attenuation of activity in hot spots is natural if particles are inhomogeneous in structure. Moreover, in natural conditions, these particles become non-uniformly covered by old foliage, grass, dust, etc.

It is important to note that the hot particles are emitting the electrons. The evaluation shows that an intensity of the α-particle radiation from hot particles does not exceed a few events per day. And the velocity of disintegration to the β-radiation channel is more than several decays per second.

Thus, hot particles constantly lose negative charges by emitting electrons. But the charge of a hot particle can partially be compensated by ionisation at the electron slowdown in the air. It creates a defined amount of ion pairs. It is possible to assume that an aerosol hot particle has a stationary positive charge. It is clear, for example, that in vacuum the charge of a hot particle tends to infinity. A velocity of β-particle radiation determines the velocity of charge increasing. The spontaneous charging of hot β-active particles in conditions close to atmosphere was investigated [5], and the measurement results of hot particle charges are shown in Figure 1. A hot particle charge in terms of an elementary charge (equal to an electron charge) is placed on y-axes and on x-axes an activity in terms of decays per second. The solid line represents calculated results with use of diffusion equations for an aerosol particle, which stands in a centre of sphere with a radius of 1 meter and emits electrons with energy 1 MeV. It is clear that the experiment and the calculation show the presence of a charge at a hot particle. It is possible to make a conclusion that this charge can has 130-300 elementary charges for a hot particle having more than 50 decays per second.

The behaviour of a hot particle charge in an exterior electrical field is considered in [6]. It is obvious that such an approach is possible to use electrical fields for purification of air streams in locations. It is shown that the exterior electrical field promotes magnification of a charge. The behaviour of a hot particle charge was investigated theoretically and experimentally [7] in conditions when it has a displacement velocity (for example, by gravity) and is lost in a gas medium containing extraneous aerosols. The charge of a hot particle grows with the increase of outside impurity concentration, and at a low concentration the charge is increased with a raise of a displacement velocity.

The results represented in [5-7] for theoretical and experimental researches of the charged aerosol β-active hot particles have a simple physical explanation. The hot particle charge, as a result of an electron radiation, increases by unit if one decay per one second is assumed. An electron loses the energy at slowdown in the air as a result of ionisation that produces a few pairs of ions. A cloud of positive ions and electrons accompanies the hot particle like this. The electrons are the most mobile cloud and partially compensate a positive charge of hot particle installing an equilibrium value. The external reasons break this equilibrium and promote magnification of a charge. The outside impurities swallow primary electrons and thus reduce concentration of ions in a cloud that also promotes magnification of an equilibrium charge.

It is possible to make a conclusion that the aerosol hot particles in natural conditions have a positive charge because of an electron irradiation. This charge is sufficient for a hot particle to reject moving the necessary direction even against gravity force on account of an electrical field with tension in some tens volt per centimetre and to deposit on the same collector. This task is easily solved locally; however, it becomes practically inapplicable to a large scale such as the Chernobyl accident.

To protect an environment from contamination in case of accident in the future, some devices should be equipped in applicable on NPP to a large scale, which are used atmospheric electrical appearances for sedimentation of particles in a defined place. However, it is unknown what charge of hot particles in conditions of the high levels of radiation and an air ionisation at the moment of a radiation ejection, and how to operate by atmospheric electrical appearances (for example, by a thunderstorm mesh). These problems are open and require solutions. It is possible to state that the Earth (as the planet) contains abundance of electrons. The electric field tension is rather high. The all aerosol hot particles under operation in this field under force gravity fall on ground and any subjects. These particles are positively charged and their density is great. In this connection, it is important to know the behaviour of a charge of a hot particle on a surface with a dielectric and the possibilities to speak about an equilibrium charge.

According to works [5-7] at a settling of a hot particle on a dielectric with zero conductivity, the charge should be increased approximately twice, as about the half decay electrons stick in a dielectric and do not participate in creation of a cloud. The experiments connected to selection of hot particles and a release from motes indirectly specify that the charge of a hot particle is high and it involves some energy of the particle interaction with dielectric. It is easy to reach trivial recommendations concerning preparation of locations, equipment, engineering, etc. to work in conditions of prospective contamination by hot particles and following activity for deactivation. A surface should be covered by a material with small dielectric constant $\varepsilon_1$ to decrease of adhesion of a hot particle with a dielectric. And if there was a contamination, it is possible to recommend wiping of a surface by a material made of a large dielectric constant $\varepsilon_2$. It is possible to note a relation $\varepsilon_2 \gg \varepsilon_1 \gg \varepsilon_0$, where $\varepsilon_0$ is connected to an air.

## Conclusion

In summary we shall mark basic moments, necessary researches and methodical works:

- Development of the two co-ordinate $\beta$-counter for a measurement of hot particles distribution on a surface of ground and their activity.

- Measurement of the ground contamination by hot particles around NPP with a realisation of continual monitoring of the soil.

- Study of a possibility to equip devices at NPP of devices for sedimentation of particles in a determined place in case of a hypothetical accident.

- Investigations of a hot particle charge on a surface of a dielectric and passive deactivation method connected to a covering of devices and engineering by painting with a small dielectric constant.

# REFERENCES

[1]    V.P. Mironov, L.E. Grushevich, M.A. Drugachenok, *et al.*, Vesti of Belorussian Academy of Science, No. 4, 1991, pp. 39 (in Russian).

[2]    M.D. Bondarkov, V.A. Zheltonozhskij, Yu.A. Izrael, T.N. Lashko *et al.,* Annual Scientific Conference of the Institute for Nuclear Researches, Ukraine (Kyiv, January 1996), pp. 114-117 (in Russian).

[3]    V.I. Gavrylyuk,, I.P. Dryapachenko, T.N. Lashko, M.V. Sokolov, *et al.,* Vesti of Belorussian Academy of Science, No. 4, 1991, pp. 35-39 (in Russian).

[4]    I.P. Dryapachenko, M.V. Sokolov, Patent of the USSR 1533520.

[5]    V.D. Ivanov, V.N. Kirichenko, *Doklady AN USSR*, 1969, Vol. 188, No. 1, pp. 65-68.

[6]    V.N. Kirichenko, V.D. Ivanov, *Doklady AN USSR*, 1969, Vol. 188, No. 2, pp. 315-317.

[7]    V.D. Ivanov, V.N. Kirichenko, V.M. Berezhnoj, I.V. Petryanov, *Doklady AN USSR*, 1972, Vol. 203, No. 4, pp. 806-809.

**Figure 1. Spontaneous charging of hot particles.**
**Sizes of particles are specified at calculated curves.**

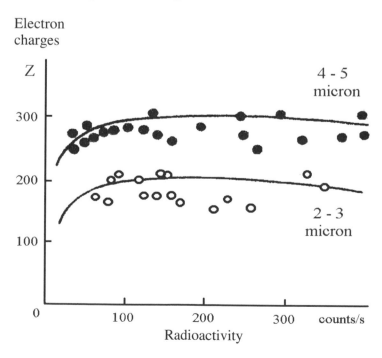

# APPLICATION OF X-RAY AND LOW ENERGY GAMMA-RAY SPECTROSCOPY TO DETERMINATION OF ACTINIDE CONCENTRATION IN REPROCESSING SOLUTION

**Toshihide Asakura, Gunzo Uchiyama,**
**Mutsumi Sawada, Hiromichi Hagiya and Sachio Fujine**
Department of Fuel Cycle Safety Research
Japan Atomic Energy Research Institute
Tokai-mura, Naka-gun, Ibaraki-ken 319-1195 Japan

**Paul J.A. Howarth**
Product Development Centre
British Nuclear Fuels plc.
Capenhurst, Chester, Cheshire CH1 6ER, England

## Abstract

X-ray and low energy gamma-ray spectroscopy was investigated as a technique to determine the concentrations of actinides within reprocessing streams produced at the NUCEF advanced reprocessing experimental research facility. The main actinides for which assay is required are U, Pu, Np and Am. The research presented here shows that passive measurements of X-rays and low energy gamma-rays from solution samples provides an accurate and non-destructive means for assaying the concentration. The experimental system described here can also be used for environmental assay of liquid samples to a minimum concentration of typically $10^{-6}$ M for Pu, $10^{-10}$ M for Am and $10^{-6}$ M for Np.

## Introduction

During these 40 years of industrial reprocessing, there has been a continual need to assay U and Pu as well as other nuclides for process control, safety, environmental monitoring and safeguards that are important factors associated with reprocessing facilities [1]. At JAERI, advanced reprocessing research is focused on the PARC (Partitioning Conundrum Key) refinement [2] of the PUREX process, using a small scale reprocessing research facility at Nuclear Fuel Cycle Safety Engineering Research Facility (NUCEF). The PARC process scopes Np separation prior to U and Pu partitioning, and Am and Cm separation from the high level waste raffinate. This research requires new detection systems capable to assay minor actinides in addition to U and Pu. Speciation of these elements is substantially important to confirm the control of the behaviour of those elements by the process. Although much attention has not been paid to minor actinide monitoring, some detection systems have been considered [3-5].

This work describes developing a suitable analysis system for use in the NUCEF facility based on X-ray and low energy gamma-ray detection of actinides in solution. Principle of the measurement technique is outlined first. Next, calibration results are presented. Then, there is the description of the results of assaying samples from the simulated active feed.

## Principal of the technique

X-ray and low energy gamma-ray emissions have been used previously for solution measurements [6-8] as well as wound monitors [9,10] and soil sample monitors [11-14]. Table 1 lists the specific emission activities for the low energy gamma-ray and L X-ray emissions from nuclides of interest. For assaying Np, the gamma-ray emission of $^{237}$Np at 29.4 keV was used since it is free of interference from other gamma-ray emissions. Am assay was achieved using the 59.5 keV gamma-ray emissions of $^{241}$Am, although interference from $^{237}$U needs to be accounted for, if it exists. For assaying Pu, the U L series X-ray emission from the decay of Pu was used since it is more intense than the characteristic gamma-ray emission, especially for $^{238}$Pu. Since product streams will contain up to 200 grams/litre of U, it also is necessary to determine the attenuation of X-ray and gamma-ray emissions. The ratio of the $^{241}$Am 26.3 keV and 59.5 keV gamma-ray emissions was used as a measurement of the self-attenuation. This ratio was also utilised as a useful means for assaying U. The gamma-ray emission at 63.3 keV from $^{234}$Th, the daughter of $^{238}$U, was used for U assay, too.

Detection of low energy photon emission also implies that the thickness of the active detector can be kept to a minimum, for example the intrinsic efficiency of a 5 mm thick SiLi detector for 15 keV X-ray radiation is ~ 100%. The advantage of using a thin detector is that the detection efficiency for higher energy gamma-rays from fission products is significantly less than for the L series X-rays, which will hopefully yield sufficient signal to noise ratio at low energy.

A procedure for separating the U L X-ray emission (from Pu decay) and the Np L X-ray emission (from $^{241}$Am decay) has been reported by Nitsche [6] and Gatti [7]. The technique makes use of relating the Np L X-ray emission to a characteristic gamma-ray line of $^{241}$Am such as 59.5 keV. Eq. 1 outlines use of the technique:

$$C_{Pu} = C_{LX} - C_\gamma k \qquad (1)$$

Here, $C_{Pu}$ is the L X-ray emission associated with the decay of Pu (the U L X-rays), $C_{LX}$ is the total L X-ray emission, and $C_\gamma$ is the characteristic gamma-ray emission from the interfering actinide, e.g. 59.5 keV emission from $^{241}$Am, k is a coefficient which relates $C_\gamma$ to $C_{LX}$ when no Pu is present. Here, the total L X-ray emission region is the background corrected count rate within the energy range of 12.5 keV to 23.0 keV. The coefficient k must be obtained for a pure [241]Am sample, which enables the 59.5 keV gamma-ray emission to be used as a measure of Np L X-ray emission. The coefficient is given by Eq. 2 where $C_{LX}$ and $C_\gamma$ are the count rates from the L X-ray region and the 59.5 keV region, respectively.

$$k = \frac{C_{LX}}{C_\gamma}$$ (2)

A similar procedure can also be used to correct interference from [237]Np by using the 29 keV gamma-ray emission or indeed any other nuclide for which a characteristic gamma-ray emission line can be used for reference.

## Calibration measurements and data interpretation

Measurement of the X-ray and gamma-ray emission from known U, Pu, Am and Np calibration samples was conducted. The detection system consists of a Series 7300 Canberra SiLi semiconductor detector (model type no. SL80175) 5 mm thick crystal with an active surface diameter of 80 mm$^2$ and a Be window of 1.0 mm thickness. Calibration sample solutions of U, Pu, Np and Am in 3 M nitric acid of 2 ml active volume in small glass vials were placed in close proximity to the detector. The amount of attenuating material between the sample and the detector was minimised to ensure maximum transmission of the X-rays. Shielding around the sample and detector consisted of a few centimetres of lead with an inner lining of 2 mm copper to attenuate Pb X-rays. The absolute efficiency of the detection system is approximately 0.2% below an energy of 30 keV.

The resulting data, corrected for self-attenuation, are shown in Figure 1. The U L X-ray emission data for the assay of Pu have been corrected for interfering X-ray emission from the decay of [237]Np and [241]Am (Np L X-rays). Typically, the linear regression for each series plotted in Figure 1 is accurate to approximately 3% for one standard deviation. Errors associated with determining the actinide concentration will depend on the linear regression uncertainty as well as statistical error analysis of the regions of interest, which will vary depending on the signal to background ratio.

## NUCEF simulated feed solution measurements

Prior to full active phase commissioning of the reprocessing research facility in NUCEF, a simulated feed solution was fed through the facility and separated. This feed solution consisted of un-irradiated uranium combined with stock Pu solution, some non-radioactive fission product simulants as well as a small concentration of their radioactive tracer counterparts [15,16]. This solution was fed to extractive equipment consisting of miniature mixer-settlers. The results for measurements of each flow stream are shown in Tables 2 to 5 for Np, Am, Pu and U together with comparisons obtained by previous techniques based on gamma-ray spectroscopy of tracers or chemical titration. The errors quoted are to one standard deviation.

For the Np data, previous measurements were based on detecting the 277.6 keV gamma-ray emission from [239]Np tracer, which was added to the feed solution. The results tend to show good agreement for all streams except the uranium product stream where the concentration was below the limit of detection of the system. The assay of Am was achieved using the 59.5 keV gamma-ray emission for both the previous and the new measurements reported here. The difference is that the previous data were obtained using a HPGe semiconductor detector uncalibrated for [241]Am assay.

Previous measurements of Pu were not possible. In order to check the X-ray technique for assaying Pu, there also was performed determination of the concentration by using the 45.4 keV gamma-ray emission from [240]Pu. Although errors associated with the latter technique are largely due to the relative weakness of the gamma emission and thus poor signal to noise ratio, the results tend to compare well. There is, however, a large difference in the results for the feed solution, which is possibly due to the self-attenuation correction factor applied to the L X-ray region of the spectrum. This is apparent in the 20% error associated with this measurement resulting from a large self-attenuation correction as well as a large correction to the X-ray region due to Np X-rays from [241]Am.

Previous assay of the U concentration was based on potentiometric titration. Table 5 shows a comparison of the titration results with the [234]Th assay technique and the 26.3 keV/59.5keV ratio technique from [241]Am. The large attenuation correction factor that was applied to the feed solution does mean that the assay of U by the 26.3 keV to 59.5 keV ratio is relatively easy and yields an accurate result. The product streams containing a low concentration of uranium are clearly difficult to measure.

## Conclusions

The results given above confirm the suitability of using X-ray and low energy gamma-ray emission to assay the concentrations of actinides in reprocessing streams. However, further work is required to determine the effect of gamma-ray emissions from reprocessed irradiated fuel, which has not yet been measured, using the instrumentation described here. In addition, the correction for the presence of [237]U will have to be investigated. Potentially, an additional application for the detection system described here is the assay of liquid environmental samples. With the current system, it is possible to determine actinide concentrations to typically $10^{-6}$ M for Pu, $10^{-10}$ M for Am and $10^{-6}$ M for Np, depending on the composition of the sample and the presence of other radionuclides, which might compromise the measurement.

## REFERENCES

[1]     C.M. Armatys, et al., ENICO-1126, Feb. 1983, Exxon Nuclear Idaho Company.

[2]     S. Fujine, et al., Global'97 Conference, Yokohama. Vol. 1, pp. 255-259.

[3]     J.K. Aaldijk, V.A. Wichers and G. Nikolaou, Proceedings of the 17th ESARDA Annual Symposium on Safeguards and Nuclear Material Management Aachen, 9-11 May 1995.

[4]     J.E. Stewart, et al., LA-13054-MS, January 1996.

[5]     W.D. Stanbro, LA-13248-MS, March 1997.

[6]     H. Nitsche, R.C. Gatti and Sh.C. Lee, Journal of Radioanalytical and Nuclear Chemistry Articles, Vol. 161, No.2 (1992), p. 401.

[7]     R.C. Gatti, H. Nitsche, *et al.*, Proceedings of the 5th International Conference on High Level High Waste Management, Las Vegas, 22-26 May 1994, Vol. 4, pp. 2719-2729.

[8]     T.K. Li, *et al.*, LA-UR-93-1726, June 1993.

[9]     D.G. Vasilik, R.W. Martin, C.J. Umbarger, *Health Physics*, Vol. 35, Oct. 1978, pp. 577-578.

[10]    H.E. Palmer, M.C. Rhoads, *Health Physics*, Vol. 56, No. 2, Feb. 1989, pp. 249-252.

[11]    O.J. Yaroshevich, *et al.*, *Nuclear Geophysics*, Vol. 9, No. 3, 1995, pp. 235-239.

[12]    M.G. Strauss, I.S. Sherman and C.T. Roche, *Nuclear Instruments and Methods*, A242 (1986), pp. 387-394.

[13]    R.J. Gehrke, K.D. Watts, E.W. Killian, M.H. Putnam, *Nuclear Instruments and Methods*, A353 (1994), pp.109-113.

[14]    D. Arnold, W. Kolb, *Appl. Radiat. Isot.*, Vol. 46, No. 11 (1995), pp. 1151-1157.

[15]    G. Uchiyama, T. Asakura, S. Hotoku and S. Fujine, 5th International Conference on Recycling, Conditioning and Disopal, Recod'98 Conference Proceedings, Nice, France, October 1998.

[16]    G. Uchiyama, T. Asakura, S. Hotoku and S. Fujine, World Nuclear Congress and Exhibition, ENC'98, European Nuclear Society, Nice, France, October 1998.

**Table 1. X-ray and gamma-ray emissions from the actinides of interest**

| Nuclide | Half-life | L X-ray emission (s/g) | Gamma-ray (keV) | $\gamma$ emission (s/g) | Ratio x/$\gamma$ emission |
|---|---|---|---|---|---|
| $^{238}$Pu | 87.8 y | $6.9 \times 10^{10}$ | 43.47 | $2.5 \times 10^{6}$ | 28 000 |
| $^{239}$Pu | 24.1 ky | $9.6 \times 10^{7}$ | 51.62 | $4.8 \times 10^{5}$ | 200 |
| $^{240}$Pu | 6.6 ky | $8.9 \times 10^{8}$ | 45.24 | $3.8 \times 10^{6}$ | 240 |
| $^{241}$Pu | 14.4 y | $4.1 \times 10^{3}$ | 98.43 | $6.9 \times 10^{6}$ | 0.0006 |
| $^{242}$Pu | 373.5 ky | $1.3 \times 10^{7}$ | 44.92 | $5.3 \times 10^{4}$ | 240 |
| $^{241}$Am | 432.7 y | $5.0 \times 10^{10}$ | 26.33 | $3.0 \times 10^{10}$ | 1.65 |
|  |  |  | 59.54 | $4.5 \times 10^{10}$ | 1.1 |
|  |  |  | 64.8 | $1.8 \times 10^{3}$ | $2.8 \times 10^{7}$ |
| $^{237}$U | 6.8 d | $1.7 \times 10^{15}$ | 26.33 | $5.9 \times 10^{14}$ | 2.9 |
|  |  |  | 59.54 | $9.9 \times 10^{14}$ | 1.7 |
|  |  |  | 64.8 | $3.1 \times 10^{11}$ | $5.5 \times 10^{3}$ |
| $^{237}$Np | 2.14 My | $1.4 \times 10^{7}$ | 29.38 | $3.4 \times 10^{6}$ | 4.0 |
| $^{234}$Th | 24.1 d | $9.1 \times 10^{13}$ | 63.3 | $3.9 \times 10^{13}$ | 2.4 |

**Table 2. Results for measurements of Np concentration**

| Product stream | Previous 277.6 keV $^{239}$Np measurement mg/litre | 29.4 keV $^{237}$Np measurement mg/litre |
|---|---|---|
| Feed solution | $137.5 \pm 18.5$ | $149.2 \pm 5.4$ |
| Tc Product | $13.9 \pm 1.9$ | $16.0 \pm 0.5$ |
| Np product | $62.6 \pm 8.4$ | $68.2 \pm 2.2$ |
| Pu product | $17.8 \pm 2.4$ | $14.2 \pm 1.4$ |
| U product | $9.1 \pm 1.2$ | – |

**Table 3. Results for measurements of the Am concentration**

| Product stream | Previous measurement mg/litre | 59.5 keV $^{241}$Am measurement mg/litre |
|---|---|---|
| Feed solution | 1.5 | $1.63 \pm 0.05$ |
| HAW | $0.85 \pm 0.11$ | $0.69 \pm 0.02$ |
| Tc Product | – | $0.0014 \pm 0.0005$ |
| Np product | – | $0.0047 \pm 0.0015$ |
| Pu product | $0.0049 \pm 0.0024$ | $0.011 \pm 0.007$ |
| U product | $0.0013 \pm 0.0008$ | $0.0013 \pm 0.0005$ |

**Table 4. Results for measurements of the Pu concentration**

| Product stream | L X-ray measurement mg/litre | $^{240}$Pu 45.4 keV measurement mg/litre |
|---|---|---|
| Feed solution | $3780 \pm 756$ | $2618 \pm 497$ |
| Tc Product | $80.4 \pm 3.2$ | $85.4 \pm 13.7$ |
| Np product | $263.4 \pm 8.4$ | $274.0 \pm 19.2$ |
| Pu product | $653.4 \pm 20.9$ | $653.0 \pm 131$ |
| U product | $85.0 \pm 2.7$ | $76.0 \pm 17.5$ |

**Table 5. Results for measurements of the uranium concentration**

| Product stream | Titration measurement g/litre | 63.3 keV $^{234}$Th measurement g/litre | 26.3 keV/59.5 keV ratio measurement g/litre |
|---|---|---|---|
| Feed solution | 250 | $242 \pm 29$ | $264 \pm 13$ |
| Tc Product | $9.9 \pm 1.8$ | $5.6 \pm 1.9$ | $5.6 \pm 0.5$ |
| Np product | $8.1 \pm 1.9$ | $7.0 \pm 2.5$ | $8.5 \pm 0.5$ |
| Pu product | – | – | – |
| U product | $41.1 \pm 7.5$ | $35.7 \pm 4.2$ | $42 \pm 6\%$ |

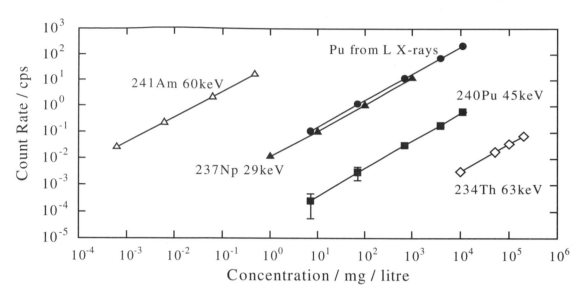

**Figure 1.** Calibration results for [241]Am, Pu, [237]Np and U. The data for [240]Pu shows the typical experimental errors (plotted for 2σ) associated with the calibration data

# CLOUD POINT EXTRACTION AND SPECIATION OF IRON(III) OF $10^{-7}$-$10^{-6}$ M LEVEL USING 8-QUINOLINOL DERIVATIVES AND TRITON X-100

**K. Ohashi, J. Ougiyanagi, S.Y. Choi, H. Ito, and H. Imura**
Department of Environmental Sciences, Faculty of Science
Ibaraki University, Mito 310-8521, Japan

## Abstract

The cloud point extraction behaviour, specification, and determination of traces of iron(III) with 8-quinolinol derivatives (HA), such as 8-quinolinol (HQ), 2-methyl-8-quinolinol (HMQ), and 2-methyl-5-octyloxymethyl-8-quinolinol (HMO$_8$Q) were investigated. Above pH 4.0, more than 95% of iron(III) was extracted with $5.00 \times 10^{-2}$ M HQ, HMQ, and HMO$_8$Q in 4 (v/v)% Triton X-100. The proposed method was applied to the determination of iron(III) in the Riverine Water Reference (JAC 0031 and JAC 0032) by graphite furnace atomic absorption spectrometry. The results agreed well with the certified values within 2% of the RSD.

# Introduction

HQ was used as the extractant for the characterisation and speciation of iron(III), which occurred during a hydrolysis reaction of iron(III) [1]. The dinuclear iron(III) was hardly extracted with HQ, but mononuclear iron(III) species was. It was also reported that the extraction rates of $Al^{3+}$, $AlOH^{2+}$, $AlSO_4^+$, and $Al_2(OH)_2CO_3^{2+}$ with HQ are larger than those of aluminium bonded with organic ligands, such as fumic acid [2]. The speciation of aluminium was done based on the difference in the extraction rates. Through the use of a solvent extraction technique, the valence recognition of metal ions is also possible. For example, chromium(VI) is extracted with $n$-octylamine from an acidic solution, but chromium(III) is not. Thus, solvent extraction can be a potential technique for not only elemental separation or pre-concentration but also the speciation of metal ions.

Recently, the cloud point extraction of organic and inorganic compounds using non-ionic surfactants has been of concern in analytical chemistry for minimising the use of toxic organic solvents [3,4]. Upon heating a surfactant solution over a critical temperature, the solution easily separates into two distinct phases. One contains a surfactant at a concentration below, or equal to, a critical micelle concentration. The other is a surfactant rich phase. An analyte can be pre-concentrated with a large degree of the concentration into a surfactant rich phase. Watanabe, *et al.* [5] have extensively investigated the cloud point extraction of several metal ions with chelating reagents into non-ionic surfactants, such as Triton X-100 and poly(oxyethylene)-4-nonylphenyl ether.

In this work, the cloud point extraction behaviour of iron(III) with HA using Triton X-100 and determination of iron(III) by absorption spectrophotometry and graphite furnace atomic absorption spectrometry (GF-AAS) were investigated to obtain fundamental information concerning the separation and speciation of traces metal ions. The solubility of HA into Triton X-100 aqueous solution was also determined at 25°C.

# Experimental

## Reagents

HQ and HMQ (Kanto Chemical Co. Ltd.) were recrystallised twice from ethanol. 2-Methyl-5-alkyloxymethyl-8-quinolinols (HMO$_n$Q, $n$ = 1, 2, 4, 6, 8, 10) were prepared by the method previously reported [6]. Triton X-100 was purchased from Aldrich. Water was distilled with a Milli-Q system (Milli-Q SP. TOC., Millipore). An acetate buffer was used to adjust the pH.

All other chemicals were of reagent grade and were used without further purification.

## Apparatus

The absorption spectra and absorbances were measured by a JASCO V-560 UV/VIS spectrometer, using a quartz cell with an optical path length of 10 mm. A GF-AAS (Shimadzu AA-646) using an auto sampler was used to determine iron(III) in an aqueous solution and iron(III)-HA complexes in a surfactant rich phase. The operating condition was as follows: wavelength, 248.3 nm; monochromator band pass, 1 nm; lamp current, 10 mA; sample uptake rate, 10 µl.

## Cloud point extraction procedure

A two ml of an iron(III) solution, 2 ml of an acetate buffer solution, and 0.5 ml of HA in a 4 (v/v)% Triton X-100 aqueous solution were taken in a test tube with a glass stopper. A solution was heated at 75°C for 30 min in a thermostated water bath. The resulting turbid solution was cooled to about -10°C for 20 min in a cooling bath and then an aqueous solution was poured off. The aqueous solution was subjected to the analysis of iron(III) by GF-AAS.

For measurements of the absorption spectra, the surfactant rich phase was diluted with water. Then, the resulting solution was supplied to absorption spectral measurements of iron(III)-HA complexes.

## Results and discussion

### Solubility of HA into 4 (v/v)% Triton X-100

The solubility of HA in a 4 (v/v)% Triton X-100 aqueous solution at 25°C increased in the following order: HMQ < 2-methyl-5-ethoxymethyl-8-quinolinol (HMO$_2$Q) < 2-methyl-5-decyloxy-methyl-8-quinolinol (HMO$_{10}$Q) < 2-methyl-5-methoxymethyl-8-quinolinol (HMO$_1$Q) < 2-methyl-5-octyloxymethyl-8-quinolinol (HMO$_8$Q) < 2-methyl-5-hexyloxymethyl-8-quinolinol (HMO$_6$Q) < 2-methyl- 5-butyloxymethyl-8-quinolinol (HMO$_4$Q) < 2-methyl-5-propyloxymethyl-8-quinolinol (HMO$_3$Q) (Figure 1). When the number of ethylene groups in HMO$_n$Q molecule is larger than 3, the solubility decreased along with an increase in the ethylene group. Though the reason why the solubility of HMO$_2$Q is the smallest among these HA molecules has not been clearly explained, the variation in the solubility may be ascribed to a difference in the interaction of the alkyl group in the HMO$_n$Q molecule with the polyethylene group of Triton X-100.

### Absorption spectra of Fe(III)-HA complexes

The iron(III)-HA complexes extracted into the surfactant rich phase showed an absorption maximum at about 600 nm. The molar extinction coefficients of iron(III)-HA complexes at 600 nm slightly increased along with an increase in the HA concentration.

### Effect of the pH on the cloud point extraction

The pH effect on the extraction percentage of iron(III) ($2.50 \times 10^{-4}$ M) with HA ($1.25 \times 10^{-3}$ M) using a 4 (v/v)% Triton X-100 aqueous solution was investigated. Iron(III) was completely extracted with HQ, HMQ and HMO$_8$Q into the surfactant rich phase at pH 4.95. The extraction percentage - pH curve for HMO$_8$Q shifted to a slightly lower pH region than those for HQ and HMQ. Above pH 4.0, more than 95% of the iron(III) was extracted with HA, HMQ, and HMO$_8$Q (Figure 2).

Speciation of hydrolysed iron(III) was also performed according to the following procedures. The pH of an iron(III) solution was adjusted to 4-12 with a sodium hydroxide solution. Then, the solution was heated at 70°C for 2 h. After the pH of the solution had been adjusted to 5.0, iron(III) was extracted into a surfactant rich phase. The absorbance at 600 nm of a iron(III)-HMQ complex extracted at pH 12 was same as that at pH 5. However, that of the iron(III)-HMQ complex extracted at pH 10 was very small.

Iron(III) speciation can be achieved by combing the cloud point extraction and high performance liquid chromatography.

### Effect of the HA concentration on the extraction of Fe(III)-HA complexes

In Figure 3, the extraction percentage of iron(III) ($2.50 \times 10^{-4}$ M) with HA ($1.25 \times 10^{-3}$ M) using 4 (v/v)% Triton X-100 is shown to increase in the following order: HMO$_8$Q < HQ < HMQ at pH 4.95. The extractability of iron(III) with HMQ was not very different from that of HQ and HMO$_8$Q. As will be mentioned below, HQ is not adequate for the absorption spectrometric determination of iron(III) because of serious interference from vanadium(V). Consequently, HMQ was used for the subsequent cloud point extraction experiments.

### Effect of the Triton X-100 concentration on the cloud point extraction

The concentration factors, $F$ ($= C_{S,Fe}/C_{Fe}$), were determined in a Triton X-100 concentration range of 0.1-8 (v/v)%, where $C_{S,Fe}$ and $C_{Fe}$ are the iron(III) concentration in a surfactant rich phase and the total iron(III) concentration, respectively. In the cases of 1, 2, and 4 (v/v)% Triton X-100, the extraction percentage was about 98%.

### Determination of iron(III) in a spiked water sample by absorption spectrometry

According to the procedure mentioned above, iron(III)-HA complexes were extracted into a surfactant rich phase. After the surfactant rich phase was diluted with an appropriate amount of water, the absorbances were measured at 600 nm. Based on the calibration curves for iron(III), $10^{-6}$ M level of iron(III) could be spectrometrically determined. The presence of diverse ions, such as vanadium(V), copper(II), nickel(II), cobalt(II), zinc(II), aluminium(III) equal to $5.0 \times 10^{-6}$ M iron(III) did not interfere with the determination of iron(III). In the presence of copper(II), the surfactant rich phase was turbid. Therefore, the solution was filtered by filter paper before measuring the absorbance.

### Application to the determination of iron(III) in the Riverine Water Reference by GF-AAS

The proposed method was applied to the determination of iron(III) in the Riverine Water Reference (JAC 0031 and JAC 0032) by GF-AAS. A 10 ml or 5 ml of the reference samples were taken in a tube. A 500 µl of $5.00 \times 10^{-2}$ M HMQ in 4 (v/v)% Triton X-100, 2 ml of an acetate buffer solution (pH 5), and 2 ml of 8 (v/v)% Triton X-100 were successively added to the solution. The total volume was made up to 20 ml with water. The solution was kept at 40°C for 60 min and then heated at 75°C for 35 min and cooled. After complete separation of the two phases, 500 µl of the surfactant rich phase was taken into 0.1 M HCl. The total volume was made up to 1.0 ml. The resulting solution was used for a GF-AAS analysis. Based on the calibration curve, iron(III) was determined. The analytical results are given in Table 1. When the Triton X-100 concentration was 1 (v/v)%, the result showed a good agreement with the certified values within 2% of the RSD.

# REFERENCES

[1]   T. Nishikawa, K. Hiraki, S. Gouda, K. Nakagawa, and M. Tamaki, *Nippon Kagaku Kaishi,* 1975, 1479.

[2]   N. Clarke and L. Danielsson, *Anal. Chim. Acta.*, 306, 5 (1995).

[3]   H. Watanabe and N. Yamaguchi, *Jpn. Anal.*, 33, E211 (1984).

[4]   M.F. Silva, L. Fernandez, R.A. Olsina, and D. Stacchiola, *Anal. Chim. Acta*, 342, 229 (1997).

[5]   H. Watanabe, T. Kamidate, S. Kawamorita, K. Haraguchi, and M. Miyajima, *Anal. Sci.*, 3, 433 (1987).

[6]   K. Ohashi, R. Iwata, S. Mochizuki, H. Imura, K. Hiratani, and H. Sugiura, *Talanta*, 43, 1481 (1996).

**Table 1 Determination of iron(III) in the Riverine water reference material by GF-AAS**

| Reference material | Sample taken/ml | Found | % RSD | Recovery (%) |
|---|---|---|---|---|
| JAC 0031* | 10.0 | $8.4\pm0.7^a$ | 8.0 | 121 |
|  | 10.0 | $7.0\pm0.1^b$ | 1.1 | 101 |
| JAC 0032** | 5.0 | $58.0\pm1^a$ | 1.9 | 101 |
|  | 5.0 | $58.0\pm1^b$ | 1.2 | 101 |

Each sample was analysed three times.
* Certified values, $6.9\pm0.5$ µg/l.
** Certified values, $57\pm2$ µg/l.
[a] 4 (v/v)% Triton X-100
[b] 1 (v/v)% Triton X-100.

# Figure 1. Solubility of HMO$_n$Q in 4 (v/v)% Triton X-100 at 25°C

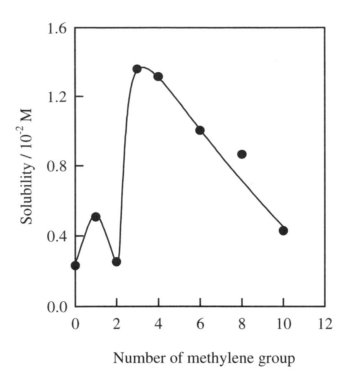

Number of methylene group

# Figure 2. Effect of pH on the cloud point extraction of Fe(III)

Fe(III), $2.50 \times 10^{-4}$ M; Triton X-100, 4 (v/v)%

■ – 8-quinolinol; ● – 2-methyl-8-quinolinol, ▲ – 2-methyl-5-octyloxymethyl-8-quinolinol

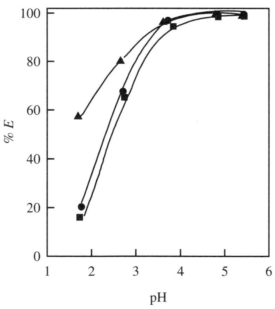

pH

# Figure 3. Effect of HA concentration on the cloud point extraction of Fe(III) at pH 4.95

Fe(III), $2.50 \times 10^{-4}$ M; Triton X-100, 4 (v/v)%
■ – 8-quinolinol; ● – 2-methyl-8-quinolinol, ▲ – 2-methyl-5-octyloxymethyl-8-quinolinol

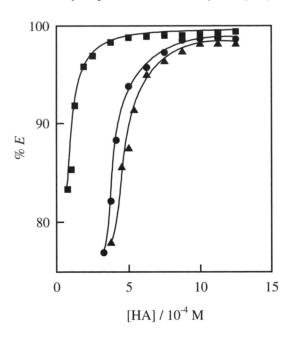

# SPECIATION OF ENVIRONMENTAL RADIONUCLIDES IN THE CHERNOBYL 30 KM ZONE

**H. Amano, Y. Hanzawa, M. Watanabe, T. Matsunaga,**
**T. Ueno, S. Nagao, N. Yanase and Y. Onuma**
Department of Environmental Sciences
Japan Atomic Energy Research Institute
Tokai-mura, Naka-gun, Ibaraki-ken 319-1195 Japan

## Abstract

This paper is a summary from previously published ones [1,2]. Speciation of environmental substances with radioactive Cs, Sr and transuranic isotopes has been examined about surface soil and water leachate sampled from the exclusion zone (30 km zone) around the Chernobyl nuclear power plant (NPP). Sequential selective extraction was carried out on surface soil samples. $^{90}$Sr in the soil was estimated to be highest in the water soluble and exchangeable fractions which were easily accessible to river and ground water as a dissolved fraction. Pu isotopes and $^{241}$Am are major radionuclides in free humic and free fulvic acid fractions. Ultra filtration has been used for water samples as another technique for speciation. Most Pu and Am exist in the molecular weight fractions of beyond 10 000 Dalton. This means that transuranic elements are associated with mobile high molecular weight materials like fulvic acids in water leacheates.

# Introduction

It has been widely recognised that migration behaviour of radionuclides is strongly affected by their chemical and physical forms [3,4]. Also recognised is the role of organic substances in the environment for migration of toxic materials [5,6]. The importance of humic substances (HS) in the natural environment for the migration of long lived radionuclides has widely been recognised, especially for transuranic elements [7,8].

Speciation of natural environmental substances with radioactive Cs, Sr and transuranic isotopes has been examined about surface soils and river water sampled from surface environment in the exclusion zone (30 km zone) around the Chernobyl Nuclear Power Plant (NPP) [1,2,9,10]. This work also aimed to investigate a secondary migration of radionuclides released and deposited from the Chernobyl accident, especially for the role of organic materials. Radionuclides determined were $^{238,239,240,241}$Pu, $^{241}$Am, $^{244}$Cm, $^{90}$Sr and gamma-ray emitting nuclides including $^{137}$Cs.

For the purposes of speciation of environmental radionuclides, sequential selective extraction techniques has been widely used (for example, Ref. [11]), but have not been standardised to date. The specificity and reproducibility of the methods depends on the chemical properties of the elements and the chemical composition of samples. Criticisms have in particular focused on two major experimental problems, selectivity and redistribution (for example, Ref. [12]). Even though, careful experiments with sequential selective extraction bring informations on speciation of trace materials in environmental samples.

Ultra filtration techniques has also been widely used for the purposes of speciation of trace materials in environmental water to estimate molecular weights (for example, Ref. [13]). These techniques themselves have brought no information on the chemical properties of target elements, but brought information on sizes of target materials.

Here we report some results on speciation of environmental radionuclides in the Chernobyl 30 km zone, using a sequential selective extraction for soils and an ultra filtration technique for water samples.

This paper is a review of previously published ones by the authors [1,2], as well as by other research groups.

# Experimental procedure

Undisturbed surface core soils were sampled at a pine forest near the Sahan River, a tributary of the Pripyat River in 1995, 5 km west from the NPP. This soil is sandy and covered with organic layers (AoL, AoF, and AoH horizon). The characteristics of the soil were as followed, for AoH and 0-1 cm mineral layer, pH ($H_2O$): 4.31 and 3.86, cation exchange capacity (CEC, meq/100 g): 10.4 and 4.7, $K^+$ (meq/100 g): 68.9 and 22.6, $Ca^{2+}$ (meq/100 g): 0.8 and 0.3, and ignition loss (%): 73.2 and 4.0, respectively.

For the purposes of speciation of environmental radionuclides, sequential selective extraction was carried out on the surface soil sample (0-1 cm mineral layer) in the same manner as reported before [1]. The speciation results about the radionuclides for the organic layers have been presented in previous paper [1]. This uses distilled water (water soluble fraction), 1 M $CH_3COONH_4$ (exchangeable fraction by adjusting pH to 7 and then to 4.5, respectively), 0.5M NaOH (to divide humin plus insoluble fraction as a precipitate) followed by centrifugation. Then, after adjusting supernatant solution to pH 1~2 followed by centrifugation, supernatant solution (fulvic acid fraction) and precipitate (humic acid fraction) were separated. This procedure is shown in Figure 1. Radionuclides separated and determined were $^{90}$Sr (radioactive half life; $T_{1/2}$: 28.8 y), $^{137}$Cs ($T_{1/2}$: 30.1 y), $^{238}$Pu ($T_{1/2}$: 87.7 y), $^{239,240}$Pu ($^{239}$Pu:$T_{1/2}$: 24 110 y; $^{240}$Pu:$T_{1/2}$: 6 550 y), $^{241}$Pu ($T_{1/2}$: 14.4 y) and $^{241}$Am ($T_{1/2}$: 433 y). Strontium-90 was determined by cation exchange and liquid scintillation counting [14]. Gamma emitters such as $^{137}$Cs were determined using Ge-detectors. $^{241}$Pu, which is a parent nuclide of $^{241}$Am, was determined by liquid scintillation techniques, following a radiochemical method [15]. Transuranic elements except for $^{241}$Pu were determined by radiochemical methods and alpha-ray spectrometry [1].

Ultra filtration was used for water leachates from soil to estimate molecular weights of dissolved materials in water phase. Surface soil (AoH horizon) which showed the highest concentration of radionuclides along the Sahan River was extracted with distilled water. The extracted water was examined for the estimation of dissolved fraction in run off components. After a filtration using a 0.2 μm Millipore filter, extracted water was treated with ultra filtration techniques separating the molecular weight fractions of greater and less than 10 000 Dalton. Each fraction was measured for the radioactivities and stable elements such as calcium. Stable elements in each fraction were measured by an inductivity coupled plasma atomic emission spectrometer (ICP-AES). The dissolved organic materials in each fraction were measured for the dissolved organic carbon (DOC).

## Results and discussion

### Depth profile and speciation of deposited radionuclides in surface soil

Figure 2 shows a depth profile of deposited radionuclides in surface soil sampled from a pine forest near the Sahan River in 1995. As seen in Figure 2, the highest amount of radionuclides exists in the surface organic layer (AoH horizon). Almost all of the radioactivities exist in the surface organic layer plus 0-3 cm depth horizon in the soil layer, as shown in Figure 2.

Figure 3 shows a result from speciation of radionuclides in the 0-1 cm horizon (including organic materials) of surface soil. As shown in Figure 3, the majority of $^{137}$Cs, Pu isotopes and $^{241}$Am in the surface soil were in the humic plus insoluble fraction. About 90% of $^{90}$Sr was determined in the water soluble and the exchangeable fraction. Plutonium isotopes and $^{241}$Am were comparable in the free fulvic acid and the free humic acid fractions. $^{90}$Sr and $^{137}$Cs were less than 4% in the free fulvic and humic acid fractions. Among these radionuclides, $^{90}$Sr was highest in the water soluble and exchangeable fractions, which were accessible to river water as a dissolved fraction to the total fractions in surface soil for all layers. Water soluble and exchangeable fractions of $^{137}$Cs, Pu isotopes and $^{241}$Am were also found, as shown in Figure 3. Pu isotopes and $^{241}$Am are major radionuclides in free humic and free fulvic acid fractions, which have possibility to be dissolved in water. Among them, $^{241}$Am is present to some extent in free fulvic acid fraction, Pu isotopes is present in free humic acid fraction more than in the free fulvic acid fraction.

### Water leachates of surface soil along a river watershed

As shown in Figure 2, the majority of radioactivities exist in the organic AoH horizon. So, water leachate from the AoH horizon was to estimate a dissolved fraction as one of main run off components of radioactivities from the river watershed.

Figure 4 shows the species distribution result of water leachates of surface soil along the river watershed. As shown in Figure 4, most Pu (79%) and Am (68%) exist in the molecular weight fractions of beyond 10 000 Dalton, in spite of the fact that most of the dissolved organic carbon fractions (69%) exist in the molecular weight below 10 000 Dalton. This means that transuranic elements, such as Pu and Am, are associated with high molecular weight materials containing carbon such as humic substances in water leachates. On the contrary, most of [137]Cs (85%) and [90]Sr (63%) exist in the molecular weight fraction of below 10 000 Dalton, reflecting the fact that most of them exist in lower molecular forms or ionic forms or associated with low molecular materials. However, some of [137]Cs and [90]Sr exist in the molecular weight of beyond 10 000 Dalton. This means that some parts of [137]Cs and [90]Sr are associated with higher molecular materials. Stable calcium shows almost the same result as [90]Sr. There is an indication that considerable amounts of radiocesium may be carried with water-soluble organic substances [16].

The results reported here were compared with the one of ultra filtration conducted in the same manner as river water of the Sahan River sampled on March 1996 [17]. It shows that most of [239+240]Pu (80%) and [241]Am (62%) are associated with colloidal and higher molecular weight fractions above 10 000 Dalton. On the contrary, about 80% of [90]Sr is present in lower molecular weight form. Most of the dissolved organic fractions (64%) in the river water exist in the molecular weight below 10 000 Dalton. Almost all Fe (99%) exists in the fraction greater than 10 000 Dalton, but almost all Ca (97%) exists in the fraction below 10 000 Dalton. These features are in good agreement with the results from our research except for [90]Sr and calcium.

Concerning the transuranic elements, they are likely associated with dissolved humic substances both in river water and in water leachate from soil, and that they are associated much more with humic substances in higher molecular fractions greater than 10 000 Dalton. The differences for [90]Sr and calcium may be attributed to the differences of the change of liquid phase from soil water leachates to river water, such as pH. The pH in the soil solution and the Sahan River was 4.3 and 7.5, respectively. This range of pH is quite sensitive to chemical reactions of precipitation and dissolution. This may be one of the reasons of the slight difference of [90]Sr and Ca between the soil water and the river water. Another possible reason is that the sources of these nuclides between soil leachate and river water might be different. In a river system, major sources of these nuclides come from groundwater mainly in ionic forms, but in water leachates most of them might come from organisms in the form of higher molecular sizes.


## Conclusion

Analysis of depth profiles of long-lived Chernobyl radionuclides in soil showed that most of the deposited radionuclides exists in the surface centimetres. Analysis of speciation of radionuclides in surface soil by means of selective sequential extraction method showed that the majority of [90]Sr was found in the exchangeable fraction which is accessible to river water. Water soluble and exchangeable fractions of [137]Cs, Pu isotopes and [241]Am were also found. Pu isotopes and [241]Am are major radionuclides in free humic and free fulvic acid fractions. While the differences of the fractions between Pu isotopes and [241]Am are not large, [241]Am tends to be associated with in the free fulvic acid fraction.

Dissolved long lived radionuclides transferred from surface soils to river water in the exclusion zone (30 km zone) near the Chernobyl Nuclear Power Plant was analysed by means of ultra filtration technique. Concerning the transuranic elements, it showed that they are associated with dissolved humic substances both in river water and in water leachate from soil. They are associated much more with humic substances in higher molecular fractions above 10 000 Dalton in spite of the fact that DOC concentration, both in river water and in water leachate from soil, is much less in the fraction above 10 000 Dalton.

*Acknowledgements*

This research has been conducted under a research co-operation between Japan Atomic Energy Research Institute and Chernobyl Scientific Centre for International Relations, Ukraine. The persons who are related to this co-operative research program are greatly acknowledged.

# REFERENCES

[1]    H. Amano. M. Watanabe, Y. Onuma, T. Ueno, T. Matsunaga and N.D. Kuchma, "Speciation of Cs, Sr and Transuranic Elements in Natural Organic Substances of Surface Soil Layers", Proc. 8th Meeting of the Int. Humic Substances Society, the Role of Humic Substances in the Ecosystems and in Environmental Protection, p. 709 (1997).

[2]    H. Amano, T. Matsunaga, S. Nagao, Y. Hanzawa, M. Watanabe, T. Ueno and Y. Onuma, "The Transfer Capability of Long-lived Chernobyl Radionuclides from Surface Soil to River Water in Dissolved Forms", *Organic Geochemistry*, 30, 437.

[3]    Bulman and Cooper (1985).

[4]    H.R. Gunten and P. Benes, "Speciation of Radionuclides in the Environment", *Radiochim. Acta*, 69, 1 (1995).

[5]    Griffith, *et al.*, (1994).

[6]    Jones and Beck (1994).

[7]    Choppin (1988): Humics and radionuclide migration. Radiochim. Acta 44/45, 23-28.

[8]    S.P. Sparkes T.W. Sanders and S.E. Long, "A Literature Survey of the Organic Speciation of Radionuclides", AERE R 12487, Harwell Laboratory (1987).

[9]    Matsunaga, *et al.* (1998).

[10]   N. Yanase, T. Matsunaga, H. Amano, H. Isobe and T. Sato, "Highly Radioactive Hot Particles as a Source of Contamination Around the Chernobyl NPP", Proceeding of the 7th International Conference on Radioactive Waste Management and Environmental Remediation, ICEM'99, Nagoya, Japan (1999).

[11]   A. Tessier, P.G.C. Campbell, "Comments on the Testing of the Accuracy of an Extraction Procedure for Determining the Partitioning of Trace Metals in Sediments", *Anal. Chem.*, 60, 1475 (1988).

[12]   T. Qiang, S. Xiao-quan, N. Zhe-ming, "Evaluation of a Sequential Extraction Procedure for the Fractionation of Amorphous Iron and Manganese Oxides and Organic Matter in Soils", *The Science of Total Environ.*, 151, 159 (1994).

[13]    B. Salbu and H.E. Bjornstad, "Analytical Techniques for Determining Radionuclides Associated with Colloids in Waters", *J. Radioanal. Nucl. Chem.*, 138, 337 (1990).

[14]    H. Amano and N. Yanase, "Measurement of $^{90}$Sr in Environmental Samples by Cation Exchange and Liquid Scintillation Counting", *Talanta*, 37, 585-590 (1990).

[15]    M. Watanabe and H. Amano, "Determination of $^{241}$Pu by Liquid Scintillation Counting Method and its Application to Environmental Samples", JAERI-Res 97-016 (1997).

[16]    U. Passeck, G. Lindner and W. Zech, "Distribution of $^{137}$Cs in Water Leachates of Forest Humus", *J. Environ. Radioactivity*, 28, 223 (1995).

[17]    Matsunaga, *et al.*, unpublished data.


*Additional references*

[18]    "Modelling and Study of the Mechanisms of the Transfer of Radioactive Materials from Terrestrial Ecosystems to and in Water Bodies around Chernobyl", EC Report, U. Sansone and O.V. Voitsekhovitch, eds., EUR 16529 (1996).

[19]    "Sediment and Pollution in Waterways – General Considerations", IAEA-TECDOC-302, IAEA (1994).

[20]    T. Matsunaga, H. Amano and N. Yanase, "Discharge of Dissolved and Particulate $^{137}$Cs in the Kuji River", *Japan. Applied Geochemistry*, 6, 159 (1991).

[21]    T. Matsunaga, H. Amano, T. Ueno, N. Yanase and Y. Kobayashi, "The Role of Suspended Particles in the Discharge of $^{210}$Pb and $^{7}$Be Within the Kuji River Watershed", *Japan. J. Environ. Radioactivity*, 26, 3 (1995).

[22]    T. Matsunaga, T. Ueno, H. Amano, Y. Tkachenko, A. Kovalyov, M. Watanabe and Y. Onuma, "Characteristics of Chernobyl-Derived Radionuclides in Particulate Form in Surface Water in the Exclusion Zone around the Chernobyl Nuclear Power Plant", *J. Contaminant Hydrology*, 35, 101 (1999).

[23]    O.V. Voitsekhovitch, M.J. Zheleznyak and Y. Onishi, "Chernobyl Nuclear Accident Hydrological Analyses and Emergency Evaluation of Radionuclide Distribution in the Dnieper River, Ukraine During the 1993 Summer Flood", PNL-9980 (1994).

**Figure 1. Speciation procedure A: extraction of soil solution and organic matter**

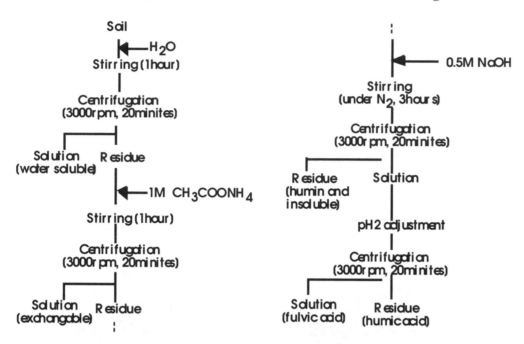

**Figure 2. Depth profile of radioactivities in surface soil at a forest near River Sahan**

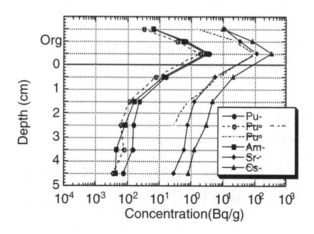

**Figure 3. Speciation of radionuclides in 0-1 cm horizon in surface soil at a forest near River Sahan**

*Upper left: $^{137}$Cs; Upper right: $^{90}$Sr; Lower left: $^{239,240}$Pu; Lower right: $^{241}$Am*

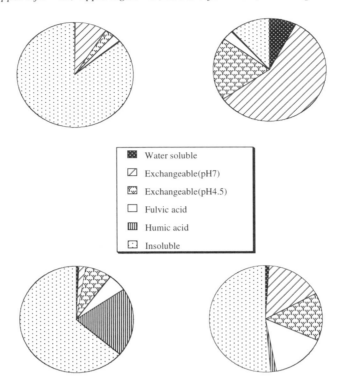

Water soluble
Exchangeable(pH7)
Exchangeable(pH4.5)
Fulvic acid
Humic acid
Insoluble

**Figure 4. Distribution of species between MW > 10 000 and MW < 10 000 in water leachate from a forest soil near River Sahan**

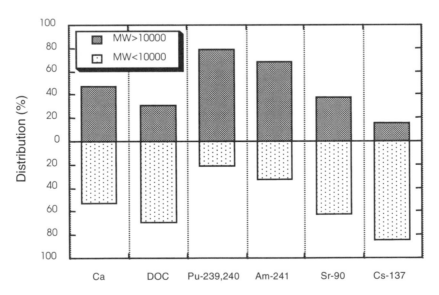

# POSTER SESSION

## *Part B: Methods for Macro Concentration Speciation ($>10^{-6}$ M)*

# Cm(III) SPECIATION IN BRINES BY TIME RESOLVED LASER FLUORESCENCE SPECTROSCOPY

**P. Paviet-Hartmann**
Los Alamos National Laboratory
Environmental Science and Waste Technology Division
Los Alamos, NM 87545, USA

**Th. Fanghänel, R. Klenze,Th. Könnecke, J.I. Kim, W. Hauser**
Forschungszentrum Karlsruhe
Institüt fur Nukleare Entsorgungstechnik
D-76021 Karlsruhe, FRG

**H. Weger**
Hemispheric Centre for Environmental Technology
Florida International University,
10555 W. Flagler St., CEAS-2100, Miami, FL 33174, USA

**K.K. Park**
Korea Atomic Energy Research Institute
Daejon, 305-600, Korea

**K.H. Chung**
Korea Electric Power Research Institute
Nuclear Power Laboratory
103-16 Munji-Dong, Yusong-Ku, Taejon 305-380, Korea

## Abstract

Time resolved laser fluorescence spectroscopy (TRLFS) is a very sensitive and selective analytical method that has been used for the understanding of actinide solution chemistry in brines which is essential for modelling that requires accurate knowledge of the interaction between actinide and different inorganic ions. Curium was selected to represent the trivalent actinides, because its excellent fluorescence spectroscopic properties enable speciation at very low concentration. Furthermore, TRLFS has proved to be a versatile tool to study hydrolysis reaction as well as complexation with sulphate, carbonate and silicate ligands in NaCl solutions. The formation of the different Cm(III) species is quantified spectroscopically in the trace concentration range by peak de-convolution of fluorescence emission spectra. Spectroscopic data (main fluorescence wavelengths, half-bandwidth, lifetime) for the different Cm(III) complexes formed as well as the complexation constants determined from the spectroscopically quantified mole fraction of each species for each particular condition are presented.

# Introduction

The possibility of disposing nuclear wastes in deep salt formations has attracted the attention of the international environmental community. Systematic investigations of actinide solution chemistry in brines are under way in the US as well as in Germany to model the behaviour of actinides in natural, multi component electrolyte solutions of high concentration [1-8]. The US's first geological repository for radioactive waste has been excavated in South-eastern New Mexico at the Waste Isolation Pilot Plant (WIPP) and opened in March 1999. Scenarios of post closure radioactive releases from the WIPP (US) or the Gorleben repository (Germany) often involve hydrologic transport of radionuclides in brine solutions. Consequently, a proper understanding of the migration of Cm(III) as well as U(VI), and Am(III) [1-4, 7-14], in the groundwater, and the effect of the presence of natural substances on the uptake of Cm(III) by natural solids is a top priority for modelling the transport of radionuclides. The study of the complexation of Cm(III) in brines with potential inorganic ligands such as $OH^-$, $SO_4^{2-}$, $CO_3^{2-}$, and the influence of the presence of silica colloids on its mobility and on its sorption on surrounding minerals, will help the understanding of its further potential migration in the environment. Furthermore, thermodynamic data pertinent to the interaction of actinides with groundwater are critical input to modelling programmes for nuclear waste disposal. The stability constants for the formation of complexes of trivalent actinides have been compiled previously by Fuger, et al. [15]. Complexation constants of Cm(III) hydroxide, sulfate, carbonate, and silica species have been investigated by different techniques. Numerous methods used in the literature, like solubility experiments [16-20], ion exchange and distribution equilibrium [21-24], are not capable of the direct speciation of complex species straightforwardly. The large scatter of experimental data led us to report new data concerning the complexation of Cm(III) with ligands present in ground waters as a function of NaCl concentration by the means of TRLFS. TRLFS is the method of choice for actinide speciation in the nanomolar range concentration because of its high sensitivity resulting from laser excitation, rapidity (few minutes analysis), and direct information obtained (spectral and temporal) on complexes present in solutions. The triple resolution:

- Excitation wavelength 375nm (with the correct choice of the laser dye).

- Emission selectivity since each element studied gives a characteristic fluorescence spectrum.

- Time resolution parameters permits discrimination of long lifetime fluorescence against undesirable short lifetime fluorescence.

# Experimental

A schematic layout of the TRLFS instrument is shown in Figure 1. For excitation, an excimer pumped dye laser (Lambda physics, EMG 201 and FL 3002) emitting at 375 nm (Laser dye QUI) is used. The laser light (pulse energy about 6 mJ at 10 Hz repetition rate) is coupled by a 460 µm Ø quartz fibre into the glove box. The interesting feature of TRLFS is the possibility to perform remote measurements via fibre optics. This instrumentation has been developed to carry out determinations in a glove box. The emitted light is collected by a fibre bundle and directed to a polychromator (Jabin Yvon, HR 320). Time resolution is performed by a gated micro channel plate. The data are read out by a control unit and stored in a PC. A more detailed set-up is described elsewhere [1,9,12]. All fluorescence measurements are performed at constant temperature $22 \pm 3°C$ within an inert glove box (100% Ar, or $CO_2$/Ar atmosphere). The emission spectrum of Cm(III) is recorded in the range 580-620 nm within a constant time window of 800 µs width exciting at 375 nm. For Cm(III) fluorescence measurements, time delay is set up to 1.2 µs. A spectral resolution of 0.2 nm is attainable for a bandwidth of 40 nm.

Detailed descriptions of the experimental procedures are given elsewhere [1-4]. The curium stock solution consisted of 97.2% $^{248}Cm$, 2.8% $^{246}Cm$, and less than 0.01% $^{244}Cm$ in weight. Concentration and pH values are given in the molal scale ($m_i$ = mol $i$/kg $H_2O$; pH = -log $m_{H+}$). The curium concentration was assayed by liquid scintillation counting (Quantulus).

## Results and discussion

The following procedure described here can be generalised to the Cm(III) species determination by peak deconvolution. In our experiments we expressed the measured spectrum for a ionic strength I for a specific ligand concentration as:

$$Sp_{exp}(\lambda) = \alpha\, Sp_{Cm(III)}(\lambda) + \beta\, Sp_{specie1}(\lambda) + \gamma\, Sp_{specie\,2}(\lambda)$$

where $Sp_{exp}$ represents the total emission fluorescence spectrum, and $Sp_{Cm(III)}$, $Sp_{specie\,n}$ the emission fluorescence spectra of the different species involved in the reaction. Since the Cm(III) pure spectrum is perfectly characterised and the last pure spectrum of the highest studied ligand concentration either $OH^-$, $SO_4^{2-}$, $CO_3^{2-}$ or silica is easily determined, the fluorescence emission spectra of the pure complexes are determined by peak deconvolution. Consequently, the total spectrum $Sp_{exp}$ is expressed as the sum of the emission fluorescence spectra of each species within the solution. A least squares fit is then applied (with $\alpha$, $\beta$, $\gamma$ as parameters) on each measured fluorescence spectrum for each ligand concentration or pH measured. This procedure allows us to obtain $\alpha$, $\beta$, $\gamma$ coefficients and the related percentage of species present in the solution. Some selected fluorescence emission spectra of Cm(III) ion and its complexation by hydroxide, sulfate, carbonate, and silica in very dilute solutions in the wavelength range from 580 to 620 nm are presented in Figure 2. At 0 molal ligand concentration, the emission spectrum represents the $Cm^{3+}$ ion. A ligand concentration increase is associated with a shift of the fluorescence emission spectrum towards higher wavelength and a change in the shape of the spectrum. These noticeable changes are attributed to the formation of Cm(III) complexes.

The fluorescence emission spectra of the individual species, i.e. $Cm^{3+}$, $Cm(OH)_n^{3-n}$, n = 1, 2, $Cm(SO_4)_n^{3-2n}$, n = 1, 2, 3, $Cm(CO_3)_n^{3-2n}$, n = 1, 2, 3, 4, as well as the measured spectra are presented in Figures 3 to 5. The lower part shows the difference between the measured spectrum and the summed spectrum of individual components. The intensities of the spectra are normalised to the same peak area. The spectra of the different Cm(III) complexes are derived by peak deconvolution knowing in all cases the $Cm^{3+}$ aquo ion emission spectrum. Table I presents the different spectroscopic data obtained (fluorescence wavelength, full width mid-height (FWMH), lifetime, fluorescence intensity factor (FI), number of co-ordinated water molecules) for the different Cm(III) species investigated under different conditions and in different media.

The relaxation of fluorescence emission is shown in Figures 6 and 7 for Cm(III) in aqueous colloidal silica solution (0.5 g/L) at various pH values as well as Cm(III) in a mixture 1.0 M $NaHCO_3$/0.01 M $Na_2CO_3$ at various pH values, respectively. The time dependence of the Cm(III) fluorescence emission in the colloidal silica investigation follows a non-monoexponential decay, which indicates that more than one Cm-species are present in the solution, and also the exchange rate between the species is low in the excited state compared to their lifetime [4]. For the carbonate complex investigation, the opposite is observed: a monoexponential decay. This suggests that the ligand exchange reaction rate is fast compared to the relaxation rate of the excited Cm(III) [2]. The insignificance of the relaxation rate of the excited Cm(III) compared to the ligand exchange reaction rate is of great importance. Since, the species distribution is governed by thermodynamic equilibrium, differences under these conditions in the absorptivity at the excitation wavelength and in the fluorescence decay rate of the different species are averaged. Furthermore, the number of co-ordinated water molecules (see Table 1) is calculated based on a linear correlation between the decay rate $k_{obs}$ [ms$^{-1}$] (reciprocal lifetime of the excited state) and the number of water molecules in the first co-ordination sphere of the Cm(III) species [25].

Based on the different experimental data described in detail in [1-4], the apparent complexation constants defined by the following equation where $\gamma_i$ is the activity coefficient, are displayed in Table 2.

$$\beta^0 = \frac{\gamma_{CmL}}{\gamma_{Cm}\gamma_L^n}\frac{[CmL_n]}{[Cm^{3+}][L]^n} = \frac{\gamma_{CmL}}{\gamma_{Cm}\gamma_L^n}\beta$$

## Conclusion

From the studies on the complexation of Cm(III) by inorganic ligands via TRLFS, a thermodynamic model based on the ion-interaction approach (Pitzer model) has been developed [1-4,7,13]. This model is of great importance for repository science since it is capable of predicting the thermodynamics of trivalent actinides in highly concentrated brines, under repository conditions.

*Acknowledgements*

The authors wish to thank Mr. Tom Baca and Dr. Caroline Mason from Los Alamos National Laboratory.

## REFERENCES

[1]   Th. Fanghänel, J.I. Kim, P. Paviet, R. Klenze, W. Hauser, *Radiochim. Acta*, 66/67, 91 (1994).

[2]   P. Paviet, Th. Fanghänel, R. Klenze, J.I. Kim, *Radiochim. Acta*, 74, 99 (1996).

[3]   Th. Fanghänel, H.T. Weger, Th. Könnecke, V. Neck, P. Paviet-Hartmann, E. Steinle, J.I. Kim, *Radiochim. Acta*, 82, 47 (1998).

[4]   K.H. Chung, R. Klenze, K.K. Park, P. Paviet-Hartmann, J.I. Kim, *Radiochim. Acta*, 82, 215 (1998).

[5]   M.R. Lin, P. Paviet-Hartmann, Y. Xu, C.D. Tait, W.H. Runde, Abstracts of Papers, American Chemical Society, 216(1-3), p. ENVR 65 (1998).

[6]   P. Paviet-Hartmann, M.R. Lin, W.H. Runde, "Spectrophotometric Investigation of the U(VI) Chloride Complexation in NaCl/NaClO$_4$ System", Mat. Res. Soc. Symp. Proc., 556 (1999)

[7]     Th. Fanghänel, Th. Könnecke, H.T. Weger, P. Paviet-Hartmann, V. Neck, J.I. Kim, *J. Chem. Sol.*, 28, 447 (1999).

[8]     J.I. Kim, R. Klenze, H. Wimmer, W. Runde, W. Hauser, *J. Alloys Compounds*, 213/214, 333 (1994).

[9]     P. Decambox, P. Mauchien, C. Moulin, *Radiochim. Acta*, 48, 23 (1989).

[10]    R. Klenze, J.I. Kim, H. Wimmer, *Radiochim. Acta*, 52/53, 97 (1991).

[11]    J.I. Kim, R. Klenze, H. Wimmer, *J. Solid State Inorg.*, 28, 347 (1991).

[12]    H. Wimmer, R. Klenze, J.I. Kim, *Radiochim. Acta*, 56, 79 (1992).

[13]    Th. Fanghänel, J.I. Kim, R. Klenze, Y. Kato, *J. Alloys Compounds*, 225, 308 (1995).

[14]    C. Moulin, P. Decambox, P. Mauchien, *J. Radioanal. Nucl. Chem.*, 226, 135 (1997).

[15]    J. Fuger, I.L. Khodakovsky, E.I. Sergeyeva, V.A. Medvedev, J.D. Navratil, "The Chemical Thermodynamics of Actinide Elements and Compounds", Part 12, The Actinide Aqueous Inorganic Complexes, IAEA, Vienna (1992).

[16]    N. Edelstein, J. Bucher, R. Silva, H. Nitsche, Report LBL-13425, Lawrence Berkeley National Laboratory, Berkeley, CA (1983).

[17]    J.I.M. Bernkopf, Ch. Lierse, F. Koppold, ACS Symposium Series, No. 246, American Chemical Society, Washington DC, 1984, pp. 115-134 (1984).

[18]    S. Stadler, J.I. Kim, *Radiochim. Acta*, 44/45, 39 (1988).

[19]    G. Meinrath, J.I. Kim, *Radiochim. Acta*, 52/53, 29 (1991).

[20]    V. Shilov, *Radiochemistry*, 40, 1 (1998).

[21]    M. Ward, G. Welch, *J. Inorg. Nucl. Chem.*, 2, 395 (1956).

[22]    I.A. Lebedev, S.V. Pirozkhov, G.N. Yakovlev, *Radiokhimiya*, 2, 549 (1960).

[23]    A. Aziz, S.J. Lyle, S.J. Naqvi, *J. Inorg. Nucl. Chem.*, 30, 1013 (1968).

[24]    M. Hussonois, S. Hubert, L. Brillard, R. Guillaumont, *Radiochem. Radioanal Lett.*, 15, 47 (1973).

[25]    T. Kimura, G. Choppin, *J. Alloys Compounds*, 213/214, 313 (1994).

**Table 1. Spectroscopic characteristics of aqueous, sulfate, carbonate, and silica-sorbed curium species**

| Species | Emission (nm) | FWHM (nm) | Lifetime ($\mu$s) | FI | n $H_2O$ | Ref. |
|---|---|---|---|---|---|---|
| $Cm^{3+}$ | 593.8 | 7.7 | $68 \pm 3$ | 1 | $8.7 \pm 0.4$ | [1] |
| $Cm(OH)^{2+}$ | 598.8 | 11.5 | $72 \pm 3$ | – | $8.1 \pm 0.4$ | [1] |
| $Cm(OH)_2^+$ | 603.5 | 11.2 | $80 \pm 10$ | – | $7.2 \pm 0.7$ | [1] |
| $CmSO_4^+$ | 596.2 | 9.5 | $88 \pm 2$ | 1.2 | $6.5 \pm 0.6$ | [2] |
| $Cm(SO_4)_2^-$ | 599.5 | 8 | $95 \pm 8$ | 2.2 | $5.9 \pm 0.7$ | [2] |
| $Cm(SO_4)_3^{3-}$ | 602.2 | 5.5 | $195 \pm 3$ | 4.2 | $2.5 \pm 0.5$ | [2] |
| $CmH(CO_3)^{2+}$ | 594.9 | 8.6 | | | | [3] |
| $Cm(CO_3)^+$ | 598.5 | 8.4 | $85 \pm 4$ | | $6.8 \pm 0.5$ | [3,8,9] |
| $Cm(CO_3)_2^-$ | 603.0 | 7.5 | $105 \pm 5$ | $1.1 \pm 0.1$ | $5.3 \pm 0.6$ | [3,8,9] |
| $Cm(CO_3)_3^{3-}$ | 605.7 | 7.1 | $215 \pm 6$ | $1.6 \pm 0.2$ | $2.1 \pm 0.5$ | [3,8,9] |
| $Cm(CO_3)_4^{5-}$ | 607.5 | 11.7 | | $3.9 \pm 0.3$ | | [3] |
| $\equiv SiOCm(I)$ | 602.3 | 8 | $220 \pm 14$ | $0.35 \pm 0.02$ | $2.1 \pm 0.5$ | [4] |
| $\equiv SiOCm(II)$ | 604.9 | 13.2 | $740 \pm 35$ | $0.82 \pm 0.03$ | $\approx 0$ | [4] |

**Table 2. Summary of the apparent complexation constants of different Cm (III) species [1-3]**

| Metal | $\mu$/medium | T (K) | Log equilibrium constant | Ref. |
|---|---|---|---|---|
| $OH^-$ | 0 | 298 | $\beta^0_{11} = 6.44 \pm 0.09$ | [1] |
| | 0 | 298 | $\beta^0_{12} = 12.3 \pm 0.2$ | [1] |
| $SO_4^{2-}$ | 3/NaCl $Na_2SO_4$ | 298 | $\beta_1 = 0.93 \pm 0.08$ | [2] |
| | 3/NaCl $Na_2SO_4$ | 298 | $\beta_2 = 0.61 \pm 0.08$ | [2] |
| $CO_3^{2-}$ | 1/NaCl | 298 | $\beta_1 = 5.90 \pm 0.1$ | [3] |
| | 1/NaCl | 298 | $K_2 = 4.37 \pm 0.2$ | [3] |
| | 1/NaCl | 298 | $K_3 = 2.91 \pm 0.15$ | [3] |
| | 1/NaCl | 298 | $K_4 = 1.0 \pm 0.2$ | [3] |

# Figure 1. Schematic layout of time resolved laser fluorescence spectrometer (TRLFS)

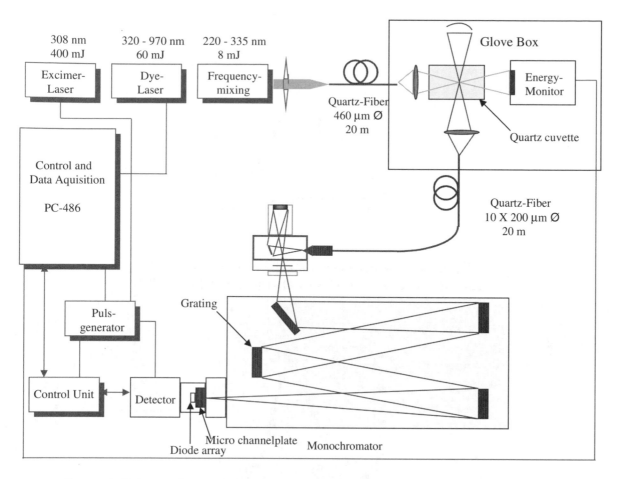

# Figure 2. Selected fluorescence emission spectra of Cm (III) in different media

*A – Hydrolysis investigation, $10^{-4}$ M $HClO_4$ - 0.1 M $NaClO_4$ at various pH*
*B – Aqueous silica colloidal solution (0.15 g/L) at various pH*
*C – $Cs_2SO_4$ solution taken in the concentration range $0 < Cs_2SO_4 < 4.7$ mol/kg $H_2O$*
*D – Mixture 0.01 M $Na_2CO_3$ -1.0 M $NaHCO_3$. Spectra are scaled to the same peak area*

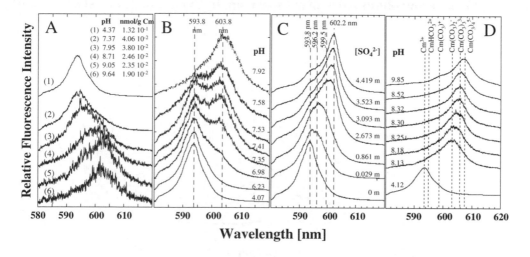

**Figure 3. Peak deconvolution of a mixed spectrum in 0.1 M NaClO₄-10⁻⁴ M HClO₄, pH = 7.88**

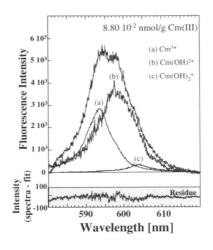

**Figure 4. Peak deconvolution of a mixed spectrum in 0.05 M NaHCO₃ /0.05 M Na₂CO₃-1.0 M NaCl, pH = 9.09**

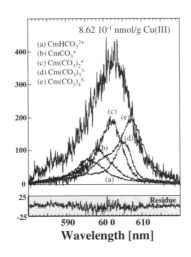

**Figure 5. Peak deconvolution of a mixed spectrum in Cs₂SO₄ solution [SO₄²⁻] = 3.093 M**

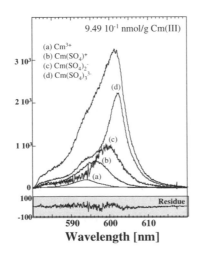

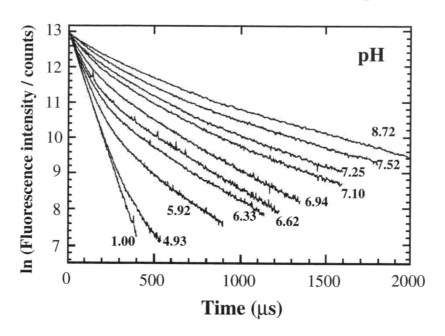

Figure 6. Time dependence of emission decay of Cm(III) in aqueous colloidal silica solution (0.5 g/L) at various pH [4]

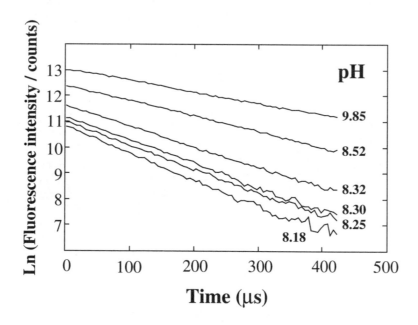

Figure 7. Time dependence of relaxation of Cm(III) carbonate species in 1.0 M NaHCO₃/0.01 M Na₂CO₃ at various pH

# SPECIATION ANALYSIS ON Eu(III) IN AQUEOUS SOLUTION USING LASER-INDUCED BREAKDOWN SPECTROSCOPY

**H. Hotokezaka, S. Tanaka**
Department of Quantum Engineering and Systems Science
The University of Tokyo
7-3-1 Hongo, Bunkyo-ku, Tokyo 113-8656, Japan

**S. Nagasaki**
Department of Environmental Studies
The University of Tokyo
7-3-1 Hongo, Bunkyo-ku, Tokyo 113-8656, Japan

## Abstract

Investigation of the chemical behaviour of lanthanides and actinides in the geosphere is important for the safety assessment of high-level radioactive waste disposal. However, determination of speciation for lanthanides and actinides is difficult, because it is too hard to distinguish between metal ion and colloidal metal in aqueous solution. Laser-induced breakdown spectroscopy (LIBS) can detect both ions and microparticles of metals in aqueous solution, especially, high sensitive to microparticles. In this study, we analysed Eu(III) ion and $Eu_2O_3$ particle in aqueous solution by LIBS, and measured the hydrolysis behaviour of Eu(III) in aqueous solution. Furthermore, we tried to detect the plasma emission of Eu(III) ions sorbed on $TiO_2$ particles, and also tried to observe the adsorption behaviour of Eu(III) ions onto $TiO_2$ particles in aqueous solution.

Plasma emission of Eu was observed at 420.505 nm both in Eu(III) aqueous solution and in $Eu_2O_3$ suspended solution. The emission line was generally used for quantitative analysis of Eu. The plasma emission intensity of $Eu_2O_3$ was higher than that of Eu(III) ions when Eu concentration was same, and absorption band of Eu(III) ions was observed in LIBS spectrum of Eu(III) aqueous solution. These results indicated Eu(III) ions can be distinguished from $Eu_2O_3$ particles in aqueous solution by LIBS.

The plasma emission intensity of Eu increased with Eu concentration. The intensity vs. concentration plot showed a good linearity in low concentration range (approximately $10^{-2}$ to $10^{-1}$ mol/dm$^3$ and $10^{-5}$ to $10^{-4}$ mol/dm$^3$ for Eu(III) ions and $Eu_2O_3$ particles, respectively).

The plasma emission intensity of Eu increased with pH of Eu(III) aqueous solution, and drastic increase of emission intensity was observed above pH = 6. The tendency was similar with solubility curve of $Sm(OH)_3$, which calculated using stability constants in literature [1]. This result suggested that LIBS is useful for measurement of hydrolysis and precipitation behaviour of Eu(III) in aqueous solution.

From plasma emission of Eu(III) ions sorbed on $TiO_2$ particles, it was found that only Eu(III) ions on $TiO_2$ may be detected by LIBS, because the emission from Eu(III) ions in solution was not observed. The plasma emission intensity of Eu(III) ions sorbed on $TiO_2$ particles increased with pH of solution at pH > 3. This tendency was similar with the result of the batch experiment using ICP spectrometry. Therefore, we considered that the plasma emission intensity of Eu(III) ions sorbed on $TiO_2$ particles corresponded to the quantity of Eu(III) ions sorbed on $TiO_2$ particles.

## REFERENCE

[1]    V. Moulin *et al.*, *Radiochim. Acta.*, 44/45 (1988) 33.

# ACTINYL(VI) SPECIATION IN CONCENTRATED SODIUM CHLORIDE SOLUTIONS

**W. Runde, M.P. Neu, C. Van Pelt, S. Reilly, and Y. Xu**
Chemical Science and Technology Division
Los Alamos National Laboratory
Los Alamos, NM 87545, USA

## Abstract

Thermodynamic parameters and sorption coefficients provide the fundamental database for the mathematical abstraction of environmental transport of actinides. We have been investigating the chemistry of the hexavalent actinides, U and Pu, in the solid state in concentrated salt (NaCl) solutions representative of conditions within geologic salt formations, sites for nuclear waste disposal. Chloride complexation is proven to play an important role for actinyl(VI) solubility and speciation. We investigated U(VI) and Pu(VI) equilibria with the predominant inorganic ligands hydroxide and carbonate in the presence of chloride. We used a number of spectroscopic techniques (UV-vis-NIR, Raman, FTIR, NMR, XAS) and X-ray diffraction to characterise solution species and solid phases as a function of pH, carbonate concentration, and ionic strength.

# Introduction

The understanding of actinide speciation and compound stability in natural aquifer systems is the basis for developing reliable site-specific solubility predictions and transport models. Most thermodynamic data on actinide solubilities and species formation constants have been determined in dilute solutions of inert electrolytes, such as $NaClO_4$ [1-3]. Natural waters, however, contain a complex mixture of inorganic and organic components that significantly affect the environmental behaviour of actinides [4]. Brines from nuclear waste repository sites in geologic salt deposits, i.e. the Waste Isolation Pilot Plant (WIPP) site in New Mexico (USA) and the Gorleben site in Niedersachsen (Germany), are highly concentrated solutions of chloride salts. Thus, for defensible risk and safety assessment calculations the understanding of actinide speciation in the presence of high chloride concentrations is crucial.

Solubility predictions for actinides in chloride solutions have been underestimated when based on thermodynamic data determined in non-interactive media [5]. We have shown previously that chloride may replace hydration water and form actinyl(VI) chloro complexes of higher stability than those of tri-, tetra-, and pentavalent actinides [6]. Building on those inner-sphere actinyl(VI) chloride interactions, we are studying the influence of chloride on the nature and stability of solution and solid actinyl(VI) compounds as a function of chloride concentration and pH in the presence and absence of carbonate. We are performing these studies using a variety of spectroscopic techniques ($^{13}$C-NMR, diffuse reflectance, conventional UV-Vis-NIR absorption, FTIR, Raman, and X-ray absorption) and X-ray diffraction.

# Experimental

## Stock solutions

Plutonium(VI) stock solutions were prepared by dissolving $^{239}$Pu metal in 7 M $ClO_4$ and fuming aliquots of this solution to near dryness with concentrated $HClO_4$, diluting with $H_2O$, then determining the total Pu concentration by liquid scintillation counting (LSC) and verifying the oxidation state purity using conventional absorbance spectrophotometry. The isotopic composition of the material was determined radioanalytically. Uranium(VI) stock solutions were prepared by dissolving uranyl nitrate (re-crystallised from nitric acid) in 1 M $HClO_4$. Sodium carbonate solutions were prepared by dissolving $^{13}$C-enriched $Na_2CO_3$ (99.9% $^{13}$C, Cambridge Isotopes) in known masses of distilled, deionised water.

## Preparation and characterisation of U(VI) hydrolysis species

Aliquots of the uranyl stock solutions were diluted in solutions of 0.1, 1, 3, and 5 M NaCl solutions to yield final U(VI) concentrations of 0.1, 0.5, 1, and 10 mM. The acidic solutions were purged with argon to minimise carbonate contamination. The pH was adjusted by addition of NaOH and the solutions were allowed to equilibrate under argon atmosphere for 24 hours. The solutions were then analysed using UV-vis, Raman, and FTIR spectroscopies.

## Preparation and characterisation of solid actinyl(VI) carbonates

Actinyl carbonates, $AnO_2CO_3$, (An = U, Pu) were prepared by bubbling $CO_2$ through stirred acidic stock solutions (pH = 4) for 3 to 5 days, washing the resulting precipitate with distilled de-ionised water, re-dissolving, and repeating precipitation. For plutonium, ozone was also bubbled through the suspension for the final 2 days to re-oxidise any reduced plutonium. The resulting pale pink-tan Pu(VI) and pale yellow U(VI) solids were characterised using powder X-ray diffraction (Inel, CPS-120) and extended X-ray absorbance fine structure (EXAFS) spectroscopy. EXAFS data were recorded at Stanford Synchrotron Radiation Laboratory (SSRL): unfocused beam line 4-2, Si-(220), double-crystal monochromator, 3.0 GeV, 60-100 mA.

## Determination of the thermodynamic constant for U(VI) Bis- and triscarbonato equilibrium

Aliquots of the U(VI) stock solutions were added dropwise to individual 30 mM $Na_2CO_3$ solutions to yield a final uranium concentration of 10 mM and NaCl concentrations between 0.05 and 5 m. A small, known amount of $D_2O$ (Cambridge Isotopes, 99.9% D) and NaCl (Baker, reagent) appropriate for the desired ionic strength were added to individual solutions. Samples were prepared with a carbonate:uranyl ratio of 3:1 to favour the formation of the monomeric triscarbonato complex. Careful titration with $HClO_4$ leads to the protonation of the carbonate ligand resulting in a decrease in the carbonate:uranyl ratio to 2:1. The p[H] of the resulting solutions was determined using a combination pH electrode and the calibration and ionic strength corrections described earlier [5]. All NMR sample solutions were loaded into pyrex NMR tubes (Wilmad 5 mm o.d. 507-PP). FT $^{13}C$ NMR spectra were recorded on a Bruker AMX500 spectrometer with a 5 mm broadband probe operating at 125.76 MHz with $^2H$ field-frequency lock. The spectral reference was set for all $^{13}C$ NMR spectra relative to the carbonyl carbon of external acetone-$d_6$ set at $\delta = 206.0$.

## Results and discussion

### Hydrolysis of U(VI) in chloride media

Uranium(VI) hydrolysis has been investigated extensively and the data reported have been evaluated recently [2]. Monomeric and polynuclear hydroxo species of general formulas $UO_2(OH)_m^{2-m}$ (m = 1-4) and $(UO_2)_m(OH)_n^{2m-n}$ (m/n = 2/2, 3/5, 4/7) are accepted U(VI) hydrolysis species which form in the absence of carbonate. At low pH, U(VI) forms the monomeric first hydrolysis product $UO_2(OH)^+$ that aggregates at high U(VI) concentrations to form the dimeric species $(UO_2)_m(OH)_n^{2m-n}$. The absorbance maximum of the dimer is observed at 421 nm (421.8 nm [7]) in $NaClO_4$ solutions, independent of the electrolyte concentration. The molar absorbance of this complex (79±5 L $Mol^{-1}$ $cm^{-1}$) is an order of magnitude greater than that of the uncomplexed $UO_2^{2+}$ ion (7.9±0.3 L $Mol^{-1}$ $cm^{-1}$). Addition of NaCl results in a significant change in the absorption spectra of the $UO_2^{2+}$ ion although the shift of the absorption maxima of up to 10 nm is far smaller than those of strong U(VI) complexes, such as the 30 nm shift observed upon carbonate complexation [8]. The maximum absorbance of $UO_2^{2+}$ in $NaClO_4$ (413 nm) is shifted to 423 nm in 5 M NaCl (Figure 1) [6]. With increasing pH, the formation of the U(VI) hydroxo species is indicated by a significant absorbance increase. However, while the shape of the spectrum is very similar to that observed in $NaClO_4$ (absorbance maximum at 421 nm) the maximum absorbance is shifted to 425 nm. These findings were confirmed by results obtained from Raman and FTIR studies. The Raman (873 $cm^{-1}$) and FTIR (961 $cm^{-1}$) bands of the uncomplexed $UO_2^{2+}$ ion are shifted in 5 M NaCl due to chloride complexation by about 10 wave numbers to 862 $cm^{-1}$ and 951 $cm^{-1}$, respectively (Figure 2). While Raman and FTIR bands of the dimeric hydroxo species in NaClO₄ appear at 852 and 941 cm⁻¹, respectively, the corresponding bands cannot be observed in NaCl solutions. However, additional peaks appear in 5 M NaCl at 837 and 922 cm⁻¹, respectively, at higher p[H]. These bands match the Raman (839 cm⁻¹) and FTIR (924 cm⁻¹) signals of the trimeric $(UO_2)_{-3}O(OH)_3^+$ observed in NaClO₄. These spectroscopic results suggest that the stability of $(UO_2)_2(OH)_2^{2+}$ is decreased significantly in the presence of chloride, such that this species is not observed spectroscopically. Clearly, the speciation of uranyl(VI) is different in concentrated chloride solutions than it is in perchlorate solutions. One explanation for the differences is the formation of mixed U(VI) chloro-hydroxo complexes, however, we have no direct evidence for such mixed ligand species. Currently, we are continuing this work, including determining formation constants for the hydroxide species formed in concentrated chloride solution which we have described qualitatively here.

**Figure 1. UV-vis absorbance spectra of U(VI) in 5 M NaClO₄ (left) and 5 M NaCl (right) as function of pH**

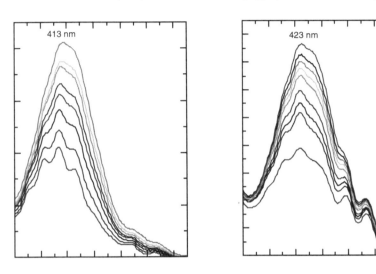

**Figure 2. FTIR (left) and Raman (right) spectra of U(VI) in 5 M NaCl as function of p[H]**

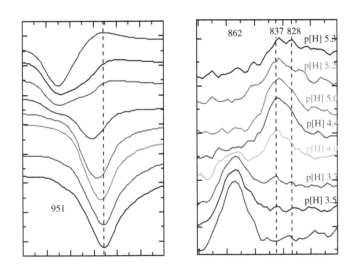

The actinyl(VI) carbonates, $UO_2CO_3$ and $PuO_2CO_3$, were precipitated from solutions (pH 4) saturated with 100% $CO_2$. The XRD powder patterns of both solids agree well with each other and with those previously reported [9,10]. To determine the local co-ordination environments of the actinides we measured the EXAFS spectra of each solid (Figure 4). The actinyl An=O distances were determined to be 1.76 for U and 1.74 for Pu. The equatorial An-OCO₂ bond distances are very similar to those in $UO_2(CO_3)_3^{4-}$ and in $(UO_2)_3 (CO_3)_6^{6-}$, i.e. 2.45 Å for An--OCO₂ with the carbonate co-ordinated in a bi-dentate fashion [11]. The Fourier transform moduli show the An---An distance of 4.2 Å, consistent with the Pmmn structure reported [12].

**Figure 3. EXAFS of UO₂CO₃ and PuO₂CO₃. The Fourier transform amplitude for each solid is show. The shells of neighbouring scatterers which comprise the fit to the data are based upon an idealised Pmmn structure and are plotted here with negative intensity.**

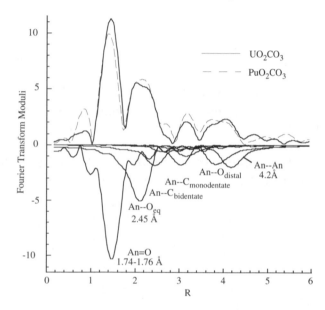

Given the large range of solubility products reported for U(VI) and Pu(VI) hydroxide phases, the correlated stability range of AnO₂CO₃ (s) is unclear. The AnO₂CO₃ (s) stability depends mainly on the $CO_2$ partial pressure and the pH, which together determine the carbonate concentration in solution, and the ionic strength. Lowering the $CO_2$ partial pressure from 1 atm (pH 4) to $10^{-3.5}$ atm favours the transformation of AnO₂CO₃(s) to a An(VI) hydroxide or oxide phase. In 5 M NaCl, UO₂CO₃ (s) transformed into a crystalline phase with significantly different Bragg reflections than the starting material. The powder pattern of the new phase compares well with that reported for the sodium uranate, $Na_2U_7O_{22}$ [13]. At lower ionic strength, the crystallinity of the final solid is lower and broad Bragg reflections match those of $UO_2(OH)_2$. Plutonium(VI) reacts sensitively to changes in ionic strength and radionuclide inventory. At low ionic strength, Pu(VI) is found to be reduced within days to increasing fractions of polymeric Pu(IV) hydroxide and $PuO_2^+$. The pale pink-tan plutonyl solid is found to be stable for only weeks under these conditions. However, in concentrated NaCl, chloride complexation and radiolysis stabilise plutonium in the hexavalent oxidation state. Radiolytically produced chlorine and hypochlorite have been reported to be responsible for creating an oxidising medium [14]. While chloride appears to stabilise Pu(VI), it does not affect the composition of the carbonate solid precipitated from concentrated NaCl solutions.

*Stability of the trimeric species, $(UO_2)_3 (CO_3)_6^{6-}$, as a function of NaCl*

Based on NMR, EXAFS and single X-ray crystal diffraction studies the trimeric species has been unequivocally characterised [11] and proven to exist in solution when U(VI) and carbonate concentrations are both in the millimolar range. However, its stability and relevance in solid-liquid phase equilibrium studies is still questionable. We studied the stability of $(UO_2)_3 (CO_3)_6^{6-}$ in 0.05 – 5 m NaCl solutions by using $^{13}C$ NMR spectroscopy. This technique allows the identification of the U(VI) carbonate species involved in the solution equilibrium via their chemical shifts and the U(VI) species concentration ratio by integrating the resonances corresponding to each species at each pH. Using the known U(VI) concentrations, measured p[H], and calculated bicarbonate concentrations, we calculated the apparent equilibrium constant relating the two species at each NaCl concentration according to Eq. (1).

$$3 \ UO_2 (CO_3)_3^{4-} + 3 \ H^+ \rightleftharpoons (UO_2)_3 (CO_3)_6^{6-} + 3 \ HCO_3^- \qquad (1)$$

Since the trimeric species is so highly charged, we anticipated a large effect of the NaCl concentration on the equilibrium between these two species. Indeed, the equilibrium constant is dependent on NaCl concentration, an example of the changes observed at a given pH as a function of ionic strength (Table 1). The values determined for the equilibrium constant range from log $K'_{eq}$ = 20.9 ± 0.2 with no NaCl in solution to 18.7 ± 0.2 in 5.0 m NaCl. The greatest changes occur for NaCl concentrations up to 1.4 m, the apparent stability constant varies little when NaCl concentrations are higher than 2 m.

**Table 1. Apparent equilibrium constants for the first protonation step of $UO_2 (CO_3)_3^{4-}$ as a function of NaCl concentrations**

| [NaCl] (m) | log $K'_{eq}$ | [NaCl] (m) | log $K'_{eq}$ |
|---|---|---|---|
| 0.000 | 20.9 ± 0.2 | 2.15 | 18.6 ± 0.2 |
| 0.496 | 19.7 ± 0.1 | 2.45 | 18.5 ± 0.2 |
| 0.597 | 19.5 ± 0.2 | 2.75 | 18.7 ± 0.1 |
| 0.938 | 19.2 ± 0.2 | 2.98 | 18.6 ± 0.2 |
| 0.998 | 19.1 ± 0.3 | 3.37 | 18.7 ± 0.1 |
| 1.380 | 18.8 ± 0.2 | 4.78 | 18.6 ± 0.2 |
| 2.000 | 18.6 ± 0.2 | 4.95 | 18.6 ± 0.1 |
| 2.036 | 18.7 ± 0.1 | 5.02 | 18.7 ± 0.1 |

**Figure 4. $^{13}C$ NMR spectra of 10 mM uranyl carbonate at [H$^+$] = 10$^{-8}$ M as a function of sodium chloride**

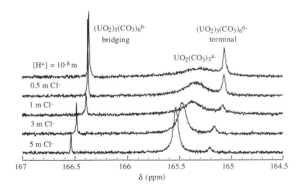

Solubility data of uranyl in carbonate media have been interpreted with the formation of only monomeric species, $UO_2(CO_3)_m^{2-2m}$ where m = 1–3 [2]. In these studies the solutions were prepared by addition of acid to the uranyl carbonate solutions to form the trimeric species without precipitation (under saturation conditions). The solubility of $UO_2CO_3$ (s) increases at $[CO_3^{2-}]$ higher than $10^{-8}$ M due to the formation of anionic U(VI) carbonate complexes. Under conditions where the biscarbonato complex is stable we observe a large scattering in the U(VI) solubility data depending on whether the experiment is performed from under saturation or over saturation. Starting the solubility experiment from the alkaline side, similar to the sample preparation for the NMR studies, allows for the stabilisation of U(VI) in solution by formation of the more soluble trimeric compound. Under saturation experiments prohibit large U(VI) concentration and favour the formation of the monomeric species $UO_2(CO_3)_2^{2-}$. Figure 5 shows the UV-Vis absorbance of U(VI) in carbonate solution taken from an oversaturation experiment compared with those of the pure trimeric species and a synthetic solution with predominantly $UO_2(CO_3)_3^{4-}$. The solution from the solubility experiment clearly matches the absorbance features of $(UO_2)_3(CO_3)_6^{6-}$ and differs significantly from those corresponding to $UO_2(CO_3)_3^{4-}$. These findings were confirmed by $^{13}C$ NMR studies. Thus far we found only the monomeric species in solution when solubility experiments are performed from under saturation.

**Figure 5. Absorbance spectra of U(VI) solution derived from a solubility experiment ([NaCl] = 2 m, pH 5.7 and 1 atm $CO_2$ partial pressure) and of known uranyl carbonate species**

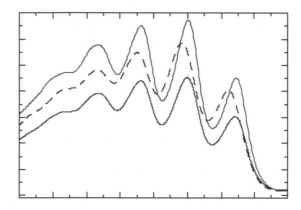

## Conclusions

We have shown that actinyl(VI) ions may form inner-sphere complexes with chloride. In brines relevant to nuclear waste repositories in geologic salt formations the dissolved chloride concentration is such that the mono-, bis- and perhaps the trischloro complexes are important. Uranyl(VI) hydrolysis reactions are more complicated and the dimeric hydroxo complex appears to be less stable in NaCl solutions. Plutonium(VI) is more stable towards reduction in highly concentrated chloride solutions than in perchlorate or dilute chloride solutions. In contrast, at low ionic strength the solubility of $PuO_2CO_3$(s) can be determined only after relatively short times of solid-liquid phase equilibration due to the redox instability of Pu(VI) and reduction to Pu(IV) colloid and Pu(V) in solution. Increasing NaCl concentrations favour the stability of the U(VI) triscarbonate complex with respect to formation of the trimeric $(UO_2)_3(CO_3)_6^{6-}$, consistent with the trend observed in sodium perchlorate [4,7]. Data from the NEA review of uranium thermodynamics yield an equilibrium constant for Eq. (1) of log K°= 17.8 [2] and Allen, *et al.* [11] reported log K' = 18.1 ± 0.5 in 2.5 m $NaClO_4$; for comparison, we have determined log K' = 18.5 in 2.5 m NaCl. We observed that the formation of the $(UO_2)_3(CO_3)_6^{6-}$ clearly depends on the details of sample preparation and its formation has to be considered in solubility experiments performed from over-saturation at $[CO_3^{2-}] \geq 10^{-8}$ M.

*Acknowledgements*

EXAFS experiments were performed at the Stanford Synchrotron Radiation Laboratory (SSRL) in collaboration with Steve D. Conradson, Materials Science and Technology Division, Los Alamos National Laboratory. SSRL is supported by the Office of Basic Energy Sciences, Division of Chemical Sciences, US Department of Energy. Full details of the EXAFS data and analysis will be reported elsewhere. This work was supported by the Waste Isolation Pilot Plant, under Contract No. AN-1756 with Sandia National Laboratory and by Los Alamos National Laboratory Directed Research Development and the Nuclear Materials Stabilisation Task Group, EM-66 of the US DOE.

# REFERENCES

[1]  R.J. Silva, G. Bidoglio, M.H. Rand, P.B. Robouch, H. Wanner, I. Puigdomenech, *Chemical Thermodynamics of Americium*, Elsevier, New York, 1995, Vol. 2.

[2]  I. Grenthe, J. Fuger, R.J.M. Konings, R.J. Lemire, A.B. Muller, C. Nguyen-Trung, H. Wanner, *Chemical Thermodynamics of Uranium*, North-Holland Elsevier Science Publishers B.V., Amsterdam, 1992, Vol. 1.

[3]  J. Fuger, I.L. Khodakovsky, E.I. Serfeyeva, V.A. Medvedev, J.D. Navratil, *Part 12, The Actinide Aqueous Inorganic Complexes*, IAEA: Vienna, 1992.

[4]  G.R. Choppin, M. Du, *Radiochim. Acta* 1992, 58/59, 101-104.

[5]  W. Runde, M.P. Neu, D.L. Clark, *Geochim. Cosmochim. Acta* 1996, 60, 2065-2073.

[6]  W. Runde, M.P. Neu, S.D. Conradson, D.L. Clark, P.D. Palmer, S.D. Reilly, B.L. Scott, C.D. Tait, in *Materials Research Society Symposium*, Boston, MA, 1996, pp. 693-703.

[7]  G. Meinrath, *J. All. Comp.,* 1998, 275-277, 777-781.

[8]  S.O. Cinneide, J.P. Scanlan, M.J. Hynes, *J. Inorg. Nucl. Chem.* 1975, 37, 1013.

[9]  J.D. Navratil, H.I. Bramlet, *J. Inorg. Nucl. Chem.* 1978, 35, 157.

[10]  I. Pashalidis, W. Runde, J.I. Kim, *Radiochim. Acta* 1993, 61, 141.

[11]  P.G. Allen, J.J. Bucher, D.L. Clark, N.M. Edelstein, S.A. Ekberg, J.W. Gohdes, E.A. Hudson, N. Kaltsoyannis, W.W. Lukens, M.P. Neu, P.D. Palmer, T. Reich, D.K. Shuh, C.D. Tait, B.D. Zwick, *Inorg. Chem.* 1995, 34, 4797-4807.

[12]  J.R. Clark, C.L. Christ, *Am. Min.* 1957, 41, 844.

[13]  JCPDS 5-132 and Wamser, *et al.*, *J. Am. Chem. Soc.* 1952, 74, 1022.

[14]  K. Büppelmann, J.I. Kim, C. Lierse, *Radiochim. Acta* 1988, 44/45, 65.

# CHARACTERISATION OF OXIDE FILMS FORMED ON STEEL SURFACE IN BWR ENVIRONMENT

**Yutaka Kameo**
Department of Decommissioning and Waste Management
Japan Atomic Energy Research Institute
Tokai, Ibaraki 319-1195, Japan

## Abstract

Oxide films formed on type 304 stainless steel and carbon steel surfaces exposed to 560 K water have been characterised. The metal oxide film consisted of two layers, comprising of lower and higher valence metal oxides, respectively; $Fe_2O_3$ as the outer layer and $Fe_3O_4$, $NiFe_2O_4$ and $Cr_2O_3$ as the inner layer. The formation of the lower valence metal oxide in the inner layer could be attributed to suppression of oxygen diffusion into the inner layer.

## Introduction

Decontamination of pipe systems and components contaminated with radioactive nuclides is one of the important subjects in the dismantlement of a nuclear power plant. It is well known that contamination of a metal surface in the primary coolant system of a nuclear reactor is caused by incorporation of radioactive nuclides into metal surface oxide films. Since chemical and physical properties of the oxide films vary with the condition of the coolant system, it is important to characterise the metal oxide films for appropriate selection and optimisation of the decontamination method.

In the present study, we characterised the oxide films formed on type 304 stainless steel and carbon steel surfaces by means of X-ray photoelectron spectroscopy (XPS), Auger electron spectroscopy (AES) and electron probe X-ray microanalysis (EPMA).

## Experimental

Small plates of type 304 stainless steel or carbon steel were oxidised for 1 000 hours in high temperature water containing 8 ppm dissolved oxygen at 560 K under a pressure of 7.4 MPa in an autoclave. The surface condition of the steel plates was assumed to be in a boiling water reactor (BWR) environment. The chemical composition of the type 304 stainless steel and carbon steel plates used for the experiment is shown Table 1.

**Table 1. Chemical composition of type 304 stainless steel and carbon steel [wt.%]**

|  | Fe | C | Si | Mn | P | S | Ni | Cr |
|---|---|---|---|---|---|---|---|---|
| **Type 304 stainless steel** | balance | 0.043 | 0.30 | 1.32 | 0.0037 | 0.022 | 8.03 | 18.83 |
| **Carbon steel** | balance | 0.17 | <0.01 | 0.36 | 0.019 | 0.026 | 0.019 | 0.020 |

The oxidised type 304 stainless steel plates were subjected to XPS or AES measurement without further pre-treatment. The XPS and AES measurement were carried out with Perkin-Elmer ESCA 5500 and JEOL JAMP-30, respectively. The XPS and AES depth profile of the oxide films were measured as a function of sputtering time. The EPMA depth profile of the oxide film formed on the carbon steel plates was examined with JEOL JXA-8621MX because the oxide film was too thick to apply XPS or AES analysis.

## Results and discussion

### Metal oxide films on type 304 stainless steel

Figure 1 shows the relative concentration of Fe, Cr, Ni, O, C in the oxide film formed on type 304 stainless steel as a function of sputtering time. The concentration of O and Fe were almost constant up to a sputtering time of 60 min, indicating that a uniform oxide film was formed. After 60 min sputtering, the concentration of O decreased, and that of Fe increased up to 120 min. Nickel and chromium were detected at a sputtering time of 50 minutes and their concentration increased up to 120 min. After 120 min sputtering, the relative concentration measured by AES conjoined to the composition of the type 304 stainless steel given in Table 1.

**Figure 1. Relative concentration of Fe, Cr, Ni, O and C
in oxide film formed on type 304 stainless steel**

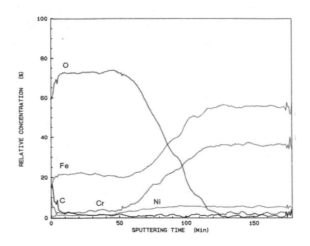

Figure 2 shows Fe(2p) and Cr(2p) XPS spectra of the oxide film on type 304 stainless steel. The Fe(2p) spectrum observed at the outermost surface of the oxide was characteristic of $Fe_2O_3$. As the sputtering time increased, the peak shape and the binding energy of the Fe(2p) transitions gradually changed; the Fe(2p) spectra characteristic of $Fe_3O_4$ and Fe metal appeared. The small peaks of the Cr(2p) transitions were also detected at the outermost surface of the oxide film. After 30 min sputtering, the clear Cr(2p) spectrum characteristic of $Cr_2O_3$ was detected. After 60 min sputtering, the Cr(2p) spectrum was deconvoluted into two peaks corresponding to $Cr_2O_3$ and Cr metal.

**Figure 2. Fe(2p) and Cr(2p) XPS spectra of oxide film formed on type 304 stainless
steel after sputtering for 30 min (A), 60 min (B), 90 min (C) and 120 min (D)**

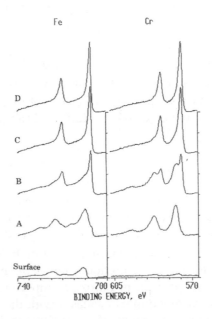

Hence, the metal oxide film formed on a type 304 stainless steel plate consisted of an inner and outer layer. The identification of chemical forms of these oxides by means of XPS and AES showed that the outer layer consisted mainly of $Fe_2O_3$ (higher valence metal oxide) and the inner layer consisted of $NiFe_2O_4$, $Fe_3O_4$ (lower valence metal oxide) and $Cr_2O_3$. These results are consistent with those observed in oxide films formed on stainless steel surfaces in similar environments [1,2,3]. In addition, the broad spectrum of the lower energy side of the Fe(2p) peak (after 30 min sputtering) implied that spinel type $FeCr_2O_4$ existed in the inner layer, as observed by Yamanaka, *et al.* in their experiment using laser Raman spectroscopy and X-ray diffraction [3]. We conclude that the inner and outer oxide layers respectively consist of lower and higher valence metal oxides.

### Metal oxide films on carbon steel

The XPS measurement was carried out for the oxide film formed on carbon steel plate after 2 min sputtering for removal of hydrocarbon contamination. The result shows that the only two elements detected were Fe and O. The Fe(2p) spectrum was characteristic of $Fe_2O_3$ as shown in Figure 3.

**Figure 3. Fe(2p) XPS spectrum of oxide film formed on carbon steel**

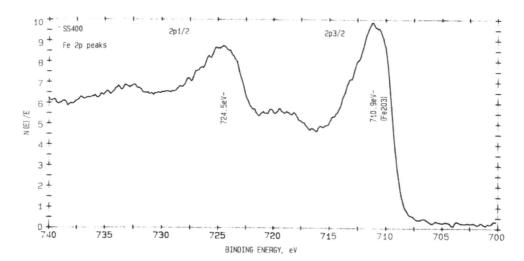

We measured the depth profile of O and Fe in the oxide film formed on the carbon steel plate. On the basis of EPMA and XPS measurements, the oxide film consisted of $Fe_2O_3$ as an outer layer and $Fe_3O_4$ as an inner layer. The structure of the oxide film consisting of lower and higher valence metal oxides is similar to that observed on the type 304 stainless steel sample.

### Structure of oxide films

The results obtained by XPS, AES and EPMA measurements in the present study indicate that the metal oxide films formed on type 304 stainless steel and carbon steel consisted of two layers. Schematic structures of the oxide films formed on the steels are shown in Figure 4. The outer layer formed on type 304 stainless steel were mainly consisted of $Fe_2O_3$, since Cr existed in the outer layer can easily be dissolved in the water. The dissolution volume of the oxide film on stainless steel is higher than that of carbon steel, so that thick metal oxide film would be formed on carbon steel [4]. It is well known that the thickness of oxide films increases with the logarithm of exposure time under the BWR environment [4]. Therefore, the growth of oxide films is dominated by diffusion of oxygen, dissolved in coolant water, into oxide films. Supposing that oxide films grow in this way, formation of the lower valence metal oxide in the inner layer could be attributed to suppression of oxygen diffusion into the inner layer.

**Figure 4. Schematic structure of oxide films formed on steels in BWR environment**

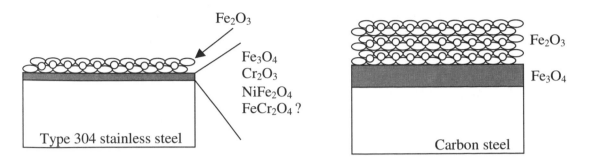

*Acknowledgements*

A part of this work was performed under a contract with the Science and Technology Agency of Japan, for which the authors express their appreciation.

**REFERENCES**

[1]   T. Fukuda, *et. al.,* Proceedings of International Conference in Stainless Steels 91, 1, pp. 168-174, Chiba, Japan (1991).

[2]   K. Yamanaka, *et al.,* 36th Fusyoku-Bousyoku Touronkai, pp. 175-178, B-110 (1989) (in Japanese).

[3]   R.L. Tapping, *et al., Corr. Sci.,* 26, pp. 563-576 (1986).

[4]   T. Honda, *et al., Boshoku Gijyutsu,* 36, pp. 267-274 (1987) (in Japanese).

# ACTINIDE SEQUESTERING AGENTS: DESIGN, STRUCTURAL AND BIOLOGICAL EVALUATIONS

**Jide Xu, Patricia W. Durbin and Kenneth N. Raymond**
Department of Chemistry, Chemical Science Division, Lawrence Berkeley Laboratory
University of California, Berkeley, CA 94720 USA

## Abstract

Multidentate actinide sequestering agents have been developed using a biomimetic approach modelled on siderophores, naturally-occurring iron(III) sequestering agents. Described are the synthesis, structure, and solution thermodynamics and animal test results of new, more effective, actinide sequestering agents.

# Introduction

The actinide elements Pu, Np, Am and U are important industrial materials and consequent environmental hazards [1,2]. Both $^{239}$Pu and $^{235}$U are nuclear fuels as well as constituents of nuclear weapons. The heavier actinides are abundant in irradiated nuclear fuel and wastes from fuel fabrication, fuel processing and weapons production. Weapons testing, accidents, and poor waste disposal practices have contaminated the environment. It seems likely that future accident and environmental contamination will have to be dealt with.

The biological and environmental behaviours of the actinides are reasonably well understood [1]. All of these hard cations form complexes with bioligands. Following human inhalation, ingestion, or deposition in wounds, the absorbed actinides circulate, bound in various degrees to the Fe transportation protein (transferring, TF). This protein binding severely inhibits renal excretion (particularly of Pu and Am) and favours deposition in tissues. Cells and structures that die or are remodelled release their actinide to the circulation, and some released Np and U and most of the released Am and Pu re-deposit at new sites within the body.

Aggressive protracted chelation therapy reduces actinide concentrations and radiation doses in the target tissues, thereby reducing chemical and radiation damage and tumour risk [1,2]. Therefore chelating agents that form stable excretable actinide complexes are the only practical treatment for reducing internal actinide contamination. The calcium and zinc salts of diethylenetriaminepentaacetate (CaNa$_3$DTPA, ZnNa$_3$DTPA) have been the only clinically accepted agents for treating internal actinide contamination. DTPA is a non-specific metal chelating agent, which does not solubilise Pu hydroxides or remove Pu from bone mineral. It is not orally effective. It is ineffective for *in vivo* chelation of Np and U. Therefore, development of ligands that are better chelating agents for actinides than DTPA remains an important component of radiation protection research.

## *Design strategy*

The actinide ions have large charge to radius ratios, they belong to the group of "hard" cations and prefer hard electron donors such as oxygen. The co-ordination properties of Fe(III) are similar to those of the actinides, especially Pu(IV), and that metal binding units with great affinity for Fe(III) also form stable Pu(IV) complexes, as is shown by the incorporation of Pu and Am into mammalian iron transport and storage systems. Siderophores are naturally occurring Fe(III) sequestering agents produced by plants and bacteria in order to obtain growth limiting iron. The siderophores have very high affinities for Fe(III). Enterobactin, a siderophore secreted by *E. coli*, has three catecholamide binding sub units attached to a macrocyclic trilactone backbone, is pre-organised for binding iron [2]. It has the highest formation constant known for ferric ion, $K_f = 10^{49}$. The binding sub units found in siderophores are catecholates, hydroxamates or hydroxypyridinonates (HOPOs), the structures of their protypes are shown below:

| Catechol | hydroxamic acid | 1.2-HOPO | 3.2-HOPO | 3.4-HOPO |
|----------|-----------------|----------|----------|----------|

These binding sub units tend to be extremely specific for Fe(III) as well as Pu(IV). Therefore, we have chosen a biomimetic approach for the design of sequestering agents for the actinide ions.

The HOPOs are of particular interest, since they selectively display a high affinity for ferric and other metal ions of charge/radius ratio comparable to Pu(IV) within the physiological pH range. The formation constants, of Fe(III)-HOPO complexes are 4-9 orders of magnitude greater than those formed with biologically essential divalent cations [2b].

| **Enterobactin** | **Fe(III)-Enterobactin complex** | **Model actinide complex** |
|:---:|:---:|:---:|
| 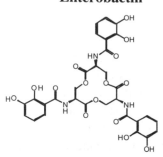 | 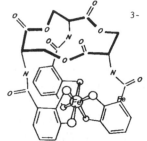 | 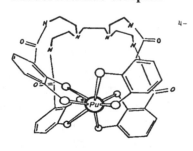 |

We have developed de-rivatised 1,2-HOPO [2c] and 3,2-HOPO [2b] binding units with a carboxy group (adjacent to the hydroxy group on the HOPO ring) which can be easily attached to an amine backbone through an amide linkage. The resultant ligand has a favourable co-ordination geometry, as shown below, stronger acidity, and therefore has increased solubility and complexation ability.

**Catecholamide complex   1,2-HOPO-6-ylamide complex   (Me-3,2-HOPO)-4-ylamide complex**

A very important feature of (Me-3,2-HOPO)-4-ylamide and 1,2-HOPO-6-ylamide ligands (abbreviated as (Me-3,2-HOPO) and (1,2-HOPO) ligands below) is that, as observed in the catcholamide complexes, strong hydrogen bonds between the amide proton and adjacent oxygen donor enhance the stability of their metal complexes [2b]. X-ray crystal structures of all the metal complexes of these ligands exhibit this hydrogen bonding.

*Plutonium sequestering agents*

The complexes of Pu(III) and Pu(V) with organic ligands are stable only in a very narrow pH range. Under biological conditions, organic sequestering agents cause disproportion to only the Pu(IV) complexes. Better understanding of the Pu(IV) co-ordination chemistry is essential for the design of potent *in vivo* Pu(IV) sequestering agents. Due to the intense $\alpha$-radiation of most of the plutonium isotopes and the decomposition of Pu(IV) complexes by self-radiation, we have chosen Ce(IV) complexes as model for studying Pu(IV) co-ordination chemistry [3a]. The ionic size of Ce(IV) is the same as Pu(IV) (0.94 Å). The aqueous co-ordination chemistry of Ce(IV) parallels that of Pu(IV), giving identical shifts in the M(IV)/M(III) redox potentials. A series of octadentate and tetradentate HOPO ligands have been designed and synthesised to match the co-ordination requirement of Pu(IV) (forming 1:1 and 1:2 complexes) [2b,2c]. These ligands are: 3,4,3-LI-(1,2-HOPO), H(2,2)-(Me-3,2-HOPO), 5-LIO-(Me-3,2-HOPO) and 5-LI-(Me-3,2-HOPO).

**5LI-(Me-3,2-HOPO (X=CH₂**
**5LIO-(Me-3,2-HOPO) (X=O)**       **3,4,3-LI-(1,2-HOPO)**      **H(2,2)-(Me-3,2-HOPO)**

The X-ray structures of Ce(5LI-(Me-3,2-HOPO))₂ (from organic solvent) and Ce(5LIO-(Me-3,2-HOPO))₂ (from water) have been determined. In each case, the central Ce(IV) is eight co-ordinated from two tetradentate ligands [2a]. The co-ordination polyhedra of the two complexes are essentially square antiprisms. Solution thermodynamic studies gave overall formation constants (log $\beta_2$) for Ce(5LI-(Me-3,2-HOPO))₂ and Ce(5LIO-(Me-3,2-HOPO))₂ of 41.9 and 41.6 respectively. From these constants, extraordinarily high pM values for Ce(IV) are obtained with the two ligands (37.5 and 37.0 respectively). The constants for Pu(IV) are expected to be essentially the same.

The great affinity of the tetradentate Me-3,2-HOPO ligands for Pu(IV) under physiological conditions has been demonstrated in animals [2b,3b]. In mice, 5LI- or 5LIO-(Me-3,2-HOPO) ligand (30 µmol/kg) injected intraperitoneally 1 hr after intravenous injection of ²³⁸Pu(IV)) removed about 82-84% of injected Pu(IV) from mice, a much better result than was obtained with an equimolar amount of CaNa₃DTPA (67%). It is notable that, unlike the catecholamide ligands, the tetradentate (Me-3,2-HOPO) ligands injected at the standard dosage promote as much Pu(IV) excretion as the octadentate ligands 3,4,3-LI-(1,2-HOPO)(86%) or H(2,2)-(Me-3,2-HOPO) (81%). Apparently, the ligand concentration established in the tissues at the standard injected dosage are large enough to allow the tetradentate ligands to complex Pu(IV) in competition with biological ligands. However, octadentate ligands have a 1:1 stoichiometry when binding Pu(IV), and when these are given at a low dosage or orally, their potency for reducing Pu(IV) in animal tissues exceeds that of tetradentate or hexadentate HOPO ligands. For example, for the octadentate ligand 3,4,3-1,2-HOPO, the same degree of efficacy is observed at injected dosages as low as 0.3 µmol/kg, and it is orally active. All of the tetradentate ligands need larger dosages (at least 10 times) and are not as orally active. Figure 1 shows the efficacy of multidentate ligands (30 µmol/kg) for removing ²³⁸Pu(IV) from mice intraperitoneally or orally.

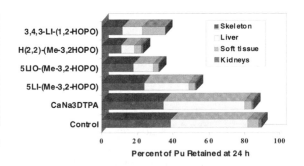

**Figure 1(a). Removal of Pu(IV) by injected ligands**       **Figure 1(b). Removal of Pu(IV) by ingested ligands**

A new octadentate, mixed HOPO, ligand was designed and synthesised recently: 3,4,3-LI-(1,2-Me-3,2-HOPO) [3]. It shows highest ability for removing Pu(IV) at physiological pH, compared with octadentate 3,4,3-LI-(1,2-HOPO) or H(2,2)-(Me-3,2-HOPO), at injected dosages of 0.01 to 0.3 µmol/kg and when the ligands are given orally. The effectiveness of this mixed ligand is remarkable when it is given orally at 10 µmol/kg, reduction of body Pu(IV) equals that obtained by injection of 10 to 30 µmol/kg of 3,4,3-LI-(1,2-HOPO). Its efficacy, when given orally at 100 µmol/kg, significantly exceeds that of any injected ligand prepared to date. It is also an effective *in vivo* chelator of Am.

## 3,4,3-LI-(1,2-Me-3,2-HOPO)

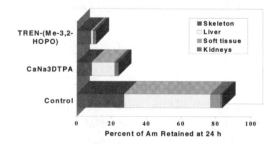

## Americium sequestering agents

Similarly to plutonium, americium generally forms eight co-ordinate complexes with ligands having hard donor atoms, but under physiological conditions only exists as Am(III). The size and co-ordination properties of Am(III) and Gd(III) are essentially identical. We have used the lanthanide ion, Gd(III), as a probe to investigate the co-ordination chemistry of Am(III) [4b]. The X-ray structure of Gd(III)-TREN-(Me-3,2-HOPO) shows a 1:1 eight co-ordinate complex with the six oxygen donors of TREN-(Me-3,2-HOPO), and two oxygen atoms of co-ordinated waters. This complex is very stable, log $\beta_1$ = 19.3, pGd = 19. Those properties are expected to be the same in Am(III) complex. Animal tests on mice show that TREN-(Me-3,2-HOPO) is very effective in removing Am(III) [4c] (Figure 2).

## TREN-(Me-3,2-HOPO)

**Figure 2. Removal of Am(III) by TREN-(Me-3,2-HOPO) from mice**

## Uranium sequestering agents

Uranium(VI) ($UO_2^{2+}$) causes chemical kidney damage, and internally deposited high specific-activity uranium isotopes can cause bone cancer. Although sought for many years, no synthetic ligand was able to reduce significantly the deposition of toxic amounts of uranyl ion in tissues, particularly kidney and bone [4a,4b]. In aqueous solution, U(VI) exists as the unique linear dioxo species, $UO_2^{2+}$. Equatorial co-ordination generally results, with $UO_2^{2+}$, complexes having pentagonal or hexagonal bipyramidal structures. Therefore, a series of tetradentate HOPO and CAM (catecholamide) ligands with 2-6 carbon linear chain backbones were designed and synthesised to match the co-ordination environment of the uranyl ion. The X-ray crystal structures of uranyl complexes with 3, 4 and 5 carbon linear chain backbones have been determined. As shown in Figure 3, the $UO_2^{2+}$ ion is equatorially co-ordinated by the tetradentate ligand 5LI-(Me-3,2-HOPO) and a solvent (DMSO) molecule. Animal testing in mice has shown that the tetradentate ligands are as effective, in some case more effective, for *in vivo* U(VI) chelation than the hexadentate or octadentate ligands containing the same binding groups [4b,4c].

**5LICAM(S)**

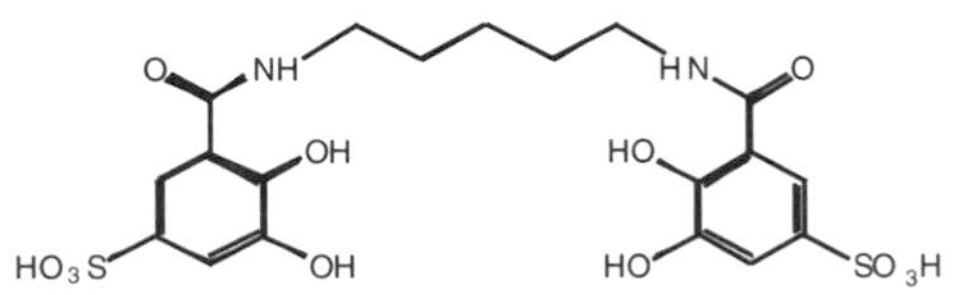

**Figure 3. X-ray structure of UO₂-5LI-(Me-3,2-HOPO)°DMSO**

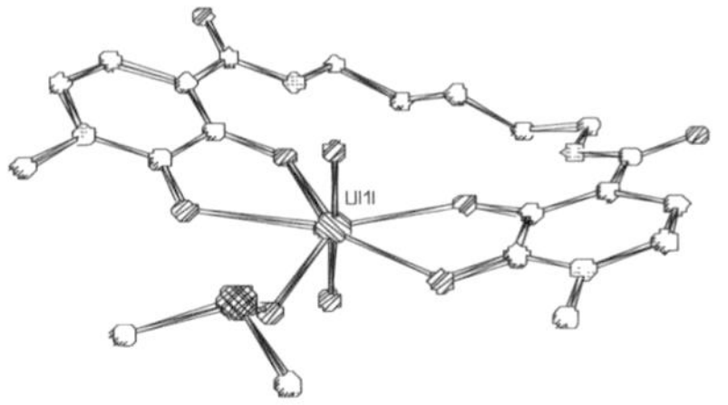

Injected U(VI) disappears from the circulating fluids of mice quickly [4b] and becomes less accessible and associated with kidneys and skeleton. The above tetradentate ligands not only chelate circulating U(VI) and divert it to excretion but they also react with and remove some of the less accessible U(VI) bound in tissues. The combination of 5LIO-(Me-3,2-HOPO) and 5LI-CAM(S) achieved better overall reduction of U(VI) in both kidneys and the skeleton than an equimolar amount of either one of them alone [4c], as shown in Figure 4.

## Neptunium sequestering agents

In aqueous solution, at physiological pH,the most stable neptunium species is Np(V) ($NpO_2^+$, a linear dioxo structure similar to that of $UO_2^{2+}$). Equatorial co-ordination with a tetradentate ligand such as 5LI-(Me-3,2-HOPO) could be expected to provide the preferred co-ordination environment. Neptunyl, is a monocharged cation with only weak binding to most ligands. For example it forms the oxine (8-hydroxy-qinoline) complex [$NpO_2^+ (OX)_2$ aq]⁻ with a stability constant of $10^{11.5}$, while that of Np(IV) complex, Np(IV)(OX)₄, is $10^{45.2}$, the difference of nearly 33.7 orders of magnitude!

## Figure 4. Removal of UO₂²⁺ from mice by a ligand cocktail

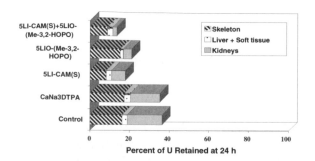

In mammals, about one half of administered NpO₂⁺ entering the blood is extracted in the urine within 24 hours, suggesting that, initially, NpO₂⁺ is the dominant species in the blood. However, prolonged retention of about 45% of the administered neptunium in the bone and 7% in the liver suggests that about one half of the circulating NpO₂⁺ is reduced to Np(IV) in those tissue in the presence of local bio-ligands that stably bind multivalent actinides [6]. Those multidentate ligands that form exceptionally stable actinide(IV) chelates should facilitate the reduction process under physiological conditions. Indeed, tetradentate 5LI-(Me-3,2-HOPO) is one of the most effective ligand for *in vivo* chelation of Np as shown in Figure 5 [6].

## Figure 5. Reduction of ²³⁷Np in mice by injected ligands

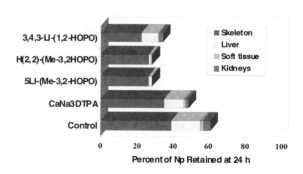

## *Conclusions*

Based on an understanding of the co-ordination chemistry of the actinide elements, a biomimetic approach has been used for the preparation of actinide sequestering agents. *In vivo* testing shows that the multidentate hydroxypyridinonate ligands are most promising. They are worthy of the continued study needed for clinical acceptance and therapeutic use.

### *Acknowledgements*

This work was supported by the Director, Office of Energy Research, Office of Basic Energy Science, Chemical Science Division, US Department of Energy under Contract Number DE-AC03-76F00098 and the National Institute of Environmental Health Science Grant Number ES02698. The authors acknowledge the invaluable assistance of Birgitta Kullgren in the conduct of animal studies.

# REFERENCES

[1]    (a) K.N. Raymond, in *Environmental Inorganic Chemistry,* K.J. Irgolic, A..E. Martell, eds., Proc. US-Italy International Workshop on Environmental Inorganic Chemistry, San Miniato, Italy, 5-10 June 1983, VCH Publishers, Inc., Deerfield Beach, Florida, 1985, 331.

   (b) P.W. Durbin, B. Kullgren, J. Xu and K.N. Raymond, *Radiation Protection Dosimetry,* 79, 433 (1998).

[2]    (a) K.N. Raymond and T.M. Garrett, *Pure and Appl. Chem.*, 60, 1807 (1988).

   (b) J.Xu, P.W. Durbin, B. Kullgren, and K.N. Raymond, *J. Med. Chem.*, 38, 2606 (1995).

   (c) D.L. White, P.W. Durbin, N. Jeung and K.N. Raymond, *J. Med. Chem.*, 31, 11 (1998).

[3]    (a) J. Xu, E. Radkov, M. Ziegler, and K.N. Raymond, submitted to *Inorg. Chem.*

   (b) P.W. Durbin, B. Kullgren, J. Xu, and K.N. Raymond, *Int. J. Radiat. Biol.*, in press.

   (c) J. Xu, P.W. Durbin, B. Kullgren, and K.N. Raymond, *manuscript in preparation.*

[4]    G.N. Stradling, *Rad. Prot. Dosimetry*, 53, 297 (1994).

   (b) J. Xu, S.J. Franklin, D.W. Whisenhunt Jr. and K.N. Raymond, *J. Am. Chem. Soc.*,117, 7245 (1995).

   (c) P.W. Durbin, B. Kullgren, J. Xu and K.N. Raymond, *Rad. Prot. Dosimetry*, 53, 305 (1994).

[5]    (a) J. Xu, and K.N. Raymond, *Inorg. Chem.*, 38, 308 (1999).

   (b) P.W. Durbin, B. Kullgren, J. Xu and K.N. Raymond, *Health Physics*, 72, 865 (1997).

   (c) P.W. Durbin, B. Kullgren, J. Xu and K.N. Raymond, submitted to *Health Physics*.

[6]    P.W. Durbin, B. Kullgren, J. Xu, and K.N. Raymond, P.G. Allen, J.J. Bucher, N.M. Edelstein and D.K. Shuh, *Health Physics*, 75, 34-50 (1998).

# TRANSURANIUM COMPOUNDS
# CHARACTERISATION FACILITY AT ITU-KARLSRUHE

**F. Wastin, E. Colineau, T. Gouder, J. Rebizant and G.H. Lander**
European Commission, Joint Research Centre, Institute for Transuranium Elements
Postfach 2340, D-76125 Karlsruhe, Germany

## Abstract

The Institute for Transuranium Elements-Karlsruhe has installed a users laboratory designed for the characterisation of transuranium elements based compounds. This facility has the capability to perform Mössbauer spectroscopy on [237]Np nucleus at low temperatures and up to 6 GPa pressure; transport measurements on small samples at temperatures ranging from 1.5 K to 1 000 K and in magnetic field up to 9 Tesla as well as magnetic measurements in fields up to 7 Tesla, in a temperature range from 2 K to 800 K, with a commercial base SQUID type apparatus.

Plans are made to complete this facility with a specific heat measurement apparatus in a near future.

Measurements of Np, Pu, Am solid compounds synthesised in the adjacent preparation laboratory are currently performed.

Surface sensitive spectroscopies as photo emission, Auger, etc. are used to characterise actinide surfaces and thin films. In particular, we are interested in the surface and interface composition, and the chemical reactions with the environment: gas adsorption, corrosion reactions, leaching and electrochemistry and solid state inter diffusion reactions are investigated on bulk actinide compounds and in-situ grown thin films.

Access to our facilities is made possible to external users either on a scientific collaboration basis or within the frame of the JRC-Secondment programme of the European Commission.

# SPECIATION OF PU(IV) COMPLEXES WITH
# WEAK LIGANDS FROM VISIBLE SPECTRA

**John M. Berg and D. Kirk Veirs**
Nuclear Materials Technology Division
Los Alamos National Laboratory
Los Alamos, New Mexico 87545, USA

## Abstract

Stoichiometries of early actinide metal ion complexes in solution equilibrium can sometimes be determined by modelling the dependence of a species-sensitive measurement on ligand concentration. Weak ligands present the additional problem that these measurements cannot be made in the simplifying limiting case of low ligand concentration relative to the background electrolyte. At high ligand concentrations, constant ionic strength no longer implies constant activity coefficients. Additional parameters must be included in the equilibrium model to account for the variation of activity coefficients with ligand concentration as well as with overall ionic strength. We present the formalism of such a model based on SIT theory and its implementation for simultaneous fitting of spectra over a wide range of ionic strengths. As a test case, we analyse a subset of the spectra we have collected on complexation of Pu(IV) by nitrate in aqueous acid solutions.

## Introduction

Most speciation techniques in common use for actinide ions in solution do not provide direct probes of species stoichiometry. They rely on modelling the functional dependence of stoichiometry sensitive observables on ligand concentration. The usual approach is to fit the changes in these observables by least squares methods to a model in which the complex-formation equilibrium constants and the values of the observables for each species are all varied simultaneously. The problem can be made sufficiently overdetermined to achieve a reliable solution by collecting enough data over a restricted range of solution conditions under which the parameters can be assumed to be constant. Visible absorption spectra of actinide ions present a particularly rich set of observables that can be interpreted through such modelling.

Solution speciation studies of actinide ions with weak ligands present problems for such methods because the functional dependence is complicated by changes in activity coefficients of the reactants and products as the ligand concentrations are increased to the levels necessary to achieve complexation. This general difficulty has been chronicled by Bjerrum [1], among others. Conclusions of studies that ignore it may be suspect, as pointed out by Spahiu and Puigdomenech in their reanalysis of several studies of actinide ion complexation [2]. They were able to show that the same data could yield different conclusions about the dominant species depending upon whether changes in activity coefficients were considered.

In an earlier paper, we demonstrated that the formation constants of the first two Pu(IV) nitrate complexes in a background electrolyte of perchloric over a broad range of ionic strengths was well described by the equations of specific ion interaction theory (SIT), provided that nitrate concentrations were calculated unconventionally [3]. We used the measured free nitrate concentrations from published Raman data on trace nitric acid in perchloric acid, rather than the stoichiometric nitrate concentrations. This modification compensated for the incomplete dissociation of $HNO_3$ in acid solutions of appreciable $H^+$ concentrations. SIT equations fit the formation constants of the first two nitrate complexes up to 19 molal ionic strength. However, that analysis approach was restricted to cases where nitrate concentration was much lower than that of the background electrolyte.

This paper reports progress in modelling equilibria in solutions where the ligand concentration varies up to the full ionic strength. Variable activity coefficients are modelled by assuming the SIT equations that were valid for the first two formation constants to high ionic strength will also be valid for high nitrate concentrations. This assumption is first applied to the dissociation of nitric acid over a wide range of nitrate concentrations then to optical spectra of Pu(IV) nitrate complexes. Free ligand concentrations and ligand activity coefficients are determined from our SIT analysis of published Raman data and are not free parameters in the model for the complex formation. An effective ionic strength is used in the equations' account for the partial dissociation of nitric acid.

This simple model may not generalise to other chemical systems. Nonetheless, it is worth exploring for the specific case of very weak actinide complexes. Methods of systematising the variation in activity coefficients at high ligand concentrations and ionic strengths are vital to being able to identify and quantify such complexes given only indirect data such as optical spectra.

## Model

A metal-ligand equilibrium system described by the overall complex formation reactions:

$$M^{q+}(aq) + nL^- \overset{\beta_n}{\rightleftharpoons} ML_n^{q-n}$$

will have an absorption spectrum that is the sum of the absorption spectra of the metal species in equilibrium:

$$A(\lambda, m_i) = c_M \epsilon_M(\lambda) + \sum_n c_{ML_n} \epsilon_{ML_n}(\lambda) \tag{1}$$

where concentrations, $c$, are in molar units and the $\epsilon$ are molar absorptivities. In this and subsequent equations, charges on the subscript ions are not shown to simplify the notation.

For a specific set of solution conditions (ionic strength, electrolyte concentrations, ligand concentration), the formation constants are related to the metal ion and metal-ligand complex concentrations by:

$$\beta_n = \frac{m_{ML_n}}{m_M m_L^{\,n}} \tag{2}$$

where concentrations, $m$, are in molal units and where $\beta_n$ is understood to mean $\beta_n(I, m_L)$. The aquated metal ion molarity can be expressed a function of the total metal ion molarity, the ligand molality, and the formation constants of all complexes that account for significant fractions of the metal.

$$c_M = \frac{c_{M_{total}} m_M}{m_{M_{total}}} = \frac{c_{M_{total}}}{m_{M_{total}}} m_M = \frac{c_{M_{total}}}{1 + \sum_{n'} m_L^{\,n'} \beta_{n'}} \tag{3}$$

The molalities of the metal-ligand complexes are also functions of the total metal ion molarity, the ligand molality and the formation constants.

$$c_{ML_n} = \frac{c_{M_{total}} m_{ML_n}}{m_{M_{total}}} = \frac{c_{M_{total}}}{m_{M_{total}}} m_M m_L^{\,n} \beta_n = c_{M_{total}} \frac{m_L^{\,n} \beta_n}{1 + \sum_{n'} m_L^{\,n'} \beta_{n'}} \tag{4}$$

It is common practice to determine formation constants by least squares fitting of the absorption spectra of metal species in a series of solutions having the same ionic strength. This relies on the approximation that activity coefficients of the reactants and products in the complex-forming reactions will not vary with ligand concentration at a constant ionic strength.

This approximation only holds if $m_L$ is small, however. This presents a practical problem for identifying and quantifying weak complexes that do not form in measurable amounts under the limiting condition of $m_L \ll I$. If $m_L$ is varied over a significant fraction of the ionic strength, the $\beta_n$ can no longer assumed to be constants at constant $I$. The usual two-step approach of determining $\beta_n$ values at a series of $I$ and then modeling their $I$ dependence cannot be used. Activity coefficient variations must somehow be explicitly incorporated into the initial fitting of the data.

The $\beta_n$ are related to the formation constants at zero ionic strength through the activity coefficients, $\gamma$ (molal scale).

$$\beta_n^0 = \frac{\gamma_{ML_n} m_{ML_n}}{\gamma_M m_M (\gamma_L m_L)^n} = \beta_n \frac{\gamma_{ML_n}}{\gamma_M \gamma_L^{\,n}} \tag{5}$$

The $\gamma$ are dependent on the concentrations of all of the other solution constituents. Specific ion interaction theory (SIT) offers a particularly simple model for this dependence under limited conditions. For solutions containing the metal ion, cationic metal ion complexes, a monoprotic acid HA acting as a non-complexing electrolyte, and a ligand introduced as a monoprotic acid HL, the relevant activity coefficients are given by SIT as:

$$\log \gamma_M(m_i) = -z_M^2 D(I) + \varepsilon_{M,L} m_L + \varepsilon_{M,A} m_A \tag{6}$$

$$\log \gamma_{ML_n}(m_i) = -z_{ML_n}^2 D(I) + \varepsilon_{ML_n,L} m_L + \varepsilon_{ML_n,A} m_A \tag{7}$$

$$\log \gamma_L(m_i) = -z_L^2 D(I) + \varepsilon_{M,L} m_M + \sum_{n=1}^{n\,\max} \varepsilon_{ML_n,L} m_{ML_n} + \varepsilon_{H,L} m_H \tag{8}$$

For cationic complexes of a metal ion at low concentration relative to the background electrolyte, Eqs. (6), (7) and (8) may be substituted into Eq. (5) to express the zero ionic strength formation constant as:

$$\log \beta_n^0 = \log\left(\frac{m_{ML_n}}{m_M m_L^n}\right) - \Delta z^2 D(I) + \left(\varepsilon_{ML_n,L} - \varepsilon_{M,L}\right) m_L + \left(\varepsilon_{ML_n,A} - \varepsilon_{M,A}\right) m_A - \varepsilon_{H,L} m_L \tag{9}$$

where terms in Eq. (8) proportional to the $M^{q+}$ and $ML_n^{q-n}$ molalities were dropped because the total concentrations of these ions is assumed to be much smaller than $m_H$, $m_A$, and $m_L$. This may be rearranged to express the phenomenological formation constants for each set of solution conditions as:

$$\log \beta_n(I, m_L, m_A) = \log \beta_n^0 + \Delta z^2 D(I) - \left(\varepsilon_{ML_n,L} - \varepsilon_{M,L}\right) m_L - \left(\varepsilon_{ML_n,A} - \varepsilon_{M,A}\right) m_A + \varepsilon_{H,L} m_H \tag{10}$$

These are the $\beta_n$ in Eqs. (3) and (4).

## Results and discussion

The validity of Eq. (10 was demonstrated for the limiting case of small $m_L$ by the observation that $\beta_1$ and $\beta_2$ for Pu(IV) nitrate complexes follow this relationship up to 19 molal total acid concentration [3]. In that study, sets of spectra at fixed ionic strengths were fit individually. The resulting $\beta_n$ for n=1,2 were fit by Eq. (10), yielding values for $\beta_n^0$ and for the combination $\varepsilon_{ML_n,A} - \varepsilon_{M,A} - \varepsilon_{H,L}$. The success of SIT theory over such a broad range of ionic strengths encourages us to explore the possibility of using the SIT equations to constrain formation constants in the more general case where $m_L$ ranges from zero to the full ionic strength.

This presents a considerably more difficult fitting problem. Because the $\beta_n$ will no longer be constant with ionic strength, they cannot be determined by fitting spectra at constant $I$. Instead, the constants to be fit are the unknown parameters in Eq. (10): $\beta_n^0$ and the $\Delta\varepsilon$ coefficients of $m_A$, $m_L$ and $m_H$. If Eq. (10) holds over a broad range of conditions, these expressions for $\beta_n$ may be substituted into Eqs. (3) and (4) and the values for the unknown parameters can be determined by fitting spectra of solutions ranging broadly in ionic strength and ligand concentration. Unlike the two-stage approach employed in our earlier work, the formation constants and the differences in ion interaction parameters would all be determined in a single fit of the global model to this larger data set.

As shown in our earlier work, Eq (10) is only valid at high $I$ for the Pu(IV) nitrate system if effective rather than stoichiometric values are used for $I$ and $m_L$ [3]. To estimate the free nitrate concentrations in mixtures of perchloric and nitric acids, we modelled the published dissociation measurements in pure nitric acid and as a trace constituent in perchloric acid using the SIT equations. The measured $K_a$ derived from published Raman data [4,5] are shown vs. ionic strength in Figure 1, along with the SIT fits. Both sets of data yielded close to the same value for $K_a^0$ (log $K_a^0 = 1.8$), suggesting that these equations may offer a reasonable approximation to $K_a$ over intermediate acid mixtures. The fits also determine the ion interaction parameters $\varepsilon_{H,L}$ and $\varepsilon_{H,A}$ ($\varepsilon_{H,L} = 0.09$ and $\varepsilon_{H,A} = 0.06$). If these were valid constants for intermediate mixtures, the free nitrate concentration could be calculated for solutions of any mixture of these two acids. For the purposes of this paper, we simply assume that they are valid without further testing. However, a series of Raman measurements of mixtures would provide valuable confirmation. Note that these $\varepsilon_{H,L}$ and $\varepsilon_{H,A}$ values differ significantly from the accepted values for the conventional application of SIT (0.07 and 0.14, respectively) [6] because they apply to a different model. Our parameters are not valid in the standard SIT model.

This model of the acid dissociation specifies $m_L$ and an effective $I$ for each sample solution, and determines the value of $\varepsilon_{H,L}$. The global model is left with three unknown parameters in Eq. (10) for each complexation equilibrium, but these three parameters are constants over all solution conditions from zero to over 10 molal ionic strength and from trace to pure nitric acid solution. Thus a single model with these as fit parameters may be applied to a data set consisting of spectra of solutions spanning multiple ionic strengths and relative nitrate concentrations.

While the parameters in this global fitting problem are overdetermined by the spectra to be fit, this does not guarantee convergence to reasonable parameter values. The increased number of parameters creates greater potential for finding false minima. We have developed a prototype software implementation to test the approach. The prototype treats $\varepsilon_{H,L}$ as a known constant and simultaneously optimises the values of $\beta_n^0$, and the $\Delta\varepsilon$ coefficients of $m_A$ and $m_L$ in Eq. (10) using unconstrained non-linear least squares. It uses unconstrained non-linear least squares on the derivative of the spectra with respect to photon energy with equal weights assigned to all data.

We tested the approach on a selection of the available data chosen because they contain only $Pu^{4+}(aq)$, $Pu(NO_3)^{3+}$ and $Pu(NO_3)_2^{2+}$. The model converged to the values $\beta_1^0 = 130$, $\varepsilon_{ML_1,A} - \varepsilon_{M,A} = -0.04$, $\beta_2^0 = 4\,600$ and $\varepsilon_{ML_2,A} - \varepsilon_{M,A} = -0.19$. These are in reasonable agreement with our earlier fits at discrete ionic strengths [3].

We have not yet succeeded in getting the fit parameters to converge reasonable values in the unconstrained case when more than two equilibria are considered. It will probably be necessary to apply constrained, weighted non-linear least squares methods to achieve reliable results for significantly more that three species fit simultaneously. We are in the process of extending the fitting method to include these capabilities.

# REFERENCES

[1]    J. Bjerrum, *Acta Chem. Scan.*, A41, 32 (1987).

[2]    K. Spahiu and I. Puigdomenech, *Radiochim. Acta*, 82, 413 (1998).

[3]    J.M. Berg, D.K. Veirs, R.B. Vaughn, M.A. Cisneros and C.A. Smith, *J. Radioanal. Nucl. Chem.*, 235, 25 (1998).

[4]    M. Sampoli, A. de Santis, N.C. Marziano, F. Pinna and A. Zingales, *J. Phys. Chem.* 89, 2864 (1985).

[5]    T.F. Young, L.F. Faranville and H.M. Smith, in *The Structure of Electrolyte Solutions*, Walter J. Hamer, ed. (Wiley: New York, 1959), p. 35.

[6]    I. Grenthe, J. Fuger, R. Konings, R.J. Lemire. A.B. Muller, C. Nguyen-Trung and H. Wanner, *Chemical Thermodynamics of Uranium* (Elsevier: Amsterdam, 1992).

**Figure 1.** Plots of log $K_a$-2 D(I) vs. the effective ionic strength show that SIT equations model nitric acid dissociation both in a nitric acid/water solution (bottom) and in trace nitric acid in perchloric acid (top). The y-intercepts are in reasonable agreement and the combination of slopes yield binary interaction parameters for $H^+$ with $ClO_4^-$ of 0.060 and of $H^+$ with $NO_3^-$ of 0.095

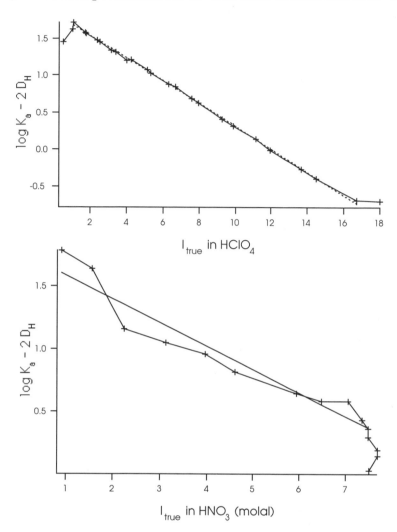

# LASER-INDUCED LUMINESCENCE LIFETIME MEASUREMENT AS AN ANALYTICAL PROBE FOR SPECIATION OF POLYCARBOXYLATES IN AQUEOUS SOLUTIONS

**Yoshio Takahashi**
Faculty of Science, Hiroshima University, Japan

**Takaumi Kimura, Yoshiharu Kato**
Advanced Science Research Centre, Japan Atomic Energy Research Institute, Japan

**Yoshitaka Minai**
Centre for Arts and Sciences, Musashi University, Japan
Nezu Institute of Chemistry, Japan

## Abstract

Luminescence from lanthanide or actinide ion is influenced by hydration structure of the ion in aqueous solution system. In particular lifetime of the luminescence has been regarded as a measure of hydration number of the lanthanide or the actinide ion based on the studies on lifetime measurement of the ion in solid and solution system. Compared with other technique like NMR to determine the hydration number, laser induced lifetime measurement is advantageous in sensitivity and selectivity. This allows us to apply this method to determining the hydration number of lanthanide or actinide ion even at low concentration.

Lifetime of luminescence emitted from europium(III) with polyacrylic acid (PAA), polymaleiic acid (PMA), polymethacrylic acid (PMAA), or poly(a-hydroxy)acrylic acid (PHAA) were determined in the aqueous solutions at various pH and ionic strength. These polyacids are regarded as analogues of humic substances which possibly controls environmental behaviour of trace amounts of lanthanide(III) and actinide(III). It is noteworthy that the measurement could be carried out even at low metal-to-ligand ratio like 1:100 appreciating intense laser light for excitation and efficient luminescence detection system.

The residual hydration number obtained from the measured lifetime indicated that the hydration number was a function of both pH and ionic strength. The hydration number for most of polycarboxylates, except PMAA, monotonously decreased with pH, indicating successive formation of europium(III)-polycarboxylate complexes and morphological change of the polymolecule. The pH dependence of the residual hydration number for PMAA showed that the increase of the number at neutral pH region where the form of the polyacid molecule changed. The pH titration curve also confirmed such morphological change of the PMAA molecule. The titration curve of humics often showed similar feature to that of PMAA. This may suggest that similar change in morphology of humics molecule may take place in particular region of pH. The dependence of the residual hydration number on ionic strength was interpreted as morphological change of the polymolecule by the shielding effect of the supporting electrolyte.

As demonstrated above, laser induced luminescence lifetime measurement is useful to elucidating the structure of polyacid in aqueous solutions if europium(III) or other metal cations can be used as luminescence probe. This technique is applicable to both solid and solution systems; furthermore, it can be used to characterising the adsorbed species on solid particles in contact with aqueous solution. Some examples of the studies on such adsorbed species would be given on site.

# POSTER SESSION

## *Part C: Methods for Empirical Formula and Molecular Structure Determination*

# DIRECT SPECTROSCOPIC SPECIATION OF ACTINIDE IONS AT THE WATER/MINERAL INTERFACE

**R. Klenze**

Forschungszentrum Karlsruhe, Institut für Nukleare Entsorgungstechnik,
D-76021 Karlsruhe, FRG

## Abstract

The reactions of actinides at the water/mineral interface constitute the most important mechanism for their retention in the geosphere. A well-founded description of the sorption mechanism necessitates identification of interface reactions (ion exchange, site specific binding, ternary surface complexes, surface precipitation, solid solution etc.) and their quantification by a thermodynamic model, e.g. a surface complexation model. However, surface models are usually parameterised by fitting experimental sorption data, without any confirmation of the species postulated.

Whereas a number of high-vacuum methods are available for the speciation of sorbed metal ions, only few methods are applicable for metal ion speciation at surfaces in contact with the aqueous phase. EXAFS has recently been applied to various surface sorbed metal ions; however, its lack of sensitivity prohibits studies at low sorption densities. For speciation at trace concentrations, more sensitive methods such as EPR, Time Differential Pertubated Angular Correlation Spectroscopy (TDPAC), and Time Resolved Laser Fluorescence Spectroscopy (TRLFS) are necessary. Applicability of these methods is restricted to only few elements.

We have used TDPAC in combination with EXAFS for the speciation of Hf(IV) (as a chemical homologue for tetravalent actinides) sorbed onto silica. TRLFS was used for the speciation of U(VI), Eu(III) and Cm(III) at various water/mineral interfaces. Advantages and disadvantages of these methods are shortly reviewed. Specific examples are presented for the speciation of the interface equilibrium reaction of Cm(III) by TRLFS with colloidal mineral phases such as silica, alumina, titania and hydrous ferric oxide.

The Cm(III) species in solution and sorbed onto the mineral surface are characterised by their excitation and emission spectra, relaxation and under certain conditions by energy transfer. The relaxation rate provides direct information about the number of water molecules within the first co-ordination shell. Changes in Cm(III) emission spectra with chemical environment is used for identification and quantification of individual species in the system by peak deconvolution of the mixed spectra. The Cm(III) speciation is determined as a function of pH, sorbent concentration, as well as equilibration time. Based on their spectroscopic and thermodynamic properties, the derived Cm(III) species are assigned to surface complexes and quantified by fitting the data to a surface complexation model.

# OXIDATION STATE AND STRUCTURE OF ACTINIDES IN ENVIRONMENTAL SAMPLES USING SYNCHROTRON-BASED TECHNIQUES

**D.T. Reed, A.J. Kropf, S.B. Aase, S. Zygmunt, and L. Curtiss**
Chemical Technology Division, Argonne National Laboratory, USA

## Abstract

X-ray synchrotron radiation (XSR) techniques are being used to determine the structure of adsorbed and solid actinide species that are of environmental importance. This research is being conducted at the Material Research Collaborative Access Team (MR-CAT) beamline at the Advanced Photon Source (APS) facility at Argonne.

Our experimental investigations are focused on the use of XANES/EXAFS to identify actinide phases and oxidation state in environmentally-relevant samples. Oxidation state trends with XANES edge position were first established for plutonium and neptunium reference solids. The results for a series of plutonium solids that differ in oxidation state are shown in Figure 1. Here a systematic ~ 2 eV shift was observed as a function of oxidation state. This is consistent with what has been reported for variable oxidation states in aqueous systems. We have also examined a series of Pu(III) solids where we have shown that even though significant differences in the geometry exist, the edge positions do not change (to an uncertainty of $\pm$ 0.3 eV). Similar experiments were done with neptunium standards but the oxidation state trends are less clear.

On the basis of the results with plutonium standards, XANES was used to establish oxidation state in actinide precipitates in WIPP brine, actinide waste forms, and biologically-induced actinide precipitation reactions. In the brine experiments, XANES analysis confirmed that iron surfaces reduced Pu(VI) to form Pu(IV) hydroxide phases and greatly reduce the overall solubility of plutonium in WIPP brine. Precipitates of uranium and neptunium in WIPP brine were shown to be U(VI) and Np(V) phases respectively. In the wasteform studies, Pu(IV) phases were shown to predominate. In microbiological interaction studies, we have shown that anaerobic methanogenic and sulfate reducing bacteria lead to biologically induced reduction of Np(V) to Np(IV) phases. All of these results have important implications to the subsurface migration and durability of actinide species.

These experimental efforts are integrated with theoretical efforts to calculate XANES/EXAFS spectra to help interpret the experimental results. Calculations of the Pu $L_{III}$ XANES of small cluster models of the local Pu environment in $PuO_2$ have been carried out to compare with the experimental results obtained at the APS. These calculations, carried out with the multiple-scattering FEFF7 code, use tangent atomic spheres and include relativistic corrections for the initial core state. The resulting spectra give relative peak energies and intensities in good agreement with experiment. Best agreement is obtained for these solid standards when the initial state potential is chosen to be $Pu^{+1}$, which accounts for core-hole screening effects in an approximate way. The result in these analysis is that we are showing that the shift in the XANES edge position can be accounted for primarily by changes in oxidation state, with possibly a much smaller dependency on geometry.

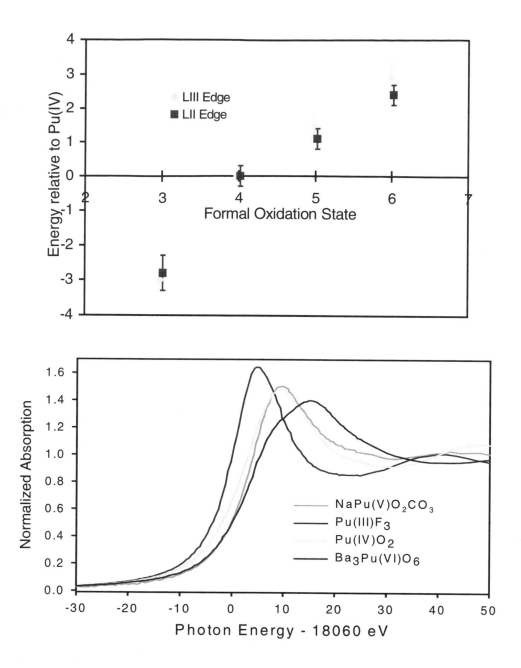

# CO-ORDINATION PROPERTIES OF DIGLYCOLAMIDE (DGA) TO TRIVALENT CURIUM AND LANTHANIDES STUDIED BY XAS, XRD AND XPS METHODS

**T. Yaita, M. Hirata, H. Narita, S. Tachimori, H. Yamamoto**
Japan Atomic Energy Research Institute, Tokai-mura, Naka-gun
Ibaraki 319-1195, Japan

**N.M. Edelstein, J.J. Bucher, D.K. Shuh, L. Rao**
Lawrence Berkeley National Laboratory
1 Cyclotron Road, MS 70A-1150
Berkeley, CA 94720, USA

## Abstract

Co-ordination properties of diglycolamide (DGA) to trivalent curium and to the trivalent lanthanides were studied by the EXAFS, the XRD and the XPS methods. The structural determinations by both the crystal XRD and the solution EXAFS methods showed that the DGA co-ordinated to the trivalent lanthanide ion in a tridentate fashion: co-ordination of three oxygen atoms of each ligand to the metal ion. The bond distances of Er-O (carbonyl) and Er-O (ether) in the Er-DGA complex were 2.35 Å and 2.46 Å, respectively, while the atom distances of Cm-O (carbonyl) and Cm-O (ether) in the Cm-DGA complex were 2.42 Å and 3.94 Å, respectively from the EXAFS data for the Cm-DGA complex. Accordingly, the DGA would behave only as a semi-tridentate in the co-ordination to trivalent curium in solution. We determined the valence band structures of the Er-DGA complex by the XPS in order to clarify the bond properties of the complex, and assigned the XPS spectrum by using the DV-DS molecular orbital calculation method.

# Introduction

In R&D on the "back end" of nuclear fuel cycle, the conversion or transmutation of the long-lived minor actinides such as americium into short-lived isotopes is done by irradiation with neutrons [1,2]. In order to complete these conversions or transmutations, it is necessary to separate trivalent minor actinides (mainly Am and Cm) from trivalent lanthanides, since some of the lanthanides have high cross sections for neutron capture and hence prevent neutron capture by the minor actinides. Many researchers, therefore, have proposed the Ln/An separation methods [12,13]. In most of the methods, co-extraction of trivalent lanthanides and actinides from high-level waste prior to Ln/An separation is assumed. Amides are one of the desired extractants for trivalent lanthanides and actinides. These compounds are completely incinerable, consisting of hydrogen, nitrogen, and oxygen, unlike organophosphorus compounds such as DIDPA and CMPO. Musikas (1987) has studied extensively the extraction of lanthanides and actinides with some amides [3,4]. The diglycolamides (DGA) are open chain polydentate ligands that can form a co-ordination structure similar to the corresponding macrocyclic ligands, and show the highest extractabilities of all the amides investigated [5,6]. However, the detailed mechanism for the high extractability of the DGA for these elements have not been clarified yet. If it becomes clear, it will provide important information for the development of a new extractant with higher extractability and selectivity.

In this study, X-ray crystallography, (XRD), and extended X-ray absorption fine structure (EXAFS) methods were utilised to clarify the co-ordination structures in both solid and solution states. Furthermore, we measured the X-ray photoelectron spectroscopy (XPS) spectrum to characterise the chemical bond properties between metal and ligand. The XPS spectrum was assigned by the preliminary results of the DV-DS molecular orbital calculation.

# Experimental

## EXAFS measurement

The sample solutions were prepared by dissolution of metal (Er, Cm)-DGA complex into ethanol. The curium samples were prepared in a glove box at LBNL. The EXAFS spectra at the $L_{III}$-edge of the metals were measured at the BL4-1 of SSRL at Stanford University, U.S.A. The Er-DGA sample was also measured at BL 27B of PF in KEK, Japan. The obtained data were analysed by using WinXAS97 program. The theoretical phase shift and back scattering amplitude parameters were calculated by FEFF 7 [14] code.

## X-ray crystallography of Er-DGA complex

A single crystal of the Er-DGA complex: $[Er(C_{18}H_{20}N_2O_3)_2(H_2O)_3]Cl_3 \cdot 6H_2O$ was prepared in mixture of ethanol and 2-pentanol. The crystal was measured by Rigaku AFC7R diffractmeter equipped with graphic monochromated Mo-$K_\alpha$ radiation. The data analysis was performed by the TEXAN crystallographic package of Molecular Structure Corporation.

## XPS spectroscopy

For the XPS analysis, a Mg $K_\langle$ X-ray (1 253.6 eV) of a VG ESCALAB-5 spectrometer were used as the excitation source, under a pressure of $<1.3 \times 10^{-7}$ Pa. The kinetic energy of the photoelectrons was calibrated such that the binding energy of the Au 4f$_{7/2}$ line of metallic gold was 84.0 eV. Its full width at half-maximum (FWHM) value was 0.85 eV. The sample was prepared by the same procedure as in the X-ray crystallography section.

## Results

Figure 1 shows the crystal structure of the Er-DGA complex, [Er(C$_{18}$H$_{20}$N$_2$O$_3$)$_2$(H$_2$O)]$^{3+}$ cation. The erbium ion is 9-co-ordinate, bound to the oxygen atoms of three waters and two diglycolamides. Diglycolamide co-ordinated to erbium in tridentate fashion. Table 1 summarises the selected bond geometric parameters of the DGA-Er complex. The Er-O (carbonyl) distances are 2.289 and 2.314 Å, and in agreement with those of Er-malonamide complex: av.2.315 Å [8]. In contrast, the Er-O (ether) distance: 2.465 Å are about 0.15 Å longer than the Er-O (carbonyl). This relationship in the Er-O distance between carbonyl and ether was also observed in the other glycolamide-lanthanide complexes [9,10].

Figure 2 shows the EXAFS oscillations, the radial structural function for the Er-DGA complexes and their fitting results, demonstrating the fit quality. The phase shift and back scattering parameters were calculated by using the crystal structure. The fitting results are listed in Table 2. The geometrical parameters: the Er-O (carbonyl) and the Er-O (ether) were 2.33 Å; N = 4 and 2.47 Å; N = 2, respectively. These parameters suggested that the complex structures in both solution and solid states were very similar, except that Er in solution was the 8-co-ordination ion, which is better than 9-co-ordination one in fitting quality. From the standpoints of the distribution ratio, the DGA is regarded as a better ligand for the trivalent lanthanide extractions than the other amides [5,6]. Accordingly, these results suggested that the role of ether oxygen in the co-ordination to lanthanide would be very important for the extraction of trivalent lanthanides.

Figure 3 shows the XPS spectrum for the valence band of the Er-DGA complex and the MO energy level structure calculated by the DV-DS calculation method [7]. The XPS spectrum mainly consists of four peaks. We tried to evaluate the contribution of atomic orbital to molecular orbital. The peak at 1 232 eV (22eV for binding energy) arose from 4f orbital of Er ion. The broad peak at about 1 222 eV (32 eV) was attributed to O 2p and O 2s for the DGA and water. The Cl 3p atomic orbital shaped the broad peaks in the range from 1 238 to 1 243 eV. The peak at 1 195 eV would be due to Er 5s. The XPS measurement for free DGA is presently being undertaken. We will discuss the detailed chemical bond properties in a future paper.

Figure 4 shows the EXAFS oscillations, the radial structural function for the Cm-DGA complexes, and their fitting results, demonstrating the fit quality. Table 3 lists the fitting results. We analysed the first peak shown in Figure 4 before the full shell fits, and confirmed that the first shell consisted of oxygen atoms for carbonyl group and water. The distances were 2.42 Å for carbonyl group and 2.45 Å for water. The distance of the Cm-O (ether) was 3.94 Å. Accordingly, it is possible that the DGA would co-ordinate to Cm in a semi-tridentate fashion: two strong bonds between Cm and oxygen of carbonyl group, and a weak bond between Cm and oxygen of ether.

## Discussion

Both the crystal XRD (crystal state) and the solution EXAFS (solution state) results showed that the DGA co-ordinated to lanthanides in a tridentate fashion. In contrast, the EXAFS data for Cm complex in solution suggested that the DGA would co-ordinate to curium in a semi-tridentate fashion: two strong bonds and a weak bond as mentioned before. The distribution ratios of americium (the distribution ratios of Am are not so different from those of Cm) in the DGA extraction system were 10 times lower than that of Eu [11]. The Eu-DGA complex also showed tridentate co-ordination similar to that of the Er complex [11]. The co-ordination of the ether oxygen to metals would stabilise the complex between the metal and the DGA by making two 5-member rings that include a metal. Accordingly, the extraction behaviours of trivalent lanthanides and actinides in the extraction with the DGA would correlate with the stabilisation of the extracted complexes.

In this study, we adopted several methods for the determinations of complex structures and the energy levels of chemical bonds. The EXAFS method is very powerful to determine the local structure of complexes in both solution and solid states, while, the X-ray single crystallography has more advantageous points than the EXAFS method for detailed structural analysis of crystal state. A small difference in complex structure between solution and solid states was observed in the distance of the Er-O (water). The distances in solution were 2.35 Å, and those in solid were 2.40 Å. This would be mainly due to an increase in total co-ordination number in the crystal state.

## Summary and conclusions

The XRD and the EXAFS determined the structures of Er and the Cm-DGA complexes in both solution and solid states. The XPS spectra for the Er-DGA complex was measured and assigned by the DV-DS calculation. The structural parameters suggested that the co-ordination of ether oxygen of the DGA to the trivalent lanthanides and actinides would stabilise their complexes. This stabilisation would be very important for the extraction of these metals.

# REFERENCES

[1]   J. Tommasi, M. Delpech, P.J. Grouillar and A. Zaetta, *Nucl. Technol.*, 111, 133 (1995).

[2]   S. Matsuura, *Nucl. Phys.*, A654, 417c-435c (1999).

[3]   C. Musikas, *Inorg. Chem. Acta*, 140, 197 (1987).

[4]   C. Musikas, H. Hubert, *Sol. Extn. Ion Exch.*, 5, 877 (1987).

[5]   H. Narita, T. Yaita, K. Tamura, S. Tachimori, *J. Radioanal Nucl.Chem.*, 239(2), 381-384 (1999).

[6]   H. Narita, T. Yaita, K. Tamura, S. Tachimori, *Radiochim. Acta*, 81, 223-226 (1998).

[7]   M. Hirata, H. Monjyushiro, R. Sekine, J. Onoe, H. Nakamatsu, T. Mukoyama, H. Adachi, K. Takeuchi, *J. Electron Spectrosc. Relat. Phenom.*, 83, 59-64 (1997).

[8]   X. Gan, Z. Liang, N. Tang, M. Tan, K. Yu, G. Tan, *Polyhedron*, 12(15), 1927-1931 (1993).

[9]   Y. Zhu, W.S. Riu, M. Tan, T. Jiao, G. Tan, *Polyhedron*, 12(8), 939-944 (1993).

[10]  E.E. Castellano, R.W. Becker, *Acta Cryst.*, B37, 61-67 (1981).

[11]  H. Narita, T. Yaita and S. Tachimori, submitted to *J. Chem. Soc. Dalton Trans.*

[12]  Z. Kolarik, U. Müllich, *Sol. Extn. Ion Exch.*, 15(3), 361-379 (1997).

[13]  Y. Zhu, *Radiochim. Acta*, 68, 95 (1995).

[14]  S.I. Zabinsky, J.J. Rehr, A. Ankudinov, R.C. Albers and M.J. Eller, *Phys. Rev. B.*, 52, 2995, (1995).

**Figure 1. The crystal structure of Er-DGA complex, [Er(C₁₈H₂₀N₂O₃)₂(H₂O)]³⁺ cation. Crystal system: Orthorhombic; Lattice parameters: α = 25.351(4) Å, β = 10.252(3) Å, γ = 18.114(2) Å; Space group: Pcca; Residuals: R = 0.045; Rw = 0.046**

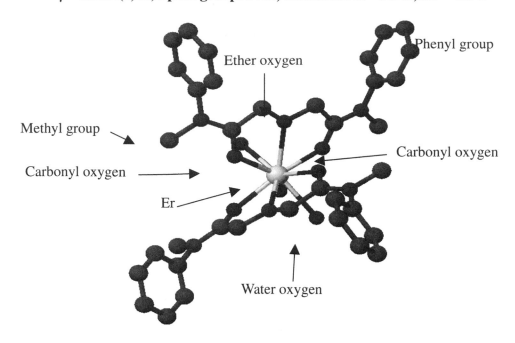

## Figure 2. EXAFS oscillation

*a) EXAFS oscillation and radial structural function*

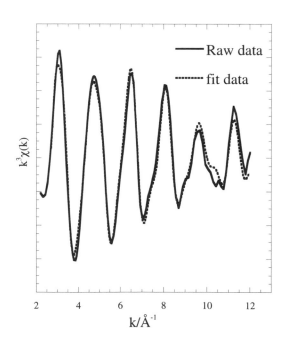

*b) EXAFS oscillation of Er-DGA complex in ethanol (phase shifts were not corrected)*

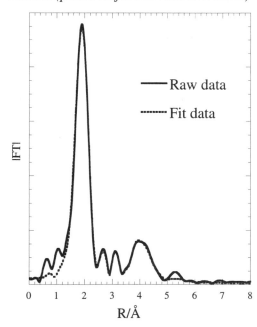

# Figure 3. Experimental XPS spectrum for Er-DGA complex and MO energy level structure calculated by DV-DS calculation

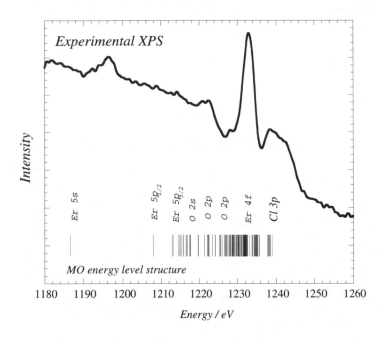

# Figure 4. EXAFS oscillation

*a) EXAFS oscillation and radial structure function*

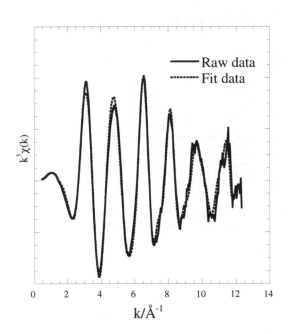

*b) EXAFS oscillation of Cm-DGA complex in ethanol (phase shifts were not corrected)*

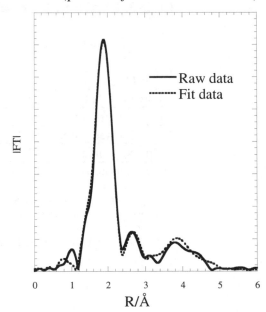

### Table 1. Selected geometrical parameters for Er-DGA complex

| Distance | N | R/Å |
|---|---|---|
| Er-O (water) | 3 | 2.399(7), 2.399(6), 2.399(6) |
| Er-O (carbonyl) | 4 | 2.289(5), 2.314(5) |
| **Er-O (ether)** | **2** | **2.465(4)** |

N: co-ordination number.
R: bond distance.
*The numbers in parenthesis are the standard deviations for the fits.

### Table 2. EXAFS fitting results for Er-DGA complex in ethanol

| Distance | N | R/Å | $\sigma^2$ |
|---|---|---|---|
| Er-O (water) | 2 | 2.35 | 0.01 |
| Er-O (carbonyl) | 4 | 2.33 | 0.005 |
| **Er-O (ether)** | **2** | **2.47** | **0.005** |
| Er-C (carbonyl) | 4 | 3.18 | 0.008 |
| Er-C (methylene) | 4 | 3.35 | 0.006 |
| Er-N (amide) | 4 | 4.44 | 0.006 |

N: co-ordination number, R: bond distance, σ: Debye-Wallar factor.

\*    N was fixed during fit.

\*\*  In addition to the distances shown above, the multiple scattering paths were also considered. The significant paths were as follows: Er-C-O (C = O); Er-N-C (amide); Er-N-O (amide).

\*\*\* The bond distances include the errors of ± 0.02.

### Table 3. EXAFS fitting results for Cm-DGA complex in ethanol

| Distance | N | R/Å | $\sigma^2$ |
|---|---|---|---|
| Cm-O (water) | 6 | 2.45 | 0.023 |
| Cm-O (carbonyl) | 4 | 2.42 | 0.005 |
| **Cm-O (ether)** | **2** | **3.94** | **0.003** |
| Cm-C (carbonyl) | 4 | 3.35 | 0.002 |
| Cm-C (methylene) | 4 | 3.49 | 0.001 |
| Cm-N (amide) | 4 | 4.62 | 0.008 |

N: co-ordination number, R: bond distance, σ:Debye-Wallar factor

\*    N was fixed during fit.

\*\*  In addition to the distances shown above, the multiple scattering paths were also considered. The significant paths were as follows: Cm-C-O (carbonyl); Cm-N-C (amide); Cm-N-O (amide).

\*\*\* The bond distances include the errors of ±0.02.

# SPECIATION OF URANIUM IN MINERALS BY SEM, TEM, μ-PIXE, XPS AND XAFS

**Toshihiko Ohnuki, Hiroshi Isobe***
Japan Atomic Energy Research Institute
Shirakata-2, Tokai, Ibaraki, 319-1195 Japan
E-mail: ohnuki@sparclt.tokai.jaeri.go.jp

**Tomihiro Kamiya, Takuro Sakai**
Japan Atomic Energy Research Institute
Watanuki 1233, Takasaki, Gunma, 370-1292 Japan

**Takashi Murakami**
The University of Tokyo
Bunkyo-ku, Tokyo, 113-8654 Japan

## Abstract

Many kinds of electromagnetic waves and particle beams are used as analytical probes for the investigation of uranium-mineralogy and speciation. The representatives of the probes are SEM, TEM, μ-PIXE, XPS and XAFS. Each of these analytical probes has its own advantages and disadvantages. We have performed studies to compare the advantages and disadvantages on the basis of the analysis of the distribution and speciation of U containing rock samples collected at the Koongarra uranium deposit, Australia. The spatial resolutions of the probes are in the order of TEM>SEM, μ-PIXE>XFAS, XPS. The lower detection limits of the probes are in the order of μ-PIXE<TEM, SEM<XFAS, XPS. Oxidation state of U was determined by XFAS and XPS. These results indicate that combination of the above probes can give us sufficient data on U speciation not only in natural rock samples but in solid samples obtained in the laboratory.

---

* Present address:  Kumamoto University
                    Kurokami 2-39-1, Kumamoto
                    Kumamoto, 860-8555 Japan

## Introduction

The two important oxidation states for uranium (U) in environment are IV and VI. The U(IV) is essentially insoluble in mildly acidic to alkaline ground waters, whereas U(IV) solubility is commonly controlled by uraninite or coffinite. On the other hand, the U(VI) is potentially much more mobile, due in part to the higher solubilities of most U(VI) minerals[1]. The mobility of dissolved U in natural water is affected by processes such as adsorption or desorption of U ions and precipitation or dissolution of U-bearing minerals. Thus, oxidation states and chemical forms of U sorbed on minerals and/or U bearing minerals should be clarified to understand the behaviour of U in environment.

The electron technology was introduced into mineralogical science in the 60s and early 70s. By employing this technique, mineralogical analysis, using electron probe microanalysis (EPMA), scanning electron microscopy (SEM) and transmission electron microscopy (TEM) have been developed. Today SEM and EPMA are traditional methods to determine U concentrations and chemical forms of U-minerals from nm to $\mu$m scale in rock samples.

Application of high-energy ion beams also goes back to the early 70s. Particle induced X-ray emission (PIXE) is one of the possible methods detecting the elements in the level of ppm [2]. If we use the probe beams of $\mu$m order or less in diameter, U distribution in a rock sample can be determined with such the spatial resolution. A light ion micro beam system with the spatial resolution of less than 1 $\mu$m was constructed on a beam line of 3 MV single ended accelerator in TIARA facility of Takasaki Establishment. And a micro-PIXE ($\mu$-PIXE) analysing system was developed on TIARA facility for various elemental analyses with a sub-micron level spatial resolution [3]. Thus, mapping of U of ppm level is available by $\mu$-PIXE.

X-ray absorption fine structure (XAFS) developed in the 80s is a powerful method for structure and electronic state analyses, which can be used not only for crystalline but also non-crystalline materials under various environments. Fluorescent XAFS has higher sensitivity, thus it is used for the materials containing low elements concentration. XAFS and X-ray photoelectron spectroscopy (XPS) were used to analyse oxidation states and chemical forms of U in the U mineral samples [4].

Using above probes, we have analysed the mineralogy, distribution and speciation of U-containing rock samples collected at the Koongarra U deposit, Australia, and have compared in terms of space resolution, detection limit in concentration, and empirical formula determination.

## Experimental

Rock samples collected at the Koongarra U deposit, Australia [5] were treated in two different ways; as thin sections and powder. Petrological thin sections of about 30 $\mu$m thick were made. The thin sections were initially checked by optical microscopy (OM), then were provided for the analyses by SEM, and EPMA. Cores of 3 mm in diameter were drilled in interesting region of some of the thin sections. The ultramicrotomy and ion-thinning methods were critical for the examinations by TEM, and analytical electron microscopy (AEM).

Cores of 6 mm in diameter were drilled in interesting region of some of the thin sections. Proton beams with the energy of 2.5 MeV from the single-ended machine was used for $\mu$-PIXE analysis. Beam spot was approximately 1 $\mu$m in diameter. X-ray detection system was similar to that used in EPMA analysis.

A powder sample of U minerals installed in an indium film was provided for XAFS and XPS measurements. Co-precipitates of uranium and iron(III) and cupper(II) oxides/hydroxides were also provided for the XPS measurement.

$L_{III}$ edge X-ray absorption spectra of U were measured in transmission mode and fluorescent mode at the BL-27 station of Photon Factory (PF) at the High Energy Accelerator Research Organisation (KEK). Synchrotron radiation from a storage ring operated at 2.5 GeV or 3.0 GeV was monochromatised with a Si(111) double crystal monochromator. XPS measurement was carried out at the same station in PF with 2 104 eV monochromatised radiation.

## Results and discussion

### SEM and TEM analysis

Back scattering electron image (BEI) of apatite, which is an accessory mineral in the host rock at Koongarra deposit shows that apatite in the sample of DDH 65-97 is surrounded with U-containing minerals (brightest area in Figure 1) [6].

**Figure 1. Back scattering image micrograph of apatite (A) surrounded with U grain (SL) [6]**

EPMA analysis of the U mineral shows that chemical component of the U mineral is expressed by $Mg(UO_2)_2(PO_4)_2 \cdot 10H_2O$. The mineral expressed by the formula is saleeite. Closer examination reveals that apatite grains are partly replaced by saleeite [7]. The cavities in the apatite grains were observed indicating that some of apatite was dissolved, and the dissolved constituents were transported downstream. Dissolution of apatite and formation of saleeite are expressed by:

$$Ca_5(PO_4)_3F + H^+ + 2UO_2^{2+} + Mg^{2+} + 10 H_2O = Mg(UO_2)_2(PO_4)_2 \cdot 10H_2O + HF + 5Ca^{2+} + PO_4^{3-}$$

Interestingly, though apatite contains calcium (Ca), the replacement of apatite by autunite $(Ca(UO_2)_2(PO_4)_2)$ was not observed. The size of saleeite aggregates was 20-100 μm.

The OM micrograph of the polished thin section from DDH 60-65 shows that U is associated with iron (Fe) mineral, which occurs midstream from the uranium deposit. EPMA analysis shows that Fe mineral contains uranium of about several wt.%. However, U mineral was not determined by EPMA. Constituents of the Fe mineral veins in sample DDH4-99, where we found U associated with phosphorous (P) by SEM-EDX, were examined by TEM. One of the U-bearing minerals was removed from the thin section and examined by TEM and AEM. The predominant crystalline phases in the Fe mineral veins are goethite and hematite, since most of the diffraction spots and rings can be attributed to these two minerals. Ferrihydrite, which is difficult to be identified using TEM or AEM, may coexist with goethite and hematite in the samples. A TEM micrograph indicates that small domains of goethite and hematite (mostly in the range 2-50 nm) are randomly oriented (Figure 2) [7].

**Figure 2. TEM micrograph of Fe minerals containing U minerals [7]**

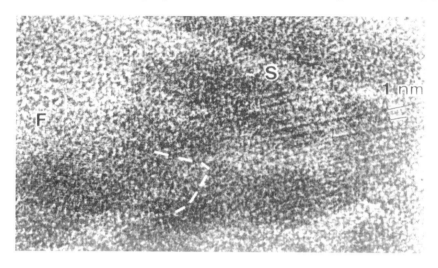

AEM indicates that U coexists with P, Mg and Cu. Phosphorous, Mg and Cu are probably derived from dissolution of apatite, chlorite and chalcopyrite of the host rock, respectively. The lattice fringes with spacing 1 nm was observed in Figure 2. AEM spectrum did not give any formula of the U minerals, because size of U-bearing minerals is quite smaller than the detection area of AEM. However, the lattice fringe indicates that the mineral is saleeite of approximately 20 nm in size.

## μ-PIXE analysis

Figure 3 shows the distribution of uranium and phosphate obtained by μ-PIXE in the thin section of the rock samples of DDH 65-97 (same position shown in Figure 1).

**Figure 3. U (left) and P (right) distributions obtained by μ-PIXE in and around apatite. Brighter spot shows higher concentration of elements.**

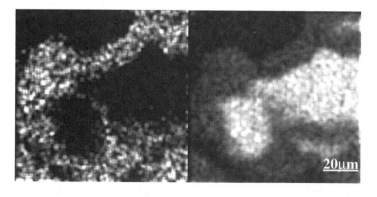

The brightest area in the right Figure indicates primary apatite. The outer rim around the primary apatite is darker than that in the primary apatite. Uranium is not present in the primary apatite, but in the rim (left figure). The primary apatite observed in the thin section from DDH 65-100, which is located slightly deeper position than DDH 65-97, does not contain any uranium. These indicate that some phosphate dissolved from the primary apatite and some uranium migrated from the primary ore react at the rim of the primary apatite to be mineralised to form saleeite. Space resolution by μ-PIXE was around 1 μm in the figure.

## XAFS and XPS analysis

The energy of the U $L_{III}$ absorption edge of uranyl minerals is higher than that of uraninite. The structure of near edge spectra of U is also dependent on the oxidation state of U(IV and VI). Minimum U content of the sample for the near edge spectra was approximately 0.1 wt.% by a fluorescent XAFS. The peak for U in the uraninite sample of ca. 0.1 mm in diameter installed in the indium foil was not detected by a transmission mode, but by a fluorescent XFAS.

XPS and XAFS analyses for U minerals show that absorption peak positions differ between $UO_2$ and $UO_3 \cdot nH_2O$. And chemical shift of U(VI) peak of $U4f_{5/2}$ and $U4f_{7/2}$ from $UO_3 \cdot nH_2O$ to metatorbernite was higher than that from uraninite to $UO_3 \cdot nH_2O$. These indicate that XPS and XAFS are effective to determine not only oxidation state of U but empirical formula of U minerals.

## Comparison among the probes

The comparison among the probes in terms of space resolution, sample preparation, detection limit and empirical formula determination are shown in Table 1. The spatial resolutions of the probes are in the order of TEM > SEM, μ-PIXE > XFAS, XPS. The lower detection limits of the probes are in the order of μ-PIXE < TEM, SEM < XFAS, XPS. Samples for XAFS and XPS are more easily prepared than for μ-PIXE, SEM and TEM. Oxidation state of U was not determined by SEM and TEM, but by XFAS and XPS.

**Table 1. Comparison between SEM, TEM, μ-PIXE, XFAS and XPS**

| Method | Resolution | Detection limit | Chemical formula | Oxidation state |
|--------|-----------|-----------------|------------------|-----------------|
| SEM | μm | 0.1 wt.% | Yes | No |
| TEM | nm | 1 wt.% | Yes | No |
| μ-PIXE | μm | 1 ppm | Yes | No |
| XFAS | mm | 1 wt.% | Yes | Yes |
| XPS | mm | 1 wt.% | Yes | Yes |

These results indicate that combination of the above probes can give us sufficient data on U speciation in not only in natural rock samples, but in solid samples obtained in the laboratory.

# REFERENCES

[1] J.K. Osmond and M. Ivanovich, in "Uranium Series Disequilibrium: Application to Earth, Marine and Environmental Science", M. Ivanovich and R.S. Harman, eds., pp. 259-289 (2nd edition), Oxford University Press, Oxford (1992).

[2] S. Sueno, *Eur. J. Mineral*, pp. 1273-1297 (1995).

[3] T. Kamiya, T. Suda and R. Tanaka, *Nuclear Inst. Method*, B 118 447-450 (1996).

[4] M.C. Duff, *et al.*, *Geochim. Cosmochim. Acta*, Vol. 61, pp. 73-81 (1997).

[5] T. Murakami, *et al.*, ARAP Final Report Vol. 9, OECD/NEA, Paris (1992).

[6] H. Isobe, T. Murakami, R. Ewing, *J. Nucl. Mat.*, 190, pp. 174-187 (1992).

[7] T. Murakami, T. Ohnuki, H. Isobe, T. Sato, *American Mineralogist*, 82, 888-899 (1997).

# POSTER SESSION

## *Part D: Methods for Redox Speciation*

# ELECTROANALYTICAL DATA ON URANIUM, NEPTUNIUM AND PLUTONIUM IONS IN ACIDIC AQUEOUS SOLUTIONS

**Sorin Kihara, Koji Maeda, Yumi Yoshida**
Department of Chemistry
Kyoto Institute of Technology
Matsugasaki, Sakyou-ku, Kyoto 606-8585, Japan

**Zenko Yoshida, Hisao Aoyagi, Yoshihiro Kitatsuji**
Advanced Science Research Centre
Japan Atomic Energy Research Institute
Tokai, Ibaraki 319-1195, Japan

**Osamu Shirai**
Department of Nuclear Energy System
Japan Atomic Energy Research Institute
Oarai, Ibaraki, Japan

## Abstract

The $E^0$s and mechanisms of the redox processes of U, Np or Pu ions in acidic aqueous solutions were evaluated taking into account the results obtained by applying modern solution chemical theories and electrochemical techniques. It has been pointed out that $E^0$s, which had been widely accepted, included considerable ambiguity arising mainly from the inaccurate activity corrections. The $E^0$s of reversible redox processes, $MO_2^{2+}/MO_2^+$ and $M^{4+}/M^{3+}$ (M: U, Np or Pu) adopted are those proposed mainly by Riglet, *et al.* on the basis of the precise correction of formal potentials, $E^{0'}$s, according to the improved theoretical approach to estimate the activity coefficient. The process of irreversible reaction of $MO_2^+$ to $M^{4+}$ has been estimated not to be simple, and new species have been proposed as the intermediates in the reaction.

## Introduction

Uranium, neptunium and plutonium are characterised by the progressive filling of the 5f electron subshells, which are more shielded than the 4f subshells of the lanthanides. The 5f and 6d electrons in these actinides are at similar energy levels because of the stabilisation of 5f electrons, compared to that of 6d electrons, with increasing atomic number. Hence, 6d electrons are still involved in chemical bonding. Because of these features of U, Np and Pu, there are various oxidation states of the elements. Five oxidation state (3+ to 6+ and sometimes 7+) have been identified for these elements in aqueous solutions. Moreover, these oxidation states are liable to change through the redox and disproportionation reactions. Deep understanding of redox behaviour of U, Np and Pu ions is one of the most important chemical subjects in the field of nuclear technology.

In the present paper, the standard potentials of U, Np and Pu ions in the acidic aqueous solutions published so far are evaluated mainly from the view point of the precise correction of formal potentials for activity coefficients by referring to the work of Riglet, *et al.* [1,2]. Electrode processes of U, Np and Pu ions in the acidic aqueous solutions are discussed by consulting the current-potential curves obtained using flow coulometry [3].

## Standard redox potential $E^0$ for U, Np and Pu ions in acidic aqueous solutions

A classic reference of $E^0$s for actinides in aqueous solutions is Latimer's *Oxidation Potentials* [4] in which most of $E^0$s were estimated from thermodynamic data. Since, in the work of Latimer, not only was the critical evaluation of the proposed data not given by taking into account, e.g. formal potentials, $E^{0'}$s, determined electrochemically; but also many thermodynamic data used were estimated from those determined for elements other than actinides, such as lanthanides. Many parts of Latimer's data have been rendered obsolete through modern publications [5,6]. The $E^0$ for U, Np or Pu redox couples was revised after Latimer's work by applying new theories for prediction or evaluation of thermodynamic data, new techniques to determine $E^{0'}$s, and a new theory for the correction of activity, which is inevitable for estimation of $E^0$ from $E^{0'}$ obtained by direct electrochemical measurements. Though the $E^0$s given in the literature were evaluated by taking into account both the thermodynamic data and $E^{0'}$s determined electrochemically, the $E^{0'}$s were related to $E^0$s by correcting for activity coefficients with the aid of empirical values or inadequate equations, without detailed discussion. Therefore, further discussion is required on the relation between $E^0$ and $E^{0'}$. In the following, focusing our attention on the $MO_2^{2+}/MO_2^+$ and $M^{4+}/M^{3+}$ (M; U, Np, Pu) redox couples which are especially important in the chemistry of M in acidic solutions, $E^0$s are evaluated by extrapolating $E^{0'}$s determined by electrochemical methods to the state of zero ionic strength ($\mu = 0$) referring mainly to the works by Riglet, *et al.* [1,2].

Rigler, *et al.* studied the $MO_2^{2+}/MO_2^+$ and $M^{4+}/M^{3+}$ couples polarographically or voltammetrically in acidic perchlorate solutions of various ionic strengths, and estimated $E^0$s of these couples at $\mu = 0$ by correcting the $E^{0'}$s obtained for activity coefficients based on the Brønsted-Guggenheim-Scatchard specific ionic interaction theory (SIT) [7]. According to the SIT, the activity, $\gamma$ of an ion, $i$, of charge, $z_i$, is given as:

$$\log \gamma_i = -z_i^2 D + \sum_j \epsilon(i,j) m_j \tag{1}$$

where D (Debye-Hückel term) $= 0.5107\mu^{1/2}/(1+1.5\mu^{1/2})$, $\epsilon(i,j)$; specific interaction coefficient between i and all the ions, *j*, of opposite charge, and $m_j$; molalities of *j*. The formal potential ($E^{0'}$) of the redox system of Eq. (2), which is measured in a $ClO_4^-$ solution, can be connected to the $E^0$ by using $\gamma_i$ and Eq. (3).

$$Ox + ne\text{-} \rightarrow Red \tag{2}$$

$$E^{0'} = E^0 + A\,[-\Delta z^2 D + \Delta\in m_{ClO4\text{-}}] \tag{3}$$

where $A = RT/nF(\log e)$, $\Delta\in\, =\, \in (Ox, ClO_4^-) - \in (Red, ClO_4^-)$, and $\Delta z^2 = z_{Ox}^2 - z_{Red}^2$.

It was confirmed that many of the existing experimental determinations of $E^0$'s of $MO_2^{2+}/MO_2^+$ or $M^{4+}/M^{3+}$ couples could be well described by the SIT correction with constant interaction coefficients. The $E^0$s of $MO_2^{2+}/MO_2^+$ and $M^{4+}/M^{3+}$ couples estimated at $\mu = 0$ by Riglet, *et al.* are listed in Table 1 and $\in$ values employed are summarised in Table 2.

## Redox reactions of U, Np and Pu in acidic aqueous solutions investigated by flow coulometry

A quantitative electrolysis can be achieved very rapidly (e.g. within 10 s) with small over voltage by using a column electrode [3], even if the electrode reaction is totally irreversible, since surface area of the working electrode of the column electrode is very large as for the solution volume in the column. Because of this unmatched advantage, the electrolysis with the column electrode in a flow system, which is called flow coulometry, is very useful for preparation of ions of the desired oxidation state as well as the rapid determination or collection of various metals. This technique is especially favourable for the preparation of unstable species for the subsequent investigation of their redox processes. The current-potential curve observed by flow coulometry is called couloupotentiogram. Flow coulometry has been applied by the authors in order to elucidate the overall redox behaviour of the actinide ions in various kinds of acidic media. Coulopotentiograms of M were measured by using multi-step column electrodes, in which 2 or 3 column electrodes were connected in series. The coulopotentiograms in perchloric acid solution are presented in Figure 1. In the figure, the number of electrons involved in the redox reaction, n, converted from the current is plotted on the ordinate.

Curves 1(U) in Figure 1 is coulopotentiogram for the reduction of $UO_2^{2+}$ which was recorded at the 1st column (CE1) of the system composed of two-step column electrodes by forcing a solution containing $UO_2^{2+}$ into the system at a constant flow rate, *f*, and scanning the potential applied to the working electrode at a constant rate, *v*. The limiting currents of curves 1(U) suggested that these curves corresponded to two-electron reduction of $UO_2^{2+}$ to U(IV). The further reduction could not be observed due to the hydrogen evolution. Curve 2(U) was recorded at the 2nd column (CE2) of the system composed of the two-step column electrodes by introducing the reduction product of $UO_2^{2+}$ at CE1, i.e. U(IV), obtained at a potential in the range available for the limiting currents in curve 1(U). The limiting currents in curve 2(U) suggested a two-electron oxidation of U(IV). The oxidation waves appeared at potentials much more positive than those for the reduction of $UO_2^{2+}$. This behaviour indicated that the redox process between $UO_2^{2+}$ and U(IV) is irreversible.

General aspects of the reduction behaviour of $NpO_2^{2+}$ and $PuO_2^{2+}$ are similar to each other. A one-electron wave for the reduction of $MO_2^{2+}$ to M(V) (M: Np or Pu ) and a two-electron wave for the further reduction of M(V) to M(III) were observed in coulopotentiograms 1(Np) and 1(Pu). The dependencies of $E_{1/2}$s on f and the logarithmic analyses of the coulopotentiograms suggested that the process of the one-electron reduction of $MO_2^{2+}$ was reversible and that of the two-electron reduction of M(V) was irreversible.

The oxidation of M(V) was investigated by introducing the ions prepared at CE1 into CE2 as shown by curves 2(Np) and 2(Pu). The $E_{1/2}$ for the oxidation of M(V) was almost the same as that for the reduction of $MO_2^{2+}$. This agreement of the $E_{1/2}$s and the results of the logarithmic analyses of the oxidation waves confirmed that the redox process of $MO_2^{2+}/M(V)$ was reversible. The oxidation of M(III) was investigated by introducing the three-electrons reduction product of $MO_2^{2+}$ at CE1 into CE2 as shown by curve 3(Np) and 3(Pu). The waves corresponding to the one-electron oxidation of M(III) to M(IV) were reversible. Further oxidation of Np(IV) to Np(VI) was observed in the coulopotentiogram recorded in perchloric acid solution. The reduction of M(IV) which had been prepared by oxidising $M^{3+}$ at CE2 was investigated by using CE3 of the three-step column electrodes system. The results are given by curves 4(Np) and 4(Pu). The one-electron reduction wave whose $E_{1/2}$ is identical with that for the oxidation of $M^{3+}$ to M(IV) suggested the reversible M(IV)/$M^{3+}$ process.

The effect of $f$ on the coulopotentiogram was elucidated. The $E_{1/2}$s for redox processes of $NpO_2^{2+} \leftrightarrows Np(V)$, $Np(IV) \leftrightarrows Np^{3+}$, $PuO_2^{2+} \leftrightarrows Pu(V)$, $Pu(IV) \leftrightarrows Pu^{3+}$ are independent of $f$, which confirms the reversible nature of these processes. The $E_{1/2}$s for processes of $UO_2^{2+} \rightarrow U^{4+}$, $Np(V) \rightarrow Np^{3+}$ and $Pu(V) \rightarrow Pu^{3+}$ depended strongly on $f$.

The effect of the hydrogen ion concentration on the redox processes was examined and it was found that the $E_{1/2}$s for redox processes of $NpO_2^{2+} \leftrightarrows Np(V)$, $Np(IV) \leftrightarrows Np^{3+}$, $PuO_2^{2+} \leftrightarrows Pu(V)$, $Pu(IV) \leftrightarrows Pu^{3+}$ were independent of the hydrogen ion concentration, which indicated that species indicated as Np(V), Np(IV), Pu(V) and Pu(IV) in the above discussion are $NpO_2^+$, $Np^{4+}$, $PuO_2^+$ and $Pu^{4+}$, respectively.

The electrochemical characteristic of GC fibre electrode depends strongly on the condition under which the fibre was prepared, and fibres prepared in different years often have different characteristics even though they were commercially available as those of the same trade names. If GC fibres of different characteristics are used properly, more detailed features of electrode reactions can be elucidated. Coulopotentiograms for the redox of Np ions in 1 M $HClO_4$ are shown in Figure 2. The electron transfer at this GC fibre electrode was relatively slow compared with that at the GC fibre used for Figure 1, and hence, the large over potential was required to observe an irreversible electrode reaction.

Two waves were observed in the coulopotentiogram for the reduction of $NpO_2^+$ recorded at the relatively high $f$ (curve 1). This coulopotentiogram was recorded at CE2 by introducing $NpO_2^+$ prepared at CE1. The two waves are attributable to the successive reductions; $NpO_2^+ \rightarrow Np(IV) \rightarrow Np(III)$, since limiting currents of both waves correspond to n = 1. When the one-electron reduction product of $NpO_2^+$ prepared at CE2 of -0.3 V vs SSE, Np(IV), was introduced into CE3, and the oxidation behaviour was investigated at CE3, an oxidation wave was observed with $E_{1/2}$ of the same as that observed in the coulopotentiogram for the oxidation of $Np^{3+}$ [curve 3(Np)]. However, the magnitude of the limiting current correspond to n = 0.45 to 0.48 indicating that the reduction product at CE2, Np(IV), was converted to $Np^{3+}$ and Np(V) almost quantitatively by the disproportionate reaction of Eq. (4) before entering into CE3.

$$2Np(IV) \rightarrow Np^{3+} + Np(V)^* \tag{4}$$

The one-electron reduction product at CE2 of -0.30 V, Np(IV), is considered to be a different species from $Np^{4+}$, taking into account that the disproportion of $Np^{4+}$ was confirmed to be slow. The reaction at potentials of the 1st wave of curve 1 in Figure 2 might be:

$$NpO_2^+ + 2H^+ + e^- \rightarrow NpO^{2+} + H_2O \tag{5}$$

292

The following consideration indicates that the $Np(V)^*$ species produced by Eq. (4), is also different from the commonly accepted form, $NpO_2^+$. The limiting current of the 1st wave of coulopotentiogram 1 in Figure 2 corresponded to n = 1. The current should be that of more than n = 1.5, however, if Np(V) is $NpO_2^+$, since $NpO_2^+$ can be reduced at -0.30 V and a half of Np(IV) must be converted to Np(V) by the quantitative reaction of Eq. (4). The species of $Np(V)^*$ has not yet been identified.

The n value for the limiting current of the 1st wave increased with a decrease of $f$, and n was 2 when $f$ was sufficiently low as 0.16 ml min$^{-1}$ (cf. curves 2 and 3 in Figure 2). When $f$ is low, the unstable $Np(V)^*$ produced in CE2 might be converted chemically into $NpO_2^+$ during its stay in CE2, and the current due to the reduction of $NpO_2^+$ might be added to the current. When the oxidation behaviour of the two-electron reduction product prepared at CE2 of -0.55 V, Np(III), was investigated at CE3, an oxidation wave identical with curve 3(Np) in Figure 1 for the oxidation of $Np^{3+}$ to $Np^{4+}$ was observed. The limiting current of the oxidation wave corresponded to n = 1.0 indicating that $NpO_2^+$ was reduced to $Np^{3+}$ quantitatively at CE2 of -0.55 V. The 2nd wave of curve 1 in Figure 2 observed at high $f$ can be explained by considering the two-electron reduction of $Np(V)^*$, of which concentration is half of that of $NpO_2^+$ introduced into CE2.

Though two-step reduction of $PuO_2^+$ has not been observed in acidic media, such species as $PuO^{2+}$ and $Pu(V)^*$, similar to $NpO^{2+}$ and $Np(V)^*$, are expected to participate in the reduction of $PuO_2^+$ to $Pu^{3+}$.

## Conclusion

It is hard to say that the redox behaviour of the U, Np and Pu ions have been understood fully even in the most widely investigated solutions, i.e. acidic aqueous solutions. The understanding of the behaviour is still less in weakly acidic or neutral solutions that might be important in the field of the underground disposal of the nuclear wastes. Therefore, the further extensive investigations leading to the more detail understanding of redox behaviours are required for the safety development of nuclear technology.

# REFERENCES

[1]  Ch. Riglet, P. Vitorge, I. Grenthe, *Inorg. Chim. Acta*, 133, 323-329 (1987).

[2]  Ch. Riglet, P. Robouch, P. Vitorge, *Radiochim. Acta*, 46, 85-94 (1989).

[3]  T. Fujinaga, S. Kihara, *CRC Critical Rev. Anal. Chem.*, 6, 223-254, (1977).

[4]  W.M. Latimer, "Oxidation Potentials", 2nd ed., Prentice-Hall, Englewood Cliffs, NJ (1952).

[5]  L. Martinot, J. Fuger, in *Standard Potentials in Aqueous Solution*, A.J. Bard, R. Parsons and J. Jordan, eds., p. 631, Marcel Dekker, New York (1985).

[6]  L.R. Morss, in *The Chemistry of the Actinide Elements*, J.J. Katz, G.T. Seaborg, L.R. Morss, eds., Vol. 2, p. 1278, Chapman and Hall, London (1986).

[7]  K.S. Pitzer, in *Activity Coefficients in Electrolyte Solutions*, R.M. Pytkowicz, ed., Vol. 1, p. 157, CRC, Boca Raton (1979).

**Table 1. $E^0$ for reversible redox process; $MO_2^{2+}/MO_2^+$ and $M^{4+}/M^{3+}$ recommended by Riglet, *et al.* [1,2]**

| Redox pair | $E^0$ (V *vs.* NHE) |
|---|---|
| $UO_2^{2+}/UO_2^+$ | $+0.089\pm0.002$ |
| $NpO_2^{2+}/NpO_2^+$ | $+1.161\pm0.008$ |
| $PuO_2^{2+}/PuO_2^+$ | $+0.954\pm0.010$ |
| $U^{4+}/U^{3+}$ | $-0.573\pm0.017$ |
| $Np^{4+}/Np^{3+}$ | $+0.218\pm0.005$ |
| $Pu^{4+}/Pu^{3+}$ | $+1.026\pm0.010$ |

**Table 2. Specific interaction coefficients; $\in(i,j)$ employed**

| i | j | $\varepsilon$ (kg/mole) |
|---|---|---|
| $MO_2^{2+}$ | $ClO_4^-$ | $0.46\pm0.02$ |
| $MO_2^+$ | $ClO_4^-$ | $0.28\pm0.04$ |
| $M^{4+}$ | $ClO_4^-$ | $0.92\pm0.24$ |
| $M^{3+}$ | $ClO_4^-$ | $0.49\pm0.04$ |

M = U, Np, Pu; $\epsilon(i,j)$ is identical for the ion of the same oxidation state. $\epsilon(M^{3+},ClO_4^-)$ is assumed to be identical to $\epsilon(Y^{3+},ClO_4^-)$; recommended by Riglet, *et al.* [1,2].

# Figure 1. Coulopotentiograms for U, Np and Pu ions in 1 M HClO₄

*Reduction of $UO_2^{2+}$, $NpO_2^{2+}$ or $PuO_2^{2+}$ – curve 1(U), 1(Np) or 1(Pu)*
*Oxidation of U(IV), Np(V) or Pu(V) – curve 2(U), 2(Np) or 2(Pu)*
*Oxidation of Np(III) or Pu(III) – curve 3(Np) or 3(Pu)*
*Reduction of Np(IV) or Pu(IV) – curve 4(Np) or 4(Pu)*
*Residual current – curve 5*
*Flow rate 1.5 ml min⁻¹, scan rate 0.2 mV s⁻¹*

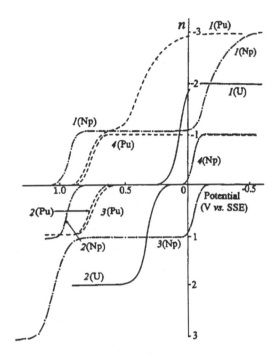

# Figure 2. Coulopotentiograms for reduction of 10⁻³ M anpO₂⁺ in 1 M HClO₄

*Flow rate 0.52 ml min⁻¹ – curve 1*
*Flow rate 0.35 ml min⁻¹ – curve 2*
*Flow rate 0.16 ml min⁻¹ – curve 3*
*Scan rate 0.2 mV s⁻¹*

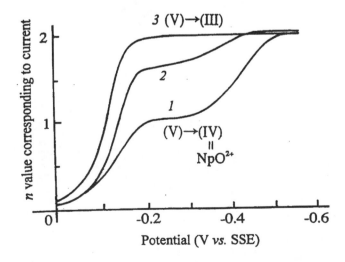

# POLAROGRAPHIC VALIDATION OF CHEMICAL SPECIATION MODELS

**John R. Duffield**
Department of Chemical and Physical Sciences
University of the West of England
Frenchay Campus, Coldharbour Lane, Bristol BS16 1QY, UK

**Jason A. Jarratt**
Centre for Advanced Micro-Analytical Systems
Faculty of Science, Technology and Design
University of Luton
Park Square, Luton, LU1 3JU, UK

## Abstract

It is well established that the chemical speciation of an element in a given matrix, or system of matrices, is of fundamental importance in controlling the transport behaviour of the element. Therefore, to accurately understand and predict the transport of elements and compounds in the environment it is a requirement that both the identities and concentrations of trace element physico-chemical forms can be ascertained. These twin requirements present the analytical scientist with considerable challenges given the labile equilibria, the range of time scales (from nanoseconds to years) and the range of concentrations (ultra-trace to macro) that may be involved. As a result of this analytical variability, chemical equilibrium modelling has become recognised as an important predictive tool in chemical speciation analysis. However, this technique requires firm underpinning by the use of complementary experimental techniques for the validation of the predictions made. The work reported here has been undertaken with the primary aim of investigating possible methodologies that can be used for the validation of chemical speciation models. However, in approaching this aim, direct chemical speciation analyses have been made in their own right.

Results will be reported and analysed for the iron(II)/iron(III)-citrate proton system (pH 2 to 10; total $[Fe] = 3$ mmol dm$^{-3}$; total $[\text{citrate}^{3-}] = 10$ mmol dm$^{-3}$) in which equilibrium constants have been determined using glass electrode potentiometry, speciation is predicted using the PHREEQE computer code, and validation of predictions is achieved by determination of iron complexation and redox state with associated concentrations.

# SEPARATION AND DETERMINATION OF URANIUM(IV,VI) BY HIGH-PERFORMANCE EXTRACTION CHROMATOGRAPHY WITH 2-THENOYLTRIFLUOROACETONE AND CITRIC ACID

**Hisanori Imura, Masayuki Michiguchi, Kiyoshi Nakano, Kousaburo Ohashi**
Department of Environmental Sciences, Faculty of Science
Ibaraki University
Mito, 310-8512 Japan

**Hisao Aoyagi, Zenko Yoshida**
Advanced Science Research Centre, JAERI

## Abstract

High-performance extraction chromatography (HPEC) for uranium(IV) and (VI) has been developed using a chelate extraction system which consists of 2-thenoyltrifluoroacetone (Htta) as an extractant and citric acid as a masking agent in the aqueous mobile phase. Octadecyl-bonded silica gel (Shiseido UG-120, 5 μm) was used as a stationary phase. Uranium(IV) and (VI) were detected by post-column derivatisation with Arsenazo III with absorbance measured at 655-665 nm. The elution order of both uranium species could be changed by adjusting pH between 2 and 3. Under the optimum conditions, i.e. 0.003 M Htta, 0.2 M citric acid, pH 2.2, and the column temperature at 40°C, the complete separation of uranium species was achieved with the resolution of 2.5. The detection limit was $1.6 \times 10^{-7}$ M for uranium(IV) and $2.4 \times 10^{c}$ M for uranium(VI) at 665 nm (S/N = 3). The present method was combined with solid phase extraction as a pre-separation method and applied to the determination of total uranium in fertiliser samples.

## Introduction

Chemical speciation of an element with different oxidation states is important because the chemical behaviour and biological activity of each species are quite different in environment. Solvent extraction is a powerful separation method of metal ions, and is applicable to the separation of the species with different oxidation states if the redox reaction does not occur during the extraction procedure. Recently, extraction chromatography, which is basically a multistage solvent extraction [1], has been greatly improved by using a HPLC technique and a reversed-phase column with or without organic solvents as a stationary phase and can be called high-performance (pressure) extraction chromatography (HPEC) [2]. In a most recent paper [3], separation of metal ions including uranium(VI) was investigated by the reversed-phase HPLC using octadecyl-silica gel or polystyrene resin dynamically coated with heterocyclic carboxylic acids in a mobile phase. The method used for the preparation of both phases and probably the partition equilibrium of metal ions are the same as in the present HPEC. The HPLC and HPEC provide the closed system for the separation and determination and are preferable for the speciation of unstable and air-sensitive species. The HPLC technique with an anion exchange column has been applied to the fast separation of uranium(IV) and (VI) formed in the nuclear reaction [4].

In this paper, the separation and determination of uranium(IV) and (VI) were studied by means of the HPEC based on the chelate extraction with 2-thenoyltrifluoroacetone (Htta) as an extractant and citric acid as a masking agent. Octadecyl-bonded silica gel used as a stationary phase is expected to play the role of the organic phase in the liquid-liquid extraction. Optimum conditions for the separation and detection of uranium(IV) and (VI) were examined in detail.

## Experimental

### Materials

A stock solution of uranium(VI) was prepared by heating an appropriate amount of uranyl nitrate (or acetate) with concentrated sulfuric acid to evolve the fumes and dissolving the residue in water, and then standardised by back titration using an EDTA and a thorium standard solution. A uranium(IV) solution was prepared by electrolytic reduction of uranium(VI) at -0.6 V vs. SSE with a column electrode [5].

Htta was purified by vacuum sublimation. Di-2-ethylhexylphosphoric acid (HDEHP) was purified by the precipitation method of the copper(II)-HDEHP complex [6]. Other reagents were of analytical grade and were used as obtained.

### Apparatus

Chromatographic equipment with a Shiseido UG-120 column (particle size 5 μm; 35 × 4.6 mm internal diameter) was of inert type. The pumping system (A) for the mobile phase was a JASCO PU-980S, and the pumping system (B) for the post-column reagent, a Shimadzu LC-10Ai. The column and a mixer (T connector and PTFE tube) for coloration reaction were thermostated at 30-55°C in a column oven. Uranium(IV) and (VI) were detected with a JASCO UV-970S spectrophotometer. A REODYNE 9125 injector with a 20-μl sample loop was connected to the outlet of the column electrode with a PTFE tube.

## HPEC procedure

An aqueous solution containing $5 \times 10^{-4}$ - $3 \times 10^{-3}$ M Htta and 0.1-0.3 M citric acid at pH 2-3 was prepared as a mobile phase and a 30 ml portion was used for the column conditioning. The sample solution of uranium(VI) ($1 \times 10^{-6}$ - $5 \times 10^{-5}$ M) was prepared by diluting the stock solution with the mobile phase and also used for the preparation of uranium(IV). The uranium(IV) solution prepared by the column electrode was directly introduced to the injector with a pump. For post-column derivatisation the effluent was mixed with $1 \times 10^{-4}$ - $4 \times 10^{-4}$ M Arsenazo III in 1-5 M perchloric acid solution with a T connector and the absorbance was measured at 655 or 665 nm.

## Results and discussion

### Separation conditions

Various chromatographic factors such as the Htta and citric acid concentration, pH, the flow rate of the mobile phase, and the column temperature were examined to optimise the separation conditions. Retention times ($t_R$) of uranium(IV) and (VI) were independent of their concentration in the sample solutions, while they were remarkably dependent on the Htta and the citric acid concentration. Figure 1 shows that the $t_R$ values for uranium(IV) and (VI) increase with an increase in the Htta concentration, and the concentration dependency in the former is larger than that in the latter. This behaviour is understandable considering the expected extraction reaction [7], i.e. $U^{4+} + 4Htta \leftrightarrows U(tta)_{4,st}$, $+ 4H^+$ and $UO_2^{2+} + 2Htta \leftrightarrows UO_2(tta)_{2,st} + 2H^+$, where the subscript (st) denotes the stationary phase. The effect of the citric acid concentration on $t_R$ is shown in Figure 2. The $t_R$ values decrease with an increase in the citrate concentration as expected by the following complexation reaction in the mobile phase; $U^{4+} + nL^{3-} \leftrightarrows UL_n^{4-3n}$ ($\beta_1 = 10^{11.53}$, $\beta_2 = 10^{19.46}$) [8]; $UO_2^{2+} + L^{3-} \leftrightarrows UO_2L^-$ ($\beta_1 = 10^{7.17}$) and $UO_2^{2+} + H_mL^{m-3} \leftrightarrows UO_2H_mL^{m-1}$ ($\beta_{HL} = 10^{9.68}$, $\beta_{H2L} = 10^{11.43}$) [9]. The stronger effect of the citric acid concentration on $t_R$ of uranium(IV) is explained by the difference in the magnitude of the formation constants ($\beta$) of citrate complexes, i.e., $\beta(U^{IV}) > \beta(U^{VI})$.

The effect of pH in the mobile phase was quite unique. As shown in Figure 3, $t_R$ of uranium(IV) ($t_{R,IV}$) gradually decreases with a rise in pH whereas that of uranium(VI) ($t_{R,VI}$) increases. These phenomena can be explicated by the extraction and complexation reactions described above. In fact, the similar trends (not the exact values) in the liquid-liquid extraction curves were obtained by calculating the distribution ratio using the extraction constant of uranium(IV) ($K_{ex} = 10^{5.3}$) and (VI) ($K_{ex} = 10^{-2.26}$) [7] and the formation constants [8,9] shown above. Furthermore, in the calculation using the hydrolysis constants of uranium(IV) ($\beta_{11} = 10^{-1.12}$ or $10^{-1.46}$) and (VI) ($\beta_{11} = 10^{-5.72}$) listed in an authentic book [10], it was confirmed that the hydrolysis of those species at pH 2 to 3.5 can be thoroughly neglected. This is attributed to the strong masking effect of citric acid. As a result, the elution order of uranium(IV) and (VI) can be changed by adjusting pH, e.g. $t_{R,IV} > t_{R,VI}$ below pH 2.7 but $t_{R,IV} < t_{R,VI}$ above pH 2.7.

The effect of the flow rate of the mobile phase and the column-temperature on $t_R$ or retention volume ($v_R$) was examined. The $v_R$ value of uranium(IV) somewhat decreased with increase in the flow rate, meaning the lowering of the distribution coefficient, while that of uranium(VI) scarcely varied. The temperature dependence of $t_R$ appeared strongly in uranium(IV); e.g., $t_R = 4.5$ min at 30°C and $t_R = 11$ min at 50°C. On the other hand, the temperature dependence in uranium(VI) was much less than in (IV). These results suggest that the extraction rate of uranium(IV) is rather low in the present HPEC system.

## Optimum conditions

The detection conditions were also optimised by varying the concentration of Arsenazo III and perchloric acid in the post-column reagent and the flow rate. The optimum conditions for the separation and detection were obtained as follows; 0.0030 M Htta, 0.20 M citric acid, pH 2.2, flow rates of 0.9 ml min$^{-1}$ for the mobile phase and 0.1 ml min$^{-1}$ for the post-column reagent containing $1.0 \times 10^{-4}$ M Arsenazo III in 1.0 M perchloric acid, and the column temperature 40°C. The complete separation of both uranium species was attained with the resolution of 2.5. The detection limit was calculated to be $1.6 \times 10^{-7}$ M for uranium(IV) and $2.4 \times 10^{-7}$ M for uranium(VI) at 665 nm on the basis of 3$\sigma$ of the background signals on the chromatogram. These are comparable to the detection limit in ICP-AES. Since the sample size to be injected is only 20 $\mu$l, the detection limit for the amount of uranium(IV) and (VI) is 0.75 and 1.2 ng respectively. Figure 4 shows the typical chromatograms after the electrolytic reduction with the column electrode under the different applied potentials. The reduction of uranium(VI) or the formation of uranium(IV) can be observed as a function of the potential.

## Application

As a simple but practical application, the determination of total uranium in fertiliser samples was made. The effect of foreign ions in the HPEC determination of uranium was examined in the presence of various metal ions commonly found in phosphate rocks and their products. No interference was observed for 3 000-fold aluminium, 850-fold magnesium, 45-fold manganese(II), 15-fold zinc, 3-fold cobalt(II) and nickel, and 2-fold cadmium. Severe interference was found for 8 700-fold iron(III), 200-fold titanium(IV), and 15-fold zirconium. Therefore, a pre-separation method of uranium from those metal ions was developed in this work. The synergistic extraction of uranium(VI) with HDEHP and trioctylphosphine oxide (TOPO) in heptane followed by the back-extraction with citric acid was tested using a solid phase extraction technique. A 10-ml portion of a sample solution in 0.2 M phosphoric acid was loaded onto the solid phase extraction column (Varian Associates, Chem Elut 10 ml), and then 12 ml of a heptane solution containing $5 \times 10^{-3}$ M HDEHP and $5 \times 10^{-3}$ M TOPO passed through the column twice. The eluate was collected, diluted with 1-octanol, and shaken with 15 ml of 0.2 M citric acid to strip uranium(VI). The quantitative recovery of uranium(VI) in this separation was confirmed by the present HPEC. This method was applied to the reference materials of fertiliser (Tokyo, 1982). A 2-g portion of the dried sample was acid-digested with conc. nitric acid, evaporated to dryness, dissolved in 1 M nitric acid, and subjected to the pre-separation followed by the HPEC. The uranium contents determined for fertiliser(A) and (B) were 17.6 and 21.5 $\mu$g g$^{-1}$ respectively and well agreed with those obtained by ICP-AES (at 367 nm), 16.1 and 20.5 $\mu$g g$^{-1}$ respectively.

In conclusion, the HPEC as modern extraction chromatography was developed for the separation and the determination of uranium(IV) and (VI). The method was based on the multistage chelate extraction of the metal ions from the aqueous mobile phase containing an extractant and a masking agent into the non-polar stationary phase. This technique is one of the instrumental methods of extraction separation and is expected to be a promising method especially for oxidation state speciation.

# REFERENCES

[1] T. Braun, G. Ghersini, "Extraction Chromatography", *Akademiai Kiado*, Budapest (1975).

[2] H. Imura, K. Yoshida, K. Ohashi, 20th Rare Earth Research Conference, Monterey (1993).

[3] M.J. Shaw, S.J. Hill, P. Jones, *Anal. Chim. Acta*, 401, 65 (1999).

[4] I. AlMahamid, J.M. Paulus, *Radiochim. Acta*, 48, 39 (1989).

[5] S. Kihara, Z. Yoshida, H. Aoyagi, *Bunseki Kagaku*, 40, 309 (1991).

[6] J.A. Partridge, R.C. Jensen, *J. Inorg. Nucl. Chem.*, 31, 2587 (1969).

[7] A.M. Poskanzer, B.M. Foreman, *J. Inorg. Nucl. Chem.*, 16, 323 (1961).

[8] D. Nebel, G. Urban, *Z. Phys. Chem.*, 233, 73 (1966).

[9] P. Vanura, L. Kuca, *Coll. Czech. Chem. Comm.*, 45, 41 (1980).

[10] E.N. Rizkalla, G.R. Choppin, "Handbook on the Physics and Chemistry of Rare Earths", K.A. Gschneidner, Jr., L. Eyring, G.R. Choppin, G.H. Lander, eds., Elsevier, Amsterdam, 1994, Vol. 18, Chap. 127.

**Figure 1. Effect of the Htta concentration on the retention times of U(IV) and (VI)**

*Mobile phase: 0.20 M citric acid, pH 2.1, flow rate $f_A$ 0.9 ml min⁻¹, $f_B$ 0.1 ml min⁻¹, temperature 50°C*

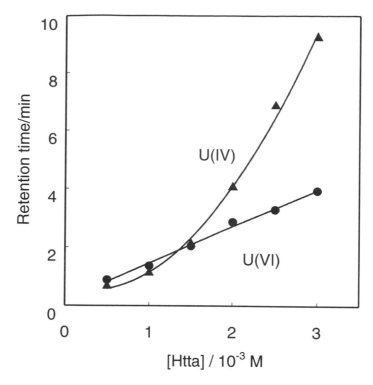

**Figure 2. Effect of the citric acid concentration on the retention times of U(IV) and U(VI)**

*Mobile phase: $3 \times 10^{-3}$ M Htta, pH 2.1, flow rate $f_A$ 0.9 ml min⁻¹, $f_B$ 0.1 ml min⁻¹, temperature 50°C*

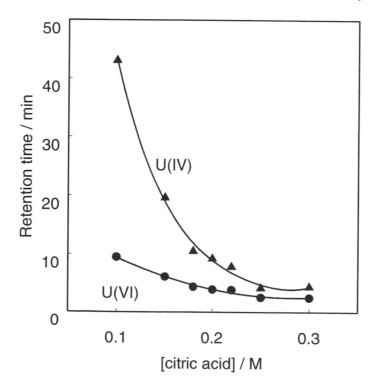

# Figure 3. Effect of pH on the retention times of U(IV) and U(VI)

*Mobile phase: $3 \times 10^{3}$ M Htta, $2.0 \times 10^{-1}$ M citric acid, flow rate $f_A$ 0.9 ml min$^{-1}$, $f_B$ 0.1 ml min$^{-1}$, temperature 50°C*

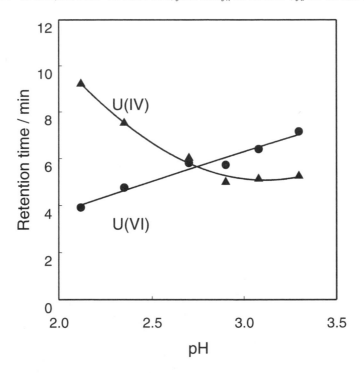

# Figure 4. Chromatograms of U(IV) and U(VI) after electrolytic reduction with the column electrode

*Reducing conditions: $1.0 \times 10^{5}$ M U(VI), $2.0 \times 10^{-1}$ M citric acid, ph 2.15, flow rate 1.25 ml min$^{-1}$, A -0.1 V, B -0.2 V, C -0.3 V (vs. SSE)*

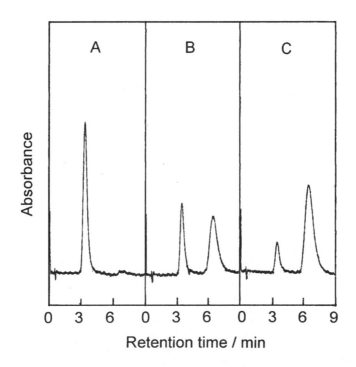

# VOLTAMMETRIC SPECIATION OF U(VI) IN THE AQUEOUS SOLUTION OF pH 2-6 UNDER CO₂ GAS/AQUEOUS SOLUTION/UO₂CO₃ SOLID PHASES EQUILIBRIUM

**Zenko Yoshida, Takaumi Kimura, Yoshiharu Kato,**
**Yosihiro Kitatsuji, Hisao Aoyagi, Yoshihiro Meguro**
Advanced Science Research Centre, Japan Atomic Energy Research Institute
Tokai, Ibaraki 319-1195, Japan
E-mail: zyoshida@popsvr.tokai.jaeri.go.jp

**Sorin Kihara**
Department of Chemistry, Kyoto Institute of Technology
Matsugasaki, Sakyo-ku, Kyoto 606-8585, Japan

## Abstract

Cyclic voltammetry using hanging mercury drop electrode (HMDE) was applied to the study of redox processes of U(VI)-carbonate species dissolved in the aqueous solution under equilibrium with $CO_2$ (1 atm ) gaseous phase and $UO_2CO_3$ solid phase. The reduction of $(1-2) \times 10^{-5}$ M $UO_2CO_3$(aq) in the solution, which corresponds to the solubility in the solution of pH 4.8-5.0, gives a well-defined wave of quasi-reversible nature. The peak potential of the reduction of $UO_2CO_3$(aq) is *ca.* 150 mV more negative than that of the reversible redox between $UO_2^{2+}$ and $UO_2^+$, which is observed in more acidic solution, e.g. of pH = 2.0. The peak current is about 15 times larger than that calculated using the theoretical equation for the diffusion-controlled voltammogram of a reversible redox reaction. It can be concluded that $UO_2CO_3$(aq) is readily adsorbed on the surface of the HMDE and U(VI) in the form of $UO_2CO_3$(ad) is reduced into U(V). The U(V)-species as the reduction product remains adsorbed on the HMDE surface, and is re-oxidised voltammetrically into $UO_2CO_3$(ad). Cyclic voltammetry is feasible for the speciation of the U(VI)-species of rather low concentration.

# Introduction

Voltammetry or polarography is very powerful for the speciation of electroactive species in a solution, since a current-potential curve observed allows simultaneous determination of a kind of the species from the potential and its concentration from the current. A limitation of the voltammetry arises when the concentration of the objective species is low, e.g. $<10^{-6}$ M, or the redox reaction is irreversible.

Voltammetric and polarographic studies of uranium ions in various media have been extensively conducted [1] and a great deal of data of standard redox potentials and the redox reaction mechanisms have been accumulated [1,2]. Most of studies so far conducted were focused on the redox of hydrated uranium ions such as $UO_2^{2+}$, $UO_2^+$, $U^{4+}$ and $U^{3+}$ in the acidic solution, e.g. $> 0.1$ M acid solutions, and on the redox of uranyl carbonate species, such as $UO_2(CO_3)_3^{4-}$ and $UO_2(CO_3)_3^{5-}$ in the alkaline solution, e.g. $0.1-1$ M $Na_2CO_3$ solutions. In these solutions, single species of enough high concentration is dissolved, which enables a measurement of the reproducible current-potential curve and a reliable data analysis. On the contrary, different hydroxide and carbonate complexes are coexisting in the solution in the pH range of 2-6 and, furthermore, the hydroxides and carbonate complexes formed are relatively insoluble in the solution. Few studies have been conducted on the redox reaction of uranyl species in the solution of weakly acidic and neutral regions.

The present work aims at the elucidation of the redox behaviour of the U(VI)-species in the aqueous solution of pH 2-6 by means of cyclic voltammetry and the development of a voltammetric speciation method of U(VI)-species. The objective U(VI)-species are carbonate complexes such as $UO_2CO_3$, $UO_2(CO_3)_2^{2-}$ and $UO_2(CO_3)_3^{4-}$ which are coexisting in the solution under equilibrium of $CO_2$ gas/aqueous solution/$UO_2CO_3$(s) solid system.

# Experimental

Three phases equilibrium system of $CO_2$ gas/aqueous solution/$UO_2CO_3$(s) was prepared by the following procedure. The aqueous solution, 100 ml, of 0.1 M (Na,H)$ClO_4$, pH = 2.0, containing $2 \times 10^{-2}$ M U(VI) was taken in a glass vessel of 200 ml volume. After the addition of a known volume, e.g. 50 ml, of 0.1 M NaOH solution to the vessel, the mixture solution was thermostated at $25.0 \pm 0.05°C$, and moistened $CO_2$ gas was bubbled continuously through the solution at a flow rate of *ca.* 50 ml/min until the equilibrium condition was established. After the measurement of the voltammogram under equilibrium at a certain pH, another aliquot of 0.1 M NaOH solution was added to the solution to increase pH further.

Cyclic voltammograms were measured by immersing hanging mercury drop electrode (HMDE), Ag-AgCl/0.1 M $NaClO_4$ + $1 \times 10^{-2}$ M NaCl (SSE) reference electrode and platinum counter electrode in the aqueous solution of three phases system as described above. All potentials in the present paper are referred to SSE. The concentration of U(VI) in the aqueous solution, which had been filtered by a membrane filter of 0.2 μm pore size, was determined by fluorescence spectrometry using the phosphoric acid medium.

The instruments for the voltammetric measurements were potentio-galvanostats, Hokuto Denko Co., HA-501, a wave function generator, Hokuto Denko Co., HB-105, and an X-Y recorder, Yokogawa Electric Co., type-3036. An electrode device for the preparation of HMDE was Yanagimoto Co. Ltd., type EA-290.

# Results and discussion

## Predictive speciation of U(VI) on the basis of solubility measurement

The solubility of U(VI) in the aqueous solution which was equilibrated with $CO_2$ gas (1 atm) and $UO_2CO_3$(s) was determined [4]. The results are plotted against pH and carbonate concentration in Figure 1(a), together with the calculated concentrations of each species. It was confirmed that the solid phase formed under the present condition was orthorhombic $UO_2CO_3$ [4]. The lowest solubility was determined to be $(1-2) \times 10^{-5}$ M U(VI) when pH of the aqueous solution was around 5. Figure 2(b) presents the relative amount of each species. The main U(VI)-species are found to be $UO_2CO_3$(aq) and $UO_2(CO_3)_3^{4-}$, respectively, in the solution of pH range 4.8-5.0 and > 6.5.

## Cyclic voltammograms of U(VI)-species

It was found that three phases equilibrium of $CO_2$ gas/aqueous solution/$UO_2CO_3$(s) was established rather slowly[4]. When $2 \times 10^{-2}$ M U(VI) solution of pH = 2.0 (100 ml) and 0.1 M NaOH solution (50 ml) were mixed and allowed to stand for a certain period, the voltammograms showed gradual change with standing time ($t_s$). Voltammograms measured at $t_s$ of 40, 65, 140 and >150 hrs are illustrated in Figure 2.

In voltammograms in Figures 2(a) and 2(b) recorded at $t_s$ shorter than 65 hrs, three cathodic peaks I, II and III and an anodic peak IV are observed. The peak currents, $i_p$, of peaks I, II and III decrease remarkably with an increase of $t_s$. It is clear that the concentration of U(VI)-species dissolved in the aqueous solution decreases with an increase of $t_s$ of this time range. The change in $i_p$ of the reduction peaks I and II are summarised in Figure 3 together with the change in peak potential, $E_p$, of these peaks and the change in pH of the solution with $t_s$. The $i_p$ of peaks I and II decrease and $E_p$ shift more negatively with $t_s$ over a period of $ca$. 140 hrs. As shown in voltammogram in Figure 2(d), $i_p$ and $E_p$ are found to remain unchanged at $t_s$ longer than 150 hrs; equilibrium being attained by standing for 150 hrs.

Peak I is assigned to the one-electron reduction of hydrated $UO_2^{2+}$ into $UO_2^+$, since the potential of peak I equals to the peak potential for the reversible reduction of $UO_2^{2+}$ to $UO_2^+$ determined in 0.1 M $(Na,H)ClO_4$ solution of pH = 2.0. Peaks II and III may correspond to the reduction of U(VI)-complexes formed in the aqueous solution.

A remarkable change in pH from 4.3 to 4.8 is observed in the range of $t_s$ from 65 to 120 hrs as shown in Figure 3. This suggests that a transformation of the U(VI)-species occurs during the standing. It can be considered that not only $UO_2CO_3$ but also relatively large portion of U(VI)-hydroxide is formed at first by mixing the solution to obtain the mixture solution of pH = 4.3-4.4, and U(VI)-hydroxide transforms into $UO_2CO_3$ releasing $OH^-$ which results in the increase of pH. Under equilibrium condition after standing for 150 hrs, the U(VI) species in the aqueous solution of pH = 4.8 is predicted to be $UO_2CO_3$(aq) (cf. Figure 1). In fact, the solution was transparent at $t_s$ shorter than 40 hrs, the solution became turbid at $t_s$ around 65 hrs, and then the precipitate of greenish yellow was developed. The colour of the precipitate changed from greenish to orangish yellow with further standing, suggesting the transformation of U(VI)-hydroxide into $UO_2CO_3$ precipitate.

Peak II in voltammograms in Figures 2(a) and 2(b) is attributable to the reduction of U(VI)-hydroxide into U(V)-species. Peak III of small $i_p$ observed at around -0.5 V may be due to the reduction of $UO_2CO_3$(aq) coexisting with U(VI)-hydroxide. Though $E_p$ of anodic peak IV observed in voltammograms in Figures 2(a)-2(c) does not coincide with that of the oxidation of $UO_2^+$ into $UO_2^{2+}$ in the solution of pH 2.0, and peak IV has irreversible characteristics, peak IV is attributed to the re-oxidation of the reduction product, $i. e.$, U(V)-species, into U(VI).

Reproducible voltammograms as shown in Figure 2(d) were obtained when the system had been allowed to stand for 150 hrs or more. A pair of main redox peaks are observed; peak III for the reduction of $UO_2CO_3(aq)$ and peak VI for the oxidation of the reduction product. The concentration of U(VI) in the aqueous solution after filtration was determined to be $2.1 \times 10^{-5}$ M U(VI). The same voltammogram as shown in Figure 2(d) was obtained even when the aqueous solution was filtered by a 0.2 μm membrane filter. This implies that such aggregates as U(VI)-hydroxide or carbonate polymer do not participate in the redox reaction at the HMDE surface.

The $i_p$ of a cyclic voltammogram for a diffusion-controlled redox reaction can be expressed by Eq. (1):

$$i_p = 269 \text{ A } n^{2/3} D^{1/2} C v^{1/2} \tag{1}$$

where A, n, D, C and $v$ denote, respectively, surface area of HMDE (0.0139 cm$^2$), number of electrons involved in the electrode reaction (n = 1), diffusion coefficient ($D_{U(VI)} = 5 \times 10^{-6}$ cm$^2$ s$^{-1}$), concentration ($2.1 \times 10^{-5}$ M) and potential scanning rate (20 mV s$^{-1}$). The $i_p$ calculated on the basis of Eq. (1) is 0.025 μA. The $i_p$ of peak III observed in voltammogram in Figure 2(d) is 0.4 μA which is about 15 times larger than that calculated. This can be explained by assuming the following electrode process; $UO_2CO_3(aq)$ in the solution under solubility equilibrium is readily adsorbed on the surface of HMDE and U(VI) as a form of $UO_2CO_3(ad)$ is reduced into U(V).

In the voltammogram in Figure 2(d), anodic peaks V and VI are observed. Peak VI is attributable to the re-oxidation of the U(V)-species as mentioned above. Since the $i_p$ of peak VI is similar to that of cathodic peak III, it is concluded that the U(V)-species produced by the reduction of $UO_2CO_3(ad)$ still remains adsorbed on the HMDE surface and is rather stable. Peak V may correspond to the oxidation of hydrated $UO_2^+$, which is generated by a dissociation of U(V)-species adsorbed on the HMDE, into U(VI)-carbonate complex.

With increasing pH higher than 5, U(VI)-carbonate complexes such as $UO_2(CO_3)_2^{2-}$ and $UO_2(CO_3)_3^{4-}$ are formed, and the equilibrium concentration of U(VI) in the solution increases with an increase of pH. Cyclic voltammograms corresponding to the reduction of these U(VI)-carbonate species are observed and show the increase of $i_p$ and the negative shift of $E_p$ with an increase of pH.

As described above, cyclic voltammetry is useful for the speciation of uranium species even at relatively low concentration. Further investigations are required to determine the concentration of each species from the peak current in the voltammogram for the electrode reaction involving the adsorption process.

# REFERENCES

[1]   G. Booman, J.E. Rein, C.F. Metz, G.R. Waterbury, in *Treatise on Analytical Chemistry*, Part 2, *Analytical Chemistry of the Elements,* I.M. Kolthoff and P.J. Elving, eds., Vol. 9, p. 1, John Wiley & Sons, New York, 1962.

[2]   L. Martinot and J. Fuger, in *Standard Potentials in Aqueous Solution*, A.J. Bard, R. Parsons, J. Jordan, eds., p. 631, Marcel Dekker, New York, 1985.

[3]   L.R. Morss, in *The Chemistry of the Actinide Elements,* J.J. Katz, G.T. Seaborg, L.R. Morss, eds., Vol. 2, p. 1278, Chapman and Hall, London, 1986.

[4]   G. Meinrath, T. Kimura, *J. Alloys Comp.*, 202, 89 (1993).

**Figure 1. Solubility of UO₂CO₃ (a) and relative amount of U(VI) species (b) as a function of pH and carbonate concentration in 0.1 M (Na,H)ClO₄ solution in equilibrium with CO₂ atmosphere at 25°C**

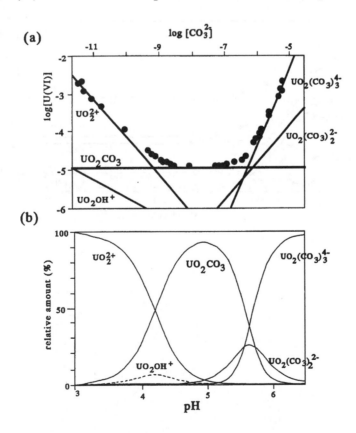

# Figure 2. Voltammograms for U(VI) species in the aqueous solution in $CO_2$ gas/aqueous solution/$UO_2CO_3$ solid phases system at various standing time, $t_s$

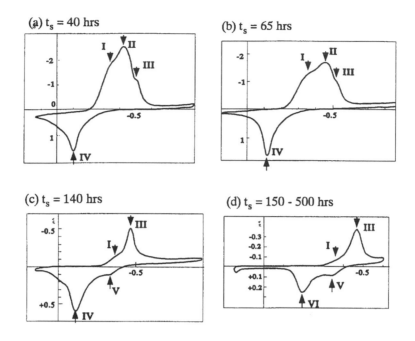

(a) $t_s$ = 40 hrs

(b) $t_s$ = 65 hrs

(c) $t_s$ = 140 hrs

(d) $t_s$ = 150 - 500 hrs

# Figure 3. $i_p$, $e_p$ and pH changes with standing time, $t_s$

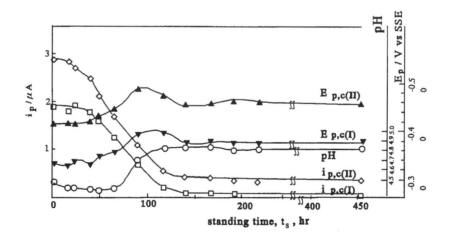

standing time, $t_s$, hr

# ION-SELECTIVE ELECTRODE METHOD FOR THE SPECIATION OF ACTINIDE IONS

**Yoshihiro Kitatsuji, Hisao Aoyagi, Zenko Yoshida**
Advanced Science Research Centre, Japan Atomic Energy Research Institute
Tokai, Ibaraki 319-1195, Japan
E-mail: kita@popsvr.tokai.jaeri.go.jp

**Sorin Kihara**
Department of Chemistry, Kyoto Institute of Technology
Matsugasaki, Sakyou-ku, Kyoto 606-8585, Japan

## Abstract

Plutonium(III)-ion selective electrode consisting of phosphine oxides such as bis-(diphenylphosphoryl)-methane as an ionophore was developed. Fundamental data, which were utilised for designing the electrode, on the ion transfer of $Pu^{3+}$ across aqueous (*w*)/organic (*org*) solutions interface facilitated by the phosphine oxide derivatives were obtained by ion-transfer polarography at the *w/org* interface. The Pu(III)-ISE prepared exhibited Nernstian response to $1 \times 10^{-7}$ - $1 \times 10^{-2}$ M $Pu^{3+}$ in $10^{-3}$ M HCl medium as the most optimum condition. It was confirmed that 10 times the concentration of $UO_2^{2+}$ did not disturb the potential for $1 \times 10^{-6}$ M $Pu^{3+}$ in $10^{-3}$ M HCl solution, and 10 times the concentration of $U^{4+}$ coexisting in the sample solution did not disturb the potential for $1 \times 10^{-3}$ M $Pu^{3+}$ in 0.2 M $H_2SO_4$ solution. A feasibility of $Pu^{3+}$-ISE developed as well as the future needs for the application of the ISE to the speciation of plutonium ions will be discussed.

## Introduction

One of the most unique characteristics of an ion-selective electrode (ISE) is its sensitivity to "activity" rather than "concentration" of the objective ionic species. It is possible to obtain information concerning speciation by means of direct potentiometry using the ISE. A rapid response time of the ISE varying usually within a few minutes allows an *in situ* measurement, and the continuous mature of the ISE measurement allows monitoring applications such as to environmental and medical samples. The ISE measurement is essentially non-destructive and non-contaminating. Sensitivity being less than 1 ppm of detection limit and a wide dynamic operating range being a reflection of the logarithmic response of potentiometric sensor are usually sufficient for most analytical applications. Despite many advantages of an analytical application of an ISE, there have been very limited studies on the ISE for actinide (An) ions other than several $UO_2^{2+}$-ISE proposed [1-3].

In principle, a potential generating at the ISE of liquid-membrane type corresponds to the potential at the aqueous solution (*w*)/organic membrane (*m*) interface, at which forward and backward ion-transfer reactions of an objective ion between two phases are in equilibrium, and therefore, the *w/m* interface is fully depolarised by the ion-transfer of the ion [4]. According to this concept on the ISE potential, the ISE can be designed on the basis of fundamental data on the ion transfer reaction across the interface between *w* and organic (*org*) phase.

In the present study, the ion-transfer reaction of Pu ions of various oxidation states in the presence of the multidentate phosphine oxides (PO) has been investigated by means of ion transfer polarography using the aqueous electrolyte dropping electrode [5]. Based on the facilitated transfer of $Pu^{3+}$ at the *w/org* interface by bis-(diphenylphosphoryl)-methane (BDPPM), $Pu^{3+}$-ISE employing BDPPM as an ionophore was developed.

## Experimental

### *Polarography for the ion transfer of actinide ions across aqueous(w)/nitrobenzene(NB) solutions interface*

An ion transfer polarogram with the aqueous electrolyte dropping electrode (AEDE) was recorded using the polarographic cell as reported previously [5,6]. For the preparation of the aqueous solution of actinide ions of a desired oxidation state, the controlled-potential electrolysis using the multi-step column electrodes [7] was employed. The effluent from the column electrode was introduced to the polarographic cell at a constant flow rate (*f*). The aqueous solution introduced to the polarographic cell was allowed to flow upward drop wise through a glass capillary into the nitrobenzene (NB) phase. The polarogram was recorded scanning the potential at a constant rate (*v*).

Crystalviolet-tetraphenylborate ($CV^+TPhB^-$) was used as a supporting electrolyte of NB phase, and the supporting electrolyte of the aqueous phase was a mixture of $HCl + NH_4Cl$ or $H_2SO_4$. The potential difference ($\Delta V$) at the *w*/NB interface was measured using Ag/AgCl electrode (SSE) in *w* and a tetraphenylborate-ion selective electrode (TPhBE) in NB [5]. The potential in this paper is referred to a standard potential denoted as TPhE, which corresponds to a transfer energy being zero. The potential of TPhBE referred to TPhE was +0.350 V at 25°C for the *w*/NB solutions couple. All measurements were carried out at $25 \pm 0.5$.

*Chemicals*

Plutonium metal (NBS-949c) was dissolved with hydrochloric acid and uranium (JAERI-U4) metal and neptunium (CEA, Fontenay-aux-Roses) dioxides were dissolved with concentrated nitric acid. The $Pu^{3+}$ solution for the ISE measurement was prepared by the controlled potential electrolysis at -0.5 V *vs*. SSE using the column electrode [7]. Hydroxylamine hydrochloride was added to the $Pu^{3+}$ solution to prevent the oxidation of $Pu^{3+}$. The concentration of more than 1 mM $Pu^{3+}$ test solutions in the absence and presence of other actinides was confirmed by the spectrophotometry.

The nitrobenzene solution containing Pu(III)-BDPPM complex was prepared by a solvent extraction method as follows. Ten millilitres of 0.1 M HCl solution containing $2 \times 10^{-3}$ M $Pu^{3+}$ + $6 \times 10^{-3}$ M $Na^+TPhB^-$ + $1 \times 10^{-2}$ M hydroxylamine hydrochloride was mixed with 10 ml of NB solution containing $6 \times 10^{-2}$ M BDPPM in a vial by shaking for 30 min. The distribution ratio (*D*) of $Pu^{3+}$ was determined to be $1.5 \pm 0.1$ by measuring the concentration of $Pu^{3+}$ in the aqueous phase by spectrophotometry. From the dependence of *D* on the concentration of BDPPM in NB, it was suggested that the extracted species was $Pu(BDPPM)_3^{3+} \cdot 3TPhB^-$.

The BDPPM ionophore was synthesised according to the procedures reported in the literature [8]. The $CV^+TPhB^-$ was prepared by purifying the precipitation obtained by mixing the methanol solutions of equivalent amounts of sodium tetraphenylborate and crystal violet [5]. All other chemicals used were of reagent grade.


## Results and discussion

### *Polarograms for the ion transfer of U, Np and Pu ions across w/NB interface facilitated by PO*

The polarograms for the transfer of $Pu^{3+}$ from *w* of 0.1 M $NH_4Cl$ + $1 \times 10^{-3}$ M HCl solution to NB solution containing $1 \times 10^{-3}$-$1 \times 10^{-2}$ M BDPPM are shown by curves 1-4 in Figure 1. Curves given by dotted line are residual polarograms obtained with *w* in the absence of $Pu^{3+}$. The anodic waves in the polarograms 1-4 correspond to the transfer of $Pu^{3+}$ from *w* to NB facilitated by BDPPM. The slope of the log-analysis, i.e. $\Delta V$ *vs*. $\log [i/(i_l-i)]$, for the anodic wave is $30 \pm 4$ mV which is larger than theoretical slope, 20 mV, for the reversible transfer of a trivalent ion. This result indicates that $Pu^{3+}$ ion transfer from *w* to NB is not reversible and includes slow reaction processes. As shown in curves 1-4, half wave potential ($\Delta V_{1/2}$) shifts 88mV more negatively with an increase of a decade of the concentration of BDPPM in NB. Consulting with the slope of the log-analysis being 30 mV, the shift of $\Delta V_{1/2}$, 88mV, indicates that species participating in the facilitated transfer reaction is $Pu(BDPPM)_3^{3+}$.

The limiting current of the anodic wave ($i_{l,a}$) was proportional to the concentration of $Pu^{3+}$ in *w* in the range of $1 \times 10^{-5}$-$1 \times 10^{-4}$ M, and $i_{l,a}$ was independent of the BDPPM concentration in NB in the range of $10^{-3}$ -$10^{-2}$ M.

Polarogram 5 in Figure 1 was measured with *w* of 0.1 M $NH_4Cl$ + $10^{-3}$ M HCl and NB of $5 \times 10^{-5}$ M $Pu(BDPPM)_3^{3+} \cdot 3TPhB^-$ + $5 \times 10^{-3}$ M BDPPM + 0.05 M $CV^+TPhB^-$. The cathodic wave corresponds to the transfer of $Pu^{3+}$ (or $Pu(BDPPM)_3^{3+}$) from NB to *w*. The limiting current of the cathodic wave ($i_{l,c}$) was proportional to the concentration of $Pu(BDPPM)_3^{3+}$ in NB, and did not depend on the concentration of BDPPM in NB.

The $\Delta V_{1/2}$ of anodic and cathodic waves for the polarogram recorded with the same BDPPM concentration ($5 \times 10^{-3}$ M) in NB are identical with each other and to be $-0.004 \pm 0.002$ V vs. TPhE. The results of the forward and backward transfer of $Pu^{3+}$ between two phases strongly suggest the possibility of $Pu^{3+}$-ISE with BDPPM ionophore.

To investigate the transfer of $U^{4+}$, $UO_2^{2+}$, $NpO_2^{+}$, $Pu^{4+}$ and $PuO_2^{2+}$ ions, identical measurements were performed. The $\Delta V_{1/2}$ values for the facilitated ion transfer reaction using $w$ of 0.1 M HCl or 0.2 M $H_2SO_4$ were listed in Table 1. Transfers of $Pu^{4+}$ and $UO_2^{2+}$ in 0.1 M HCl solution were more effectively facilitated by BDPPM than $Pu^{3+}$, and $U^{4+}$ was also facilitated. However, in 0.2 M $H_2SO_4$ solution, extents of the facilitation for the transfer of $U^{4+}$ and $Pu^{4+}$ were not so large because of a formation of the stable $An^{4+}$-sulfate complexes in $w$. The transfer of $UO_2^{2+}$ is facilitated more remarkably than $Pu^{3+}$ even with $w$ of 0.2 M $H_2SO_4$.

## $Pu^{3+}$-ISE

On the basis of the polarographic data as above, $Pu^{3+}$-ISE cell using BDPPM as an ionophore was prepared. The ISE cell configuration is:

| Ag\|AgCl 0.1 M LiCl (SSE1) | $1 \times 10^{-3}$ M $Pu^{3+}$ (HCl solution) inner solution | $1 \times 10^{-3}$ M $Pu(PO)_3 3TPhB$ + $1 \times 10^{-2}$ PO (liquid membrane) | $10^{-9}$-$10^{-1}$ M $Pu^{3+}$ (HCl, $H_2SO_4$ solution) test solution | AgCl\|Ag 0.1 M LiCl (SSE2) |
|---|---|---|---|---|

The potential, denoted as ISE potential ($\Delta V_{ISE}$), generated between SSE1 and SSE2 after the potential attained at constant was measured. The $\Delta V_{ISE}$ of ISE using BDPPM is plotted against the concentration of $Pu^{3+}$ in the test solution in Figure 2. A slope of a liner portion of the plots to be 19 mV/decade in Figure 3 agrees with the theoretical Nernstian response corresponding to the reversible transfer of a trivalent ion. With $10^{-3}$ M HCl solution of the aqueous supporting electrolyte, ISE potential indicates a Nernstian response to $Pu^{3+}$ in the concentration range of $1 \times 10^{-7}$-$1 \times 10^{-2}$ M.

### Interference of other actinide ions

Referring to the results of polarographic studies as above, it is suggested that $Pu^{4+}$, $U^{4+}$ and $UO_2^{2+}$ are interfering ions in the measurement of $Pu^{3+}$. Half-wave potentials as given in Table 1 imply that only $UO_2^{2+}$ transfers from $w$ of 0.2 M $H_2SO_4$ solution to NB preferentially to $Pu^{3+}$. The ISE potentials measured with the mixture solution of $1\times10^{-3}$ M $Pu^{3+}$ and other actinide ions in 0.2 M $H_2SO_4$ are listed in Table 2. Only $UO_2^{2+}$ interferes the measurement of $1 \times 10^{3+}$ M $Pu^{3+}$, and these results are in good accordance with the polarographic data. The selectivity coefficient ($K^{pot}$) for $Pu^{3+}$-ISE was studied by separate solutions method. The $K_{Pu(III),An}^{pot}$ were calculated using the reduced Eisenman equation:

$$\text{Log } K_{Pu(III),An}^{pot} = (E_2 - E_1)/(2.303RT/n_1F) + (1 - n_1/n_2)\log C \tag{1}$$

where $E_1$ and $E_2$ are the potentials observed with the solutions of the same concentration of $Pu^{3+}$ and interference ion, $n_1$ and $n_2$ are the charges of $Pu^{3+}$ and interference ion, respectively, and $C$ is the concentration of $Pu^{3+}$. The values of selectivity coefficients log $K_{Pu(III),An}^{pot}$ at $1 \times 10^{-3}$ M $Pu^{3+}$ toward $Pu^{4+}$, $U^{4+}$ and $UO_2^{2+}$ were found to be -1.40, -1.40 and 3.50, respectively, with 0.2 M $H_2SO_4$ supporting electrolyte solution. This means that $Pu^{3+}$-ISE response to $1 \times 10^{-3}$ M $Pu^{3+}$ is not influenced by the presence of $Pu^{4+}$ or $U^{4+}$ at least 10 times concentration of $Pu^{3+}$.

The response of $Pu^{3+}$-ISE with the mixed solutions of $Pu^{3+}$ and $UO_2^{2+}$ in $10^{-3}$ M HCl was investigated and results are plotted in Figure 3. The ISE measurements of $10^{-4}$ M and $10^{-6}$ M $Pu^{3+}$ were not influenced by the presence of $UO_2^{2+}$ of the same concentration and 10 times concentration, respectively. The selectivity coefficient log $K_{Pu(III),U(VI)}^{pot}$ for $1 \times 10^{-3}$ M of $Pu^{3+}$ toward $UO_2^{2+}$ is 1.44 in $10^{-3}$ M HCl. The interference of $Pu^{4+}$ and $U^{4+}$ are remarkable in the ISE measurement of $Pu^{3+}$ with HCl media.

## Application

Uranium(VI) disturbs the measurement of $10^{-3}$ M $Pu^{3+}$ and 10 times concentration of $U^{4+}$ does not interfere the measurement of $Pu^{3+}$ in 0.2 M $H_2SO_4$ medium. Plutonium concentration in 0.2 M $H_2SO_4$ sample solution can be determined by using $Pu^{3+}$-ISE even in the presence of uranium ion of less than 10 times concentration, if the oxidation state of uranium ion is controlled to be tetravalent prior to the ISE measurement. The controlled-potential electrolysis with the column electrode is useful for the oxidation state control. For example, the sample solution of the mixture of plutonium and uranium ions is electrolysed at -0.6 V *vs.* SSE using the column electrode electrolysis to obtain the sample solution of $Pu^{3+}$ and $U^{4+}$. The $Pu^{3+}$-ISE developed is also useful as an indicator electrode for the redox titration of plutonium ion in the sample solution.

The selectivity of the ISE developed is not sufficiently high for using to the direct speciation of $Pu^{3+}$ ion. The development of new ionophore, which has higher selectivity or sensitivity, makes possible to direct measurement of actinide solution samples by using ISE.

## REFERENCES

[1]    I. Goldberg, D. Meyerstein, *Anal. Chem.*, 52, 2105 (1980).

[2]    J. Senkyr, D. Ammann, P.C. Meier, W.E. Morf, E. Pretsch, W. Simon, *Anal. Chem.*, 51, 786 (1979).

[3]    N.V. Serebrennikova, I.I. Kukushkina, N.V. Plotnikova, *Zh. Anal. Khim.*, 37, 645 (1982).

[4]    S. Kihara, Z. Yoshida, *Talanta*, 31, 789 (1984).

[5]    S. Kihara, M. Suzuki, K. Maeda, K. Ogura, S. Umetani, M. Matsui, Z. Yoshida, *Anal. Chem.*, 58, 2954 (1986).

[6]    Y. Kitatsuji, H. Aoyagi, Z. Yoshida , S. Kihara, *Anal. Sci.*, 14, 67 (1998).

[7]    H. Aoyagi, Z. Yoshida, S. Kihara, *Anal. Chem.*, 59, 400 (1987).

[8]    S. Umetani, M. Matsui, *Anal. Chem.*, 64, 2288 (1992).

**Table 1. Half wave potential for actinide ion transfer**

| Metal ions | $\Delta V_{1/2}$(V vs. TPhE) | |
|---|---|---|
| | w; 0.1 M HCl | w; 0.2 M H₂SO₄ |
| Pu(III) | -0.038 | +0.005 |
| U(IV) | -0.034 | >+0.05[b] |
| U(VI) | <-0.040[a] | -0.080 |
| Np(V) | >+0.07[b] | >+0.05[b] |
| Pu(IV) | >+0.070[b] | >+0.05[b] |
| Pu(VI) | <-0.040[a] | >+0.05[b] |

NB; $2 \times 10^{-3}$ M BDPPM + 0.05 M CV⁺TPhB⁻
a) The polarographic wave was overlapped with the final fall of the residual current.
b) The polarographic wave was overlapped with the final rise of the residual current.

**Table 2. ISE potential for the measurement of $1 \times 10^{-3}$ M Pu³⁺ in the presence of other actinide ions**

| Actinide ions | Concentration of the ion/10⁻³ M | $\Delta V_{ISE}$ / mV |
|---|---|---|
| None | – | -29.5 |
| Pu(IV) | 1.0 | -29.0 |
| | 10 | -29.0 |
| U(IV) | 1.0 | -29.4 |
| | 10 | -29.5 |
| U(VI) | 1.0 | -68.2 |
| Np(V) | 1.0 | -28.5 |

Supporting electrolyte of the test solution; 0.2 M H₂SO₄

**Figure 1. Polarograms for the transfer of Pu³⁺ between w/NB interface**

w; 0.1 M NH₄Cl + 10⁻³ M HCl + (1-4) $5 \times 10^{-5}$ M Pu³⁺, (1'-4', 5) none, NB; 0.05 M CV⁺TPhB⁻ + (1,1') $1 \times 10^{-3}$ M, (2,2') $2 \times 10^{-3}$ M, (3,3',5) $5 \times 10^{-3}$ M, (4,4') $1 \times 10^{-2}$ M BDPPM, + (5) $5 \times 10^{-5}$ M Pu(BDPPM)₃³⁺·3TPhB⁻, $f$ = 1.0 ml min⁻¹, $v$ = 3 mV s⁻¹

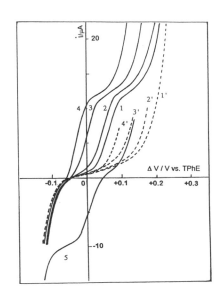

## Figure 2. Potential measured at Pu³⁺-ISE

Test solution; (1) 1 M HCl, (2) 0.1 M HCl, (3) $10^{-2}$ M HCl, (4) $10^{-3}$ M HCl

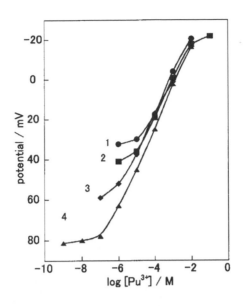

## Figure 3. Response of Pu³⁺-ISE with the test solutions of the mixture of Pu³⁺ and UO₂²⁺ in 10⁻³ M HCl

Concentration of $UO_2^{2+}$; (1) $1 \times 10^{-4}$ M, (2) $5 \times 10^{-5}$ M, (3) $2 \times 10^{-5}$ M, (4) $1 \times 10^{-5}$ M, (5) none

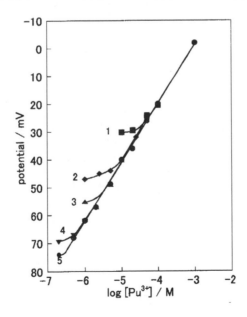

# REDOX SPECIATION OF Np IN TBP EXTRACTION PROCESS BASED ON ELECTROCHEMICAL OXIDATION STATE CONTROL OF Np

**Kwang-Wook Kim, Kee-Chan Song, Eil-Hee Lee, In-Kyu Choi, Jae-Hyung Yoo**
Korea Atomic Energy Research Institute
Taejon 305-350, Korea

## Abstract

The change of Np oxidation state in nitric acid and the effect of nitrous acid on the oxidation state were analysed by spectrophotometry, solvent extraction, and electrochemical methods. An enhancement of Np extraction to 30 vol.% TBP was carried out through adjustment of Np oxidation state by using a glassy carbon fibre column electrode system. The information of electrolytic behaviour of nitric acid was important because the nitrous acid affecting the Np redox reaction was generated during the electrolytic adjustment of the Np oxidation state. The Np solution used in this work consisted of Np(V) and Np(VI) without Np(IV). The composition of Np(V) in the range of 0.5 M ~ 5.5 M nitric acid was 32% ~ 19%. The electrolytic oxidation of Np(V) to Np(VI) in the solution enhanced Np extraction efficiency about five times higher than the case without the electrolytic oxidation. It was confirmed that the nitrous acid of less than about $10^{-5}$ M acted as a catalyst to accelerate the chemical oxidation reaction of Np(V) to Np(VI).

## Introduction

The concept of partitioning-transmutation to treat high-level radioactive liquid waste (HLLW) has been studied in several countries during the last decade, because it could diminish the potential toxicity of long-lived minor actinides (Am, Cm, Np, etc.), meet the public's acceptance, and encourage further development of nuclear power for peaceful uses [1-3]. Np, among the long-lived minor actinides, is liable to spread over the different streams in the partitioning process due to the complicated chemical behaviour of Np in nitric acid, where it can exist simultaneously in three stable oxidation states: Np(IV) ($Np^{4+}$), Np(V) ($NpO_2^+$), Np(VI) ($NpO_2^{2+}$) [4-5]. The Np ions of different oxidation states exhibit different extraction behaviours to tri-butyl phosphate (TBP). In nitric acid, Np(VI) is easily extractable, Np(IV) is less extractable, and Np(V) is nearly unextractable [5]. These oxidation states are readily inter-convertible depending on the nitric acid and nitrous acid concentrations. The nitrous acid severely affecting Np oxidation states could be produced through the reduction of the nitric acid during electrolysis for controlling the oxidation state of Np in nitric acid or through the radiolysis of nitric acid. Moreover, the nitric acid concentration has an effect on the Np valences through a disproportionation reaction as well. The ratio of the Np oxidation states in HLLW from several countries' reprocessing plants has been known to be different [4]. For example, at CEA in France, Np(V) and Np(VI) exist. At ORNL in the USA, the ratio of Np(IV):Np(V):Np(VI) is 10:80:10. At Windscale in England, Np(IV) exists in significant quantity. This discrepancy in the Np oxidation states is considered to be due to the differences of nitrous acid and nitric acid concentrations at each facility.

Accordingly, the estimation of the ratio of the Np ions of different oxidation states and the change of the oxidation states to a desired state are necessary to improve the partitioning process. Controlling the Np oxidation states electrochemically is promising. The electrochemical method does not require, in principle, the reactive reagents, which minimises generation of secondary wastes and change of the stoichiometric concentrations of the other components. Also the chemical reagents can bring out the corrosion problem in the process. As mentioned above, because the nitrous acid, which could be generated during the process of controlling Np oxidation states in nitric acid, affects the Np oxidation state, it is necessary to know the electrochemical redox behaviours of nitric acid and Np simultaneously. However, there are only a few publications about quantitative data of Np oxidation states in a wide range of nitric acid concentration, taking into consideration the effect of nitrous acid on Np oxidation states. Moreover, the electrochemical redox mechanism of nitric acid is complicated such that there are a few papers on the electrochemical behaviours of Np in nitric acid, which mostly deal with low nitric acid concentration of less than 3 M.

In this work, Np oxidation states in nitric acid of 0.5 M to 5.5 M were analysed by spectrophotometry, solvent extraction, and electrochemical methods, and then the effect of nitrous acid on the Np extraction to TBP was investigated. An enhancement of Np extraction by controlling the Np oxidation states using a glassy carbon (GC) fibre column electrode system was carried out.

## Experimental

All reagents used in this work were chemical reagent grade and used as received. Np obtained from AEA Technology in England as $^{237}$Np in 2.0 M nitric acid (total alpha per cent: 99.82%) was used without further purification. A set of two GC fiber column electrode systems was used for controlling Np oxidation state. Details of the GC column electrode system were described in other work [6]. The Np extraction was carried out by mixing vigorously the aqueous solution of Np with 30 vol.% TBP/dodecane in a batch at phase ratio of 1 for 30 minutes. Np concentrations in aqueous and organic phases were measured by a γ-spectrometer with a HP-Ge detector (Oxford Inc., Model 5000 Radiation

Analyser). Nitric acid concentrations in the solution before and after the electrolysis were measured by an auto-titrator (Kyoto Electronics, Model AT-400). The concentration of nitrous acid in the nitric acid solution after the electrolysis was determined by a calorimetric method proposed by B.E. Saltzman [13]. In this method, the nitrous acid reacts with sulfanilic acid to form a diazonium salt which is coupled with $\alpha$-naphtylamine to form a red dye. The absorbency at 547 nm was measured by an UV-VIS spectrometer (Shimadzu, Model UV-160A). The standard solution was prepared from the sodium nitrite reagent. Np absorption spectrum in the range of 350-1 300 nm was measured by an UV-IR spectrophotometer (Varian, Model Cary 5E UV-VIS-NIR). All experiments were performed at $23\pm0.5°C$.

## Results and discussion

Neptunium ions of different oxidation states give absorption peaks at specific wavelengths. For example, Np(IV) is representatively detected at 960 nm and 723 nm, Np(V) at 618 nm and 980 nm, and Np(VI) at 1 223 nm. Figure 1 shows an absorption spectrum of $8.0 \times 10^{-4}$ M Np in 2.0 M nitric acid obtained from AEA Technology. Three peaks are observed at 981 nm, 616 nm, and 1 223 nm. They are for Np(VI) and Np(V). Np(IV) is concluded not to exist in the solution because there are no peaks at 723 nm and 960 nm.

When the Np oxidation state in nitric acid is electrochemically controlled, nitrous acid can be generated, depending on the nitric acid concentration. The generated nitrous acid affects the redox of Np in two ways as follows. When the concentration of nitrous acid is below $ca.$ $10^{-5}$ M, the nitrous acid acts as a catalyst to accelerate oxidation of Np(V) to Np(VI) as Eq. (1):

$$2NpO_2^+ + 3H^+ + NO_3^- = 2NpO_2^{2+} + HNO_2 + H_2O \qquad (1)$$

When the nitrous acid is more than $ca.$ $10^{-4}$ M, the nitrous acid in nitric acid acts as a reductant so that it reduces Np(VI) to Np(V) rapidly. The reason can be explained by the fact that the backward reaction rate of Eq. (1) increases according to the Le Châtelier's principle when the concentration of the nitrous acid is higher. And Np(V) can change to Np(VI) or Np(IV) spontaneously through a disproportionation reaction as Eq. (2), depending on the nitric acid concentration.

$$2NpO_2^+ + 4H^+ = NpO_2^{2+} + Np^{4+} + 2H_2O \qquad (2)$$

The interconversion of $NpO_2^+$ to $Np^{4+}$ requires an extensive rearrangement of the primary co-ordination sphere of $NpO_2^+$ (i.e. a change in the linear dioxo structure) before electron transfer. This results in a relatively high activation barrier such that the rate of the disproportionation is slow and can be considered to be negligible in reprocessing process from which HLLW comes. This can be a reason why Np(IV) does not exist in the nitric solution as shown in Figure 1. Because the nitrous acid, which could be generated during controlling Np in nitric acid, severely affects the Np oxidation state, it is required to understand the electrochemical redox behaviours of nitric acid and Np at the same time. Some experimental results on the electrolytic reaction of nitric acid were reported in our previous paper [7] where sulfamic acid of 0.1 M was confirmed to be enough to decompose the nitrous acid generated during the electrolytic reaction of nitric acid of less than 3.5 M.

Figure 2 shows the effect of nitric acid concentration on an apparent distribution coefficient of Np from the aqueous solution of Np(VI) and Np(V) mixture into 30 vol.% TBP without controlling the oxidation state. The apparent distribution coefficients increase with an increase of nitric acid concentration and they are between those of pure Np(V) and those of Np(VI). Figure 2 shows that there is no effect of the sulfamic acid on the Np extraction. However, in the presence of more than $10^{-3}$ M nitrite ion, the distribution coefficient drops a lot. This can be explained by considering that the nitrous acid resulted from nitrite ion and proton of nitric acid acts as a reductant toward Np(VI) as Eq. (1) so that the portion of Np(V) in the system increases.

The distribution coefficients of pure Np(VI) and Np(V) are well known and the compositions of Np(VI) and Np(V) can be estimated from the apparent distribution coefficient of Np in Figure 2 by using the following Eqs. (3) and (4). The results of the estimation are shown in Figure 3.

$$X_{Np(V), i} \, D_{Np(V), i} + X_{Np(VI), i} \, D_{Np(VI), i} = D_{App.Np(VI),Np(V), I}, \text{ i : nitric acid} \tag{3}$$

$$X_{Np(V), i} + X_{Np(VI), i} = 1, \text{ i : nitric acid} \tag{4}$$

where $X_{Np(V)}$ and $X_{Np(VI)}$ are mole fractions of Np(V) and Np(VI), respectively. $D_{Np(V)}$ and $D_{Np(VI)}$ are distribution coefficients of pure Np(V) and Np(VI), respectively. $D_{App.Np(VI),Np(V)}$ is the apparent distribution coefficient of Np in the solution of Np(VI) and Np(V) coexisting.

Because the disproportionation rate increases with an increase of the concentration of nitric acid as Eq. (2), the composition of Np(VI) in Figure 3 increases and consequently that of Np(V) decreases with an increase of nitric acid concentration. From this result, it is confirmed that the composition of Np(V) in the initial Np solution is less than 30% for the nitric acid concentration from 0.5 M to 5.5 M. Figure 4 shows a comparison of mole fractions of Np(V) estimated by solvent extraction method (*cf.* Figure 3) with those estimated by spectrophotometry (*cf.* Figure 1) using an extinction coefficient of Np(V) of 390 $M^{-1}cm^{-1}$ at. They agree relatively well with a maximum error of 4%.

Figure 5 shows a voltammogram observed in the positive potential scanning with $5.1 \times 10^{-3}$ M Np in 1.5 M nitric acid being fed into Cell 1, a voltammogram in the negative scanning direction for the Np solution coming out of Cell 1 where a constant potential of +1.05 V was supplied, and their respective background voltammograms. The redox reaction of Np(VI) and Np(V) are expressed as follows.

$$NpO_2^{2+} + e^- = NpO_2^+ \quad E_o = +1.14 \text{ (V vs SHE)} \quad (+0.87 \text{ V vs SSE}) \tag{5}$$

In the voltammogram at Cell 1, a reduction limiting current observed at more negative potential than +0.8 V is considered to be due to the reduction of Np(VI) in the feeding solution to Np(V) and an oxidation limiting current at more positive than +0.95 V is considered to be due to the oxidation of Np(V) to Np(VI). On the basis of the background current at Cell 1, the current difference between the oxidation limiting current and the background current is 1.65 mA, and the one between the reduction limiting current and the background current is 4.49 mA. The current ratio is the ratio of the compositions of Np(VI) and Np(V) in the nitric acid which is fed into Cell 1. The compositions of Np(V) and Np(VI) in 1.5 M nitric acid calculated from the ratio are 27% and 73%, respectively, which are very close to those calculated by spectrometry as given Figure 4. The net current between the limiting oxidation and reduction currents at Cell 1 is 6.14 mA. From this voltammogram of Np oxidation, the potential over +0.95 is enough to oxidise Np(V) to Np(VI). Accordingly, when +1.05 V is applied constantly to Cell 1 into which $5.1 \times 10^{-3}$ M Np in 1.5 M nitric acid is fed, the solution coming out from Cell 1 contains only Np(VI) so that the voltammogram in the negative scanning direction at Cell 2 represents the reduction of pure Np(VI) to Np(V). The net current between the voltammogram and its background at Cell 2 is 6.15 mA which is very close to the net current between the limiting oxidation and reduction currents at Cell 1. From the results of Figure 1 to Figure 5, it can be said that the Np(V) composition in 0.5 M to 5.5 M nitric acid solution is 32% to 19% and Np(VI) is dominant in the solution and increases with an increase of the concentration of nitric acid. The other papers dealing with Np oxidation in nitric acid of more than 5.5 M mentioned that the degree of disproportionation of Np(V) was 76.2%, in other words, the composition of Np(V) is 23.8%, which was close to 19% as shown in Figure 4 even though there are a few errors among both values.

Figure 6 shows the distribution coefficient of Np in the solution eluting from Cell 1 of +1.05 V, after mixing with 30 vol.% TBP. The distribution coefficients obtained with the electrolytic oxidation of Np(V) to Np(VI) are about 5 times bigger than those without the electrolytic oxidation. However, they are lower than those of pure Np(VI) shown in the previous papers [5] where Np(VI) were prepared by chemical oxidants in excess. Np(VI) generated in Cell 1 is assumed to be partially re-reduced to Np(V) during the extraction. Np(VI) is known to be reduced to Np(V) easily. Figure 6 also shows the effect of sulfamic acid on the Np distribution coefficient after the electrolytic oxidation. In nitric acid containing 0.05 M sulfamic acid, the distribution coefficients deviate from those measured in nitric acid when the concentration is higher than 2 M. This could be explained as follows. The existence of a small amount of nitrous acid allow Np(V) to be oxidised to Np(VI) at a high nitric acid concentration, *cf.* Eq. (1). Without sulfamic acid as a nitrous acid scavenger, nitrous acid, which could be generated during the oxidation of Np(V) at Cell 1, is considered to help Np(V) partially converted from Np(VI) during the extraction to re-oxidise to Np(VI). However, with sulfamic acid, there is no chance for the oxidation of Np(V) to Np(VI) during the extraction. In order to confirm the existence of nitrous acid in the eluents from Cell 1 at +1.02 V and from Cell 2 at +0.72 V, nitrite ion concentrations from eluents were analysed by a colorimetric method and the results are plotted in Figure 7. The nitrous acid in the effluent from Cell 1 is less than $10^{-5}$ M, which implies that nitrous acid acts as a catalyst shown by Eq. (1). However, in the solution from Cell 2, the nitrous acid concentration is $\sim 10^{-3}$ M, which indicates the auto-catalytic reactions of the nitrous acid generated at the Cell 1 [7]. From the results of Figure 6 and Figure 7, a small amount of nitrous acid may accelerate the oxidation of Np(V) to Np(VI) and the sulfamic acid is considered not to be necessary in the feeding nitric acid solution only for the preparation of Np(VI) by the electrolytic oxidation. Figure 8 shows the distribution coefficient of Np in the eluents from Cell 2 at +0.72 V with Cell 1 at +1.02 V after mixing with 30 vol.% TBP. The nitric acid solution from Cell 2 is considered to contain only Np(V). The distribution coefficient increases exponentially with an increase of nitric acid concentration up to5.5 M. This is attributable to the re-oxidation of Np(V) through the disproportionation reaction of Eq. (4) during the extraction step. The equilibrium constant of Eq. (2) is known to be $4 \times 10^{-7}$ in 1.0 M nitric acid and 0.13 in 5.3 M nitric acid.

## Conclusions

An enhancement of Np extraction efficiency into 30 vol.% TBP was carried out through the adjustment of Np oxidation state by using GC fibre column electrode system. The Np solution used in this work consisted of Np(V) and Np(VI). The composition of Np(V) in 0.5 M ~ 5.5 M nitric acid solution was 32% ~ 19%. The electrolytic oxidation of Np(V) in the solution enhanced the TBP extraction efficiency of Np about five times higher than that without the electrolytic oxidation.

# REFERENCES

[1]    OECD Final Report : NEA/PTS/DOC(98)4 (1998).

[2]    D. Lelièvre, H. Boussier, J.P. Grouiller and R.P. Bush, EUR-17485 (1996).

[3]    SKB Technical Report, "Partitioning and Transmutation 1997 Status Report", TR-98-14 (1998).

[4]    Y. Morita and M. Kubota, "Recovery of Neptunium", JAERI-M-84-043 (1984).

[5]    M. Benedit, T.H. Pigford, and H.W. Levi, "Nuclear Chemical Engineering", 2nd ed., McGraw-Hill Book Company (1981).

[6]    S. Kihara, Z. Yoshida, and H. Aoyagi, *Bunseki kagaku*, 40, 309 (1991).

[7]    K.W. Kim, E.H. Lee, I.K. Choi, J.H. Yoo, H.S. Park, *J. Radioanal. Nucl. Chem.* to be printed (2000).

**Figure 1. Absorbance spectrum of $8.0 \times 10^{-4}$ M Np in 2.0 M nitric acid**

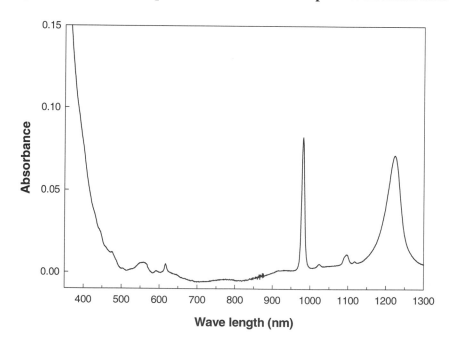

# Figure 2. Np distribution coefficient in nitric acid with and without sulfamic acid and sodium nitrite

*Initial aqueous Np: 0.0011 M, organic phase: 30 vol.% TBP*

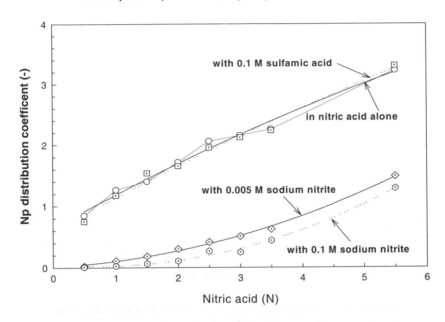

# Figure 3. Np(VI) and Np(V) in nitric acid with and without sulfamic acid

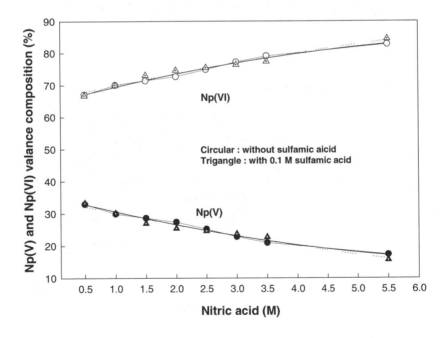

# Figure 4. Speciation of Np(V) estimated by absorbance and apparent distribution coefficient

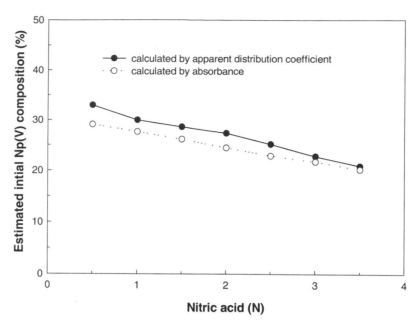

# Figure 5. Voltammograms for reduction and oxidation of Np(VI) and Np(V) in 1.5 M nitric acid at Cell 1 and Cell 2

*Feeding flow rate: 0.7 ml/min, Np concentration: $5.1 \times 10^{-3}$ M*

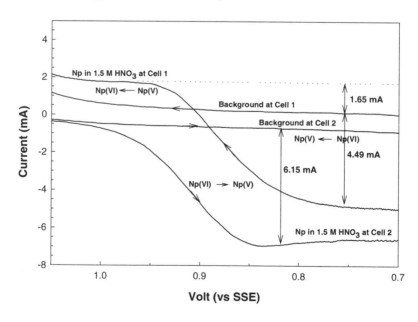

## Figure 6. Np distribution coefficient with electrolytic oxidation

*Np concentration: $1.1 \times 10^{-3}$ M, organic phase: 30 vol.% TBP, applied potential at Cell 1: 1.02 V vs. SSE*

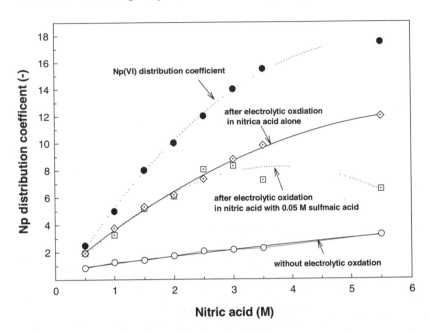

## Figure 7. Nitrite concentrations after Cell 1 of +1.02 V and Cell 2 of +0.72 V

*Flow rate: 0.6 ml/min*

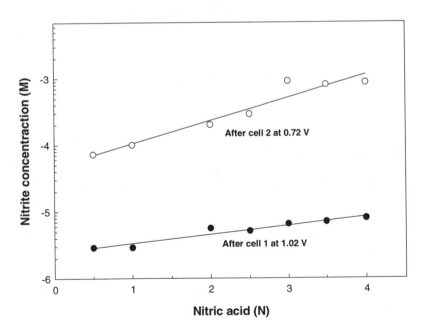

# Figure 8. Np distribution coefficient with electrolytic reduction at Cell 2

*Np concentration: $1.1 \times 10^{-3}$ M, organic phase: 30 vol.% TBP, applied potential at Cell 2: 0.72 vs. SSE*

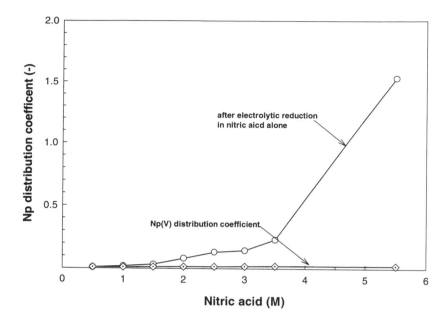

# POSTER SESSION

## Part E: Predictive Approach to Speciation

# CONTRIBUTION TO LANTHANIDE AND ACTINIDE SPECIATION: THERMODYNAMIC STUDIES OF HYDROXYCARBONATE SPECIES

**F. Rorif, J. Fuger and J.F. Desreux**

Co-ordination Chemistry and Radiochemistry, University of Liège
B16, Sart Tilman, B-4000 Liège, Belgium

## Abstract

The accurate prediction of the conditions for the formation of solid compounds, that may appear as the result of the long-term interaction of groundwater with the nuclear waste matrix, requires a better understanding of the thermodynamic parameters associated with the formation of such compounds. Studies reported by this laboratory on the enthalpy of formation of lanthanide and americium hydroxycarbonates $M(OH)CO_3 \cdot 0.5H_2O(cr)$ in their orthorhombic form [1] and their relation to the corresponding hydroxides [2] are contributions to this aim.

The authors report here some thermodynamic parameters of the hexagonal form $M(OH)CO_3(cr)$ (M = Nd, Sm). An extrapolation to the hexagonal variety of $AmOHCO_3(cr)$ which was isolated for the first time more than a decade ago from J-13 groundwater of Yucca Mountain [3] is now possible.

In this context, the authors have developed a new sealed micro calorimeter, the original version of which was previously designed in their laboratory. The new calorimetric chamber was made of 18-carat gold alloy in order to improve the overall performances.

Thermodynamically meaningful results require well-defined solids. Each compound was prepared in triplicate and thoroughly characterised by X-ray diffraction, IR spectra, TGA, and gravimetry of the sesquioxide.

The standard enthalpies of formation of the compounds were obtained from the measured heat effect associated with their dissolution in $6.00$ mol•dm$^{-3}$ HCl, using relevant auxiliary data. These results are discussed in the light of literature data on the enthalpy of formation and solubility of the corresponding $M(OH)CO_3 \cdot 0.5H_2O(cr)$ (M = Nd, Sm, Am) hydroxycarbonates in their orthorhombic form and of other related compounds.

## Introduction

In the back-end of the nuclear fuel cycle, the long-lived radionuclides, and more particularly the transuranium elements, are very important as potential long-term contaminants of the environment if the strategy is to store the waste, reprocessed or not, in proper locations such as deep geological formations. After the initial controlling-step of leaching from the waste form, the fate of each radionuclide and its migration through the geomedia will be very different depending on its own chemical properties and on the specific conditions and intrinsic properties of the chosen repository. Carbonate complexes of actinide ions will play an important role in the migration process because many kinds of natural water found in the vicinity of nuclear waste repositories contain a significant amount of carbonate and bicarbonate. In this context, lanthanide and actinide carbonates, hydroxycarbonates and hydroxides are important compounds (e.g. [4]).

Computer simulations have been increasingly used in recent years since many important decisions may depend on them, for instance, in the process of proposal and/or acceptance of a nuclear waste repository. When calculating the migration of radionuclides from an underground site by using geochemical codes, one of the important input data is the solubility of a given element from its solubility limiting phase, which then will provide limits of the concentration of this element in those types of water. Such limits are important for verifying the correctness of the radionuclide transport model used to perform the calculations since the model should be capable of predicting the resulting solubility of a particular species within a given range of conditions (engineered, natural). The contribution of the laboratory to performance assessment has been through the measurement, using solution calorimeters, of the enthalpies associated with the dissolution of relevant compounds, which leads to the accurate determination of their enthalpies of formation, and ultimately, to their solubility products.

In previous studies, the same group reported results on the thermochemistry of the orthorhombic variety $M(OH)CO_3 \cdot 0.5H_2O(cr)$ (M = Nd, Sm, Am) [1], usually noted as $A-M(OH)CO_3 \cdot 0.5H_2O(cr)$, and $M(OH)_3(cr)$ (M = La, Nd, Sm, Am) [2], and studied the relation between them. Lately the authors have been concerned with the thermochemistry of the hexagonal variety $M(OH)CO_3(cr)$ (M = Nd, Sm), noted as $B-M(OH)CO_3(cr)$. Properties of $B-AmOHCO_3(cr)$ identified in waters from well # J-13 (Yucca Mountain) [3] can be inferred from those of the lanthanide homologues.

In this paper, the essential characteristics and performances of a new sealed isoperibolic micro calorimeter are also described. The calorimetric chamber was made of 18-carat gold alloy and especially designed for the present studies.

## Calorimetric measurements on carbonate compounds

There is one major difficulty to overcome when carrying out such measurements. Indeed carbon dioxide formed upon dissolution of carbonate compounds in acidic media (typically, 0.100, 1.00, and 6.00 $mol \cdot dm^{-3}$ HCl) has a strong tendency to supersaturate in solution. This results in a slow release of $CO_2(g)$ from the solution thus causing a non-reproducible thermal heat effect, which can become a quite significant source of errors, especially when measuring small calorific effects. The alternative was thus to force the gas to remain entirely in solution. As a consequence, the authors have designed a new version of an already described semi-adiabatic (often called isoperibolic) micro calorimeter [1]. The former version consisted of a sealed gold-plated brass chamber equipped with two wells for the calibrating heating resistance and the temperature sensing element. Once completely filled with acid the chamber offers no free volume and it was estimated that more than 99.3% of the $CO_2$ produced remain in solution.

A calorimetric chamber must have other properties: a low specific heat relative to that of the contained aqueous solution in order to gain sensitivity, an excellent thermal conductivity to compensate the poor agitation of the designed system, and finally a chemical inertness against the contained solution. An 18-carat gold alloy was selected for a new calorimetric chamber while essentially conserving the previous design and system support. The new system has a heat capacity of *ca.* 50 $J \cdot K^{-1}$ when filled with 6.5 $cm^3$ of water.

The performance of the instrument was verified by using two test reactions. Firstly, the molar enthalpy of solution of TRIS [tris(hydroxymethyl)-aminomethane(cr)] in 0.100 $mol \cdot dm^{-3}$ HCl was measured. The resultant effect, $\Delta_{sln}H_m = -(29.90 \pm 0.18)$ $kJ \cdot mol^{-1}$*, is comparable to the value recommend by the National Bureau of Standards (NBS), $\Delta_{sln}H_m = -(29.769 \pm 0.029)$ $kJ \cdot mol^{-1}$ [5]. A second test with a reaction leading to $CO_2(g)$ formation was also carried out by dissolving $Na_2CO_3(cr)$ in 1.00 $mol$ $dm^{-3}$ HCl. The resultant molar energy of solution of $Na_2CO_3(cr)$ (P.A. Merck, heated at 623 K under vacuum) is $\Delta_{sln}E_m = -(51.94 \pm 0.55)$ $kJ \cdot mol^{-1}$. This value can be compared with that obtained using a suitable Hess cycle and auxiliary data**, $\Delta_{sln}E_m = -(53.3 \pm 0.8)$ $kJ \cdot mol^{-1}$ [1]. The difference could be attributed to approximations used for the calculation, such as $\Delta_f H_m^0(CO_2, sln) = \Delta_f H_m^0(CO_2, aq)$. The accuracy of the new calorimeter chamber ranges between 0.6 and 1% for chemical effects ranging from 1 to 5 Joules, which is appropriate since limited amounts of samples are handled (a few mg).

## Characterisation of lanthanide hydroxycarbonates

If orthorhombic A-form, $A-M(OH)CO_3(cr)$ containing variable amounts of water, is usually isolated from aqueous solution, the hexagonal B-form, $B-M(OH)CO_3(cr)$, is obtained best under hydrothermal conditions by decomposition of the hydrated sesquicarbonate according to Eq. (1) [6].

$$M_2(CO_3)_3 \ nH_2O(cr) \rightarrow 2 \ B-M(OH)CO_3(cr) + (n-1) \ H_2O(g) + CO_2(g) \qquad M = Nd, Sm. \qquad (1)$$

Each compound was prepared in triplicate and carefully characterised. Each prepared sample was analysed by gravimetry of the lanthanide sesquioxides. For each of them, all observed diffraction lines were indexed in the hexagonal system and intensities were consistent with the space group P(-6) (No. 174). All the result, lattice parameters agree well with literature data. They closely relate to those published for $B-Am(OH)CO_3(cr)$ [3]. The FT-IR spectra of all samples in KBr pellets show the two typical well-defined double bands of the same intensity in the region of OH-stretching ($v_{OH}$) vibration near 3 500 $cm^{-1}$ [7]. Absence of DTA signals below 400°C confirms that the samples do not contain water.

## Thermochemistry of lanthanide hydroxycarbonates

The standard molar enthalpy of formation of lanthanide hydroxycarbonates can be obtained by using the thermodynamic cycles shown in Table 1. The first reaction corresponds to the energy of solution of $B-M(OH)CO_3(cr)$ in 6.00 $mol \cdot dm^{-3}$ HCl and measured with the new calorimetric chamber. Within the uncertainty limits, no systematic differences could be detected between the various preparations of a given compound. In each cycle, the second reaction involves the molar enthalpy of dissolution of the lanthanide metals of high purity in the same medium, 6.00 $mol \cdot dm^{-3}$ HCl [9]. In this table, the last two reactions (3 and 4) represent the partial molar enthalpies of formation of $H_2O$ and

---

* Uncertainty limits are reported for the 95% confidence level using standard statistical methods [11].

** Unless specified, auxiliary thermodynamic data are those recommended by CODATA [9].

$CO_2$ in the solution and were calculated from apparent molar enthalpies. The standard molar enthalpies of formation for B-Nd(OH)$CO_3$(cr) and B-Sm(OH)$CO_3$(cr) are then obtained by combination of the four reactions: $\Delta_f H_m^0\{$B-M(OH)$CO_3$(cr)$\} = -\Delta H_1 + \Delta H_2 + 2 \cdot \Delta H_3 + \Delta H_4$.

The solubility product, $K_{sp}^0$, of the two hydroxycarbonates, corresponding to Eq: (2), at infinite dilution in water:

$$\text{B-M(OH)CO}_3\text{(cr)} = \text{M}^{3+}\text{(aq)} + \text{CO}_3^{2-}\text{(aq)} + \text{OH}^-\text{(aq)} \qquad \text{(M = Nd, Sm)} \qquad (2)$$

is then obtained using the entropies of the orthorhombic hydroxycarbonates estimated from those of corresponding trihydroxides and auxiliary data [1]. Using the relationship:

$$\log K_{sp}^0\{\text{B-M(OH)CO}_3\text{(cr)}\} = -\Delta_r G_m^0 \{\text{Eq. (2)}\}/2.303 \bullet R \bullet T \qquad (3)$$

the authors obtain -(23.5 ± 0.4) and -(23.6 ± 0.7) for the neodymium and samarium hydroxycarbonates, respectively.

In a second step, the authors approximate $\Delta_f H_m^0$ of anhydrous A-M(OH)$CO_3$(cr) by subtracting half of $\Delta_f H_m^0$($H_2O$, l) = -(285.83 ± 0.04) kJ•mol$^{-1}$ from the previously published values for $\Delta_f H_m^0$ {A-M(OH)$CO_3$•0.5$H_2O$(cr)}. From the above values the difference δ(M), expressed as follows:

$$\delta(M) = [\Delta_f H_m^0\{\text{B-M(OH)CO}_3\text{(cr)}\}] - [\Delta_f H_m^0\{\text{A-M(OH)CO}_3\text{(cr)}\}] \qquad (4)$$

is -(12.7 ± 2.8) and -(10.2 ± 4.3) kJ•mol$^{-1}$, for the neodymium and the samarium compounds, respectively. The uncertainty limits of these values, δ(Nd) and δ(Sm), are believed to be large enough to cover differences in the entropies of the two varieties. Thus, δ(M) can be taken, as a first approximation, as a measure of the difference in stability of the polymorphs.

## Extrapolation to B-Am(OH)CO$_3$(cr)

From the analogy of the properties of $Am^{3+}$ and the lanthanides with similar ionic radii one may anticipate that carbonate complex formation of these trivalent elements is comparable. For americium, the authors use the weighted average of the δ(Nd) and δ(Sm), i.e. δ(Am) = -(12.0 ± 2.3) kJ•mol$^{-1}$, to calculate $\Delta_f H_m^0\{$B-Am(OH)$CO_3$(cr)$\}$ = -(1552.0 ± 3.5) kJ•mol$^{-1}$. Using a suitable entropy estimate [1], they obtain log $K_{sp}^0\{$B-Am(OH)$CO_3$(cr)$\}$ = -(25.9 ± 1.0). The following value log was reported in [1]: $K_{sp}^0\{$A-Am(OH)$CO_3$•0.5$H_2O$(cr)$\}$ = -(23.1 ± 1.0). It thus can be seen that the orthorhombic A-form of hydroxycarbonate containing 0.5 water molecule is meta stable with regards to the hexagonal B-form.

## Conclusions and perspectives

This study has allowed improving interrelationships between the hydroxycarbonates of lanthanides and of americium. It was also demonstrated that the performances of the new calorimeter are adequate with milligram amounts of samples. The next measurements will involve carbonate compounds of Np, such as MNpO$_2$CO$_3$.x$H_2O$ and M$_3$NpO$_2$(CO$_3$)$_2$•x$H_2O$ (M = Na, K), which are particularly relevant [4].

*Acknowledgements*

The financial support of the Inter-University Institute for Nuclear Sciences (Brussels) to the Laboratory of Co-ordination Chemistry and Radiochemistry of University of Liège is acknowledged. The first author is also indebted to the Belgian National Agency for the Management of Radioactive Waste and Fissile Materials (NIRAS/ONDRAF) for a special travel grant.

# REFERENCES

[1]   L. Merli, J. Fuger, *Radiochim. Acta*, 74, 37 (1996).

[2]   L. Merli, B. Lambert, J. Fuger, *Journal of Nuclear Materials*, 247, 172 (1997).

[3]   E.M. Standifer, H. Nitsche, *Lanthanide and Actinide Research*, 2, 383 (1988).

[4]   H. Nitsche, Mat. Res. Soc. Symp. Proc., 257, 289 (1992).

[5]   E.J. Prosen, M.V. Kilday, *J. Res. Natl. Bur. Stand.*, A77, 581 (1973).

[6]   A.N. Christensen, *Acta Chem. Scand.*, 27, 2973 (1973).

[7]   T.R.N. Kutty, M.N. Viswanathiah, J.A.K. Tareen, Proc. Indian Acad. Sci. Chem. Sci., 40, No. 4, 69 (1978).

[8]   L. Merli, F. Rorif, J. Fuger, *Radiochim Acta*, 82, 3 (1998).

[9]   "CODATA Key Values for Thermodynamics", J.D. Cox, D.D. Wagman, V.A. Medvedev, eds., Hemisphere, New York (1989).

[10]  V.B. Parker, "Thermal Properties of Aqueous Uni-Univalent Electrolytes", Report NSRDS NBS-2 (1965).

[11]  W.J. Youden, "Statistical Methods for Chemists", Wiley, New York, pp. 18-20 (1967).

# Table 1. Thermochemical cycles for the calculation of $\Delta_f H_m^o\{B\text{-}M(OH)CO_3(cr)\}$, M = Nd, Sm

*All cycles at T = 298.15 K and $p^0$ = 101.325 kPa. Sln = 6.00 mol•dm$^{-3}$ HCl.*

|  | $\Delta_r H_m^o$ (kJ•mol$^{-1}$) |
|---|---|
| ♦ **B-Nd(OH)CO$_3$(cr)** | |
| 1. B-Nd(OH)CO$_3$(cr) + 3 HCl(sln) = (NdCl$_3$ + 2 H$_2$O + CO$_2$)(sln) | -(67.5 ± 0.5) [a] |
| 2. Nd(cr) + 3 HCl(sln) = NdCl$_3$(sln) + 1.5 H$_2$(g) | -(695.7 ± 1.8) [b] |
| 3. 2 H$_2$(g) + O$_2$(g) = 2 H$_2$O(sln) | -2 × (286.646 ± 0.041) [c] |
| 4. C(cr) + O$_2$(g) = CO$_2$(sln) | -(413.26 ± 0.20) [d] |
| | |
| **Nd(cr) + C(cr) + 2 O$_2$(g) + 0.5 H$_2$(g) = B-Nd(OH)CO$_3$(cr)** | $\Delta_f H_m^o\{$**B-Nd(OH)CO$_3$(cr)**$\}$ = **-(1614.7±1.9)** |
| | |
| ♦ **B-Sm(OH)CO$_3$(cr)** | |
| 1. B-Sm(OH)CO$_3$(cr) + 3 HCl(sln) = (SmCl$_3$ + 2 H$_2$O + CO$_2$)(sln) | -(68.3 ± 0.7) [a] |
| 2. Sm(cr) + 3 HCl(sln) = SmCl$_3$(sln) + 1.5 H$_2$(g) | -(693.2 ± 3.5) [e] |
| 3. 2 H$_2$(g) + O$_2$(g) = 2 H$_2$O(sln) | -2 × (286.646 ± 0.041) [c] |
| 4. C(cr) + O$_2$(g) = CO$_2$(sln) | -(413.26 ± 0.20) [d] |
| | |
| **Sm(cr) + C(cr) + 2 O$_2$(g) + 0.5 H$_2$(g) = B-Sm(OH)CO$_3$(cr)** | $\Delta_f H_m^o\{$**B-Sm(OH)CO$_3$(cr)**$\}$ = **-(1611.4±3.6)** |

[a] Measured in this study: 10 and 9 measurements for B-Ln(OH)CO$_3$(cr) with Nd and Sm, respectively.

[b] Measured by [8].

[c] From [9] plus relative partial molar enthalpy of H$_2$O (l) calculated from [10].

[d] $\Delta_f H_m^o$ (CO$_2$, aq), [9]. Assuming the same value as in H$_2$O(l) at $p^0$ = 101.325 kPa.

[e] Calculated from value in 1.00 mol•dm$^{-3}$ HCl, i.e. -(689.8 ± 1.7) kJ•mol$^{-1}$, corrected with the term (3.4 ± 3.1) kJ°mol$^{-1}$ estimated from results of Nd and Gd [8].

338

# DATA PROCESSING FOR THE DETERMINATION OF STABILITY CONSTANTS

**Osamu Tochiyama**
Department of Quantum Science and Energy Engineering
Graduate School of Engineering, Tohoku University

## Abstract

In the process of extracting values for stability constants from experimental measurements, two types of variables, logarithmic (such as $\log[X]$ and $\log K$) and linear (such as $[X]$ and $C_x$), must be handled in the chemical speciation calculation. This makes it difficult for us to assess the reliability of complex speciation distribution and to evaluate errors in the stability constants propagated from the errors in the measurements. To overcome this difficulty, two calculation tools, (i) non-linear least-squares program and (ii) tableau-solver formulation, were developed and implemented in commercial spreadsheet software. The former directly correlates the random and statistical errors in the data with those in the estimated stability constants, whereas the latter simulates the equilibrium for a set of given total concentrations of the metal and ligand with the values of stability constants.

## Introduction

The essence of data processing to extract values of stability constants from experimental measurements is the comparison of the experimental data, obtained from the measurements, with calculated values of a model function by using appropriate parameter values. In this process, complications arise from the fact that it must handle two types of variables, logarithmic and linear, in the chemical speciation calculation. That is, equilibrium relationships are described in terms of variables that are proportional to logarithms of concentrations such as $\log[X]$, $\log K$ and $\Delta G$, while mass balance relationships are described in terms of variables that are proportional to concentrations such as $[X]$ and $C_X$. The experimentally measured values may also be logarithmic (e.g. $\log D$ and pH) or linear (e.g. $[X]$, % and absorbance), and are usually described by a non-linear function of some independent variable(s) and given parameters (such as assumed stability constants). For this reason, it is common practice in this field to mathematically transform experimental data in order to obtain algebraically simple functional relationships. However, the complicated mathematical transformation, including conversions of linear values to logarithmic values and vice versa, makes it difficult for us to assess the reliability of the speciation of the complexes and the errors in their stability constants. In order to facilitate the calculation which correlates the errors in the measurements with those in the estimated stability constants, two simple, portable and user-friendly calculation tools, (i) a non-linear least-squares program and (ii) tableau-solver formulation were developed and implemented in a commercial spreadsheet software. The former enables us to directly correlate the random and statistical errors in the data with those in the estimated stability constants, whereas the latter enables us to carry out a sensitivity analysis by simulating the equilibrium for a set of given total concentrations of the metal and ligand, and the values of stability constants. These tools are available as Excel workbook files on request to osamu.tochiyama@qse.tohoku.ac.jp by E-mail.

## Non-linear least-squares fitting in Excel

When the experimentally obtained data can be modelled by an explicit function of a given set of control variables and parameters (assumed stability constants), it is preferable to obtain values of parameters by a non-linear least-squares curve fitting, since it enables us to fit all experimental data simultaneously while allowing us to evaluate the errors in the stability constants obtained, propagated from the errors in the data. To overcome the reluctance to use a non-linear fitting method, a non-linear least-squares fitting program based on a Marquardt algorithm [1,2] has been incorporated as a Visual Basic Application (VBA) macro in spreadsheet software (Microsoft Excel).

Table 1 shows an example of the user worksheet, which contains data to be analysed. In cells from B5 to B24, the measured values of logarithms of distribution ratios, $\log D$, of Np(V) obtained by solvent extraction [3] are listed as a function of the logarithm of the concentration of dissociated malonate anion, $\log[L]$. This example assumes that the distribution ratio, $D$, is expressed by

$$D = \frac{D_0}{1 + \beta_1[L] + \beta_2[L]^2} \tag{1}$$

where $D_0$, $\beta_1$ and $\beta_2$ are the constant parameters. Usually in solvent extraction, the concentrations of some metal ion in both organic and aqueous phases are measured to obtain $D$ values. If absolute errors in the determination of metal ion concentration are nearly equal, the relative errors in $D$ ($|\Delta D / D| = |\Delta \log D|$) will be nearly constant. In this case, the best set of parameters will be obtained by minimising the weighted residual sum of squares:

$$S = \Sigma \frac{\left(\log D_{obs,i} - \log D_{calc,i}\right)^2}{\sigma_i^2} \qquad (2)$$

where $\log D_{calc,i}$ is the calculated value by Eq. (1) and $= \Delta \log D_{obs,i}$ is the estimated standard deviation of the observed $\log D_{obs,i}$. Since the variable(s) and parameters should be selected so that their uncertainties will contribute equally to the residual sum, the variable and parameters should be taken as $\log[L]$, $\log\beta_1$ and $\log\beta_2$ (not $[L]$, $\beta_1$ and $\beta_2$). Thus, the best form that should be taken for least-squares method is:

$$\log D = \log D_0 - \log\left(1 + 10^{\log[L]+\log\beta_1} + 10^{2\log[L]+\log\beta_2}\right) \qquad (3)$$

Therefore, the formulae of this equation are written in cells from D5 to D23. For example, the following formula is written in cell D5.

$$[D5] = \$F\$4 - LOG(1 + 10^\wedge(\$F\$5 + \$A4) + 10^\wedge(\$F\$6 + 2*\$A4)) \qquad (4)$$

where cells F5, F6 andF7 contain initial estimates of $\log D_0$, $\log\beta_1$ and $\log\beta_2$, respectively.

In order to use the non-linear program, the workbook (LeastSqr_E.xls) which contains this program should be opened. After returning to the user workbook, the program (Non-linear LSQ) can be used as a macro. The program asks for the inputs as listed in (b) of Table 1, where bold face indicates the user inputs and italics are the buttons. After specifying the cell ranges for each request, one can execute the program to obtain the best set of the values in cells F5, F6 andF7 to minimise $S$ expressed by Eq. (2). In the resultant sheet, the standard deviations of the parameters, multiple correlation coefficient and chi-square probability will be also given, the meanings of which are given in standard text [1,2]. When the errors (standard deviations) of the data are not specified (by removing the check mark), the program assumes that the errors in the data ($\sigma_i$) are equal and will give the estimated standard deviation. Since a user can define any kind of non-linear function in the spreadsheet as far as it can be calculated explicitly by a given set of control variables and parameters (assumed stability constants), this program can be used flexibly and easily, without requiring any knowledge of computer programming. The only demands for users are setting up of the appropriate function and pointing the initial estimates of the parameters. As shown by the above example, a strategy of taking logarithms for the independent variable and fitting parameters usually works well, since the changes contributed to the function are comparable in magnitude.

**Tableau-solver method for the calculation of chemical systems**

Although non-linear least-squares curve fitting is a powerful tool for evaluating errors in the estimated stability constants, there still are some cases where the data obtained experimentally cannot be formulated appropriately to easily calculate with a given set of control variables and parameters. When the process requires calculation of the concentration(s) of some species from mass balance relationship(s), such as the calculation of free ligand concentration from the difference between its total added concentration and its concentration in the complex formed, the error in the total concentration cannot be handled appropriately. Since the difference cannot be negative, this kind of error would introduce serious systematic errors in the result. In such a case, the best way may be to reproduce species distribution and experimental measurements from the obtained parameters under a given condition. One can look into the contribution of such changes by changing the assumed complex species, values of obtained parameters or the total initial concentration of a metal ion or ligand (sensitivity analysis). In order to facilitate the calculation of equilibrium concentrations of all species in the solution from a given set of initial (total) concentrations and relevant stability constants, a "tableau-solver" method is devised. This is a systematic approach to organising chemical equilibrium problems and calculating the concentrations of all species in equilibrium.

In order to estimate the equilibrium concentrations of all species in the solution from given initial concentrations of some chemical materials and relevant formation constants, we will adopt a systematic approach to organise chemical equilibrium problems using the tableau format advocated by Morel and Hering [4]. Table 2(a) is an example of the tableau prepared for the titration of an equimolar mixture of acetic acid and carbonic acid.

As illustrated in Table 2(a), a tableau is a matrix of stoichiometric coefficients in which the components define the columns and the species present in the system (top part) and those in its recipe (bottom part) define the rows. In the bottom block of the leftmost column, a recipe is given to express how the system is constructed from reagents. In this example case, the solution is made from initial concentrations of carbonic acid (0.01 M) and acetic acid (0.01 M) and the added NaOH (0.01 M). Their concentrations are given in the third column in the same row. In the top block of the leftmost column, a list of the all species at equilibrium is given. In the top row, a list of components (a set of chemical entities that permits a complete description of the stoichiometry of a system) is given. In this example, $H^+$, $H_2CO_3$ and $CH_3COOH$ are enough to describe the stoichiometry of all species in the system. As discussed in [4], predominating species (species with higher concentrations) are preferred as components to avoid failure in the following numerical calculation. Although water is always a component in tableaux, it is omitted for simplicity, the concentration of $H_2O$ being effectively constant for dilute solutions. By giving the stoichiometric coefficients in the cells where the component and equilibrium or recipe species cross, all species are represented stoichiometrically in terms of the components, such as:

$$(H^+) = (H^+)_1 \tag{5}$$

$$(OH^-) = (H_2O)_1(H^+)_{-1} = (H^+)_{-1} \tag{6}$$

$$(H_2CO_3) = (H_2CO_3)_1 \tag{7}$$

$$(HCO_3^-) = (H_2CO_3)_1(H^+)_{-1} \tag{8}$$

$$(CO_3^{2-}) = (H_2CO_3)_1(H^+)_{-2} \tag{9}$$

and so on. By giving the relevant formation constants (should be given as concentration constants at the ionic strength concerned) in the rightmost columns, the concentrations of equilibrium species are expressed by the concentrations of component species such as:

$$\log[H^+] = 1 \times \log[H^+] + 0 \times \log[H_2CO_3] + 0 \times \log[CH_3COOH] + 0 \tag{10}$$

$$\log[OH^-] = -1 \times \log[H^+] + 0 \times \log[H_2CO_3] + 0 \times \log[CH_3COOH] - 14 \tag{11}$$

$$\log[HCO_3^-] = -1 \times \log[H^+] + 1 \times \log[H_2CO_3] + 0 \times \log[CH_3COOH] - 6.35 \tag{12}$$

$$\log[CO_3^{2-}] = -2 \times \log[H^+] + 1 \times \log[H_2CO_3] + 0 \times \log[CH_3COOH] - 16.68 \tag{13}$$

Each column expresses the mass balance (conservation equation) as:

$$[H^+] - [OH^-] - [HCO_3^-] - 2[CO_3^{2-}] - [CH_3COO^-] = -[NaOH]_T \tag{14}$$

$$[H_2CO_3] + [HCO_3^-] + [CO_3^{2-}] = [H_2CO_3]_T \tag{15}$$

$$[CH_3COOH] + [CH_3COO^-] = [CH_3COOH]_T \tag{16}$$

Since concentrations of all species in Eqs.(14)-(16) are expressed in terms of the concentrations of three independent components ($[H^+]$, $[H_2CO_3]$ and $[CH_3COOH]$), the problem is organised to find $[H^+]$, $[H_2CO_3]$ and $[CH_3COOH]$ which satisfy Eqs.(14)-(16). Since equilibrium species distribution is realised by minimising Gibbs free energy of the system, the true variables are not $[H^+]$, $[H_2CO_3]$ and $[CH_3COOH]$ but $\log[H^+]$, $\log[H_2CO_3]$ and $\log[CH_3COOH]$, which are proportional to free energy. To solve this problem, the tableau in Table 2(a) is prepared in an Excel worksheet. For example, when the leftmost top cell is A1, the initial guesses of log[H], log[L] and log[M] are written in the second row (in cells D2, E2 and F2). Formulae to express the logarithmic concentrations of equilibrium species are written in the second column, for example, the following formulae are written in cells B3 and B4 to express Eqs. (10) and (11).

$$[B3] = SUMPRODUCT(D\$2:F\$2*D3:F3) + G3 \tag{17}$$

$$[B4] = SUMPRODUCT(D\$2:F\$2*D4:F4) + G4 \tag{18}$$

where, for example, Eq. (10) is equivalent to:

$$[B3] = (D\$2*D3 + E\$2*E3 + F\$2*F3) + G3 \tag{19}$$

In the third column, formulae to give linear concentrations of equilibrium species are written, such as:

$$[C3] = 10\wedge B3, \quad [C4] = 10\wedge B4, \tag{20}$$

Then in cells D14 to F14 (just below the recipe rows), the following formulae are written to express mass balances.

$$[D14] = SUMPRODUCT(\$C3:\$C10*D3:D10) \tag{21}$$
$$-SUMPRODUCT(\$C11:\$C13*D11:D13)$$

$$[E14] = SUMPRODUCT(\$C3:\$C10*E3:E10) \tag{22}$$
$$-SUMPRODUCT(\$C11:\$C13*E11:E13)$$

$$[F14] = SUMPRODUCT(\$C3:\$C10*F3:F10) \tag{23}$$
$$-SUMPRODUCT(\$C11:\$C13*F11:F13)$$

To satisfy Eqs. (14)-(16) simultaneously, the values estimated by the formulae (21)-(23) must all be zero. Thus, in cells D15 to F15:

$$[D15] = D14\wedge 2, \quad [E15] = E14\wedge 2, \quad [F15] = F14\wedge 2 \tag{24}$$

are written, and the sum of these squares is calculated in cell G15.

$$[G15] = LOG(SUM(D15:F15)) \tag{25}$$

Here, the problem is organised to find log[H], log[L] and log[M] in order to minimise the value of cell G15. To solve this implicit problems, a "solver" prepared in Excel can be used conveniently. By running the solver to minimise the value of the target cell (G15) by changing the values of cells D2:F2, the approximate solution (set of log[H], log[L] and log[M]) can be obtained. The concentrations of all equilibrium species are obtained simultaneously. This procedure can be repeated by changing the recipe (the concentration of $[H^+]_T$, $[H_2CO_3]_T$ or $[CH_3COOH]_T$) to give simulated curves of titration, extraction etc. The workbook, Tableau_E.xls, provides the macro (Repeating Tableau) for this simulation. This macro requests the inputs as listed in Table 2(b). By running this macro and saving the results of each step, equilibrium concentrations of any species can be expressed as a function of the total concentration added of any material. Although appropriate initial guesses are necessary to avoid erroneous estimations, this tableau formulation provides us with a flexible and easy to use means of simulating equilibrium.

**REFERENCES**

[1]   W.H. Press, B.P. Flannery, S.A. Teukolsky, W.T. Vetterling, "Numerical Recipes in Pascal", Cambridge University Press, New York, 1989.

[2]   T. Nakagawa, Y. Koyanagi, "Experimental Data Analysis by Least-Squares Method-Programme SALS", Tokyo Daigaku Shuppankai, 1982 (in Japanese).

[3]   O. Tochiyama, Y. Inoue, S. Narita, *Radiochim. Acta*, 58/59, 129 (1992).

[4]   F.M.M. Morel, J.G. Hering., "Principles and Application of Aquatic Chemistry", John Wiley & Sons, Inc., New York,1993.

# Table 1. Spreadsheet arrangement of non-linear least-squares program

*a) An example of the data worksheet*

|    | A | B | C | D | E | F | G |
|----|------|------|------|------|------|------|------|
|    | log[L] | logD(obs) | logD(calc) | ΔlogD |  | parameters | StdDev |
| 4  | log[L] | logD(obs) | logD(calc) | ΔlogD |  | parameters | StdDev |
| 5  | -0.3010 | -2.7310 | -2.7386 | 0.01 | logDo= | 1 |  |
| 6  | -0.4260 | -2.4856 | -2.4956 | 0.01 | logB1= | 3 |  |
| 7  | -0.6021 | -2.1901 | -2.1569 | 0.01 | logB2= | 4 |  |
| 8  | -0.9031 | -1.5829 | -1.5934 | 0.01 |  |  |  |
| 9  | -1.3010 | -0.8994 | -0.8999 | 0.01 |  |  |  |
| 10 | -1.4260 | -0.6985 | -0.7001 | 0.01 |  |  |  |
| 11 | -1.6021 | -0.4228 | -0.4374 | 0.01 |  |  |  |
| 12 | -1.9031 | -0.0575 | -0.0470 | 0.01 |  |  |  |
| 13 | -2.3010 | 0.3327 | 0.3433 | 0.02 |  |  |  |
| 14 | -2.4260 | 0.4402 | 0.4361 | 0.02 |  |  |  |
| 15 | -2.6021 | 0.5458 | 0.5443 | 0.02 |  |  |  |
| 16 | -2.9031 | 0.6657 | 0.6751 | 0.02 |  |  |  |
| 17 | -3.3010 | 0.7729 | 0.7698 | 0.02 |  |  |  |
| 18 | -3.4260 | 0.7853 | 0.7872 | 0.02 |  |  |  |
| 19 | -3.6021 | 0.8128 | 0.8052 | 0.02 |  |  |  |
| 20 | -3.9031 | 0.8289 | 0.8237 | 0.02 |  |  |  |
| 21 | -4.3010 | 0.8354 | 0.8352 | 0.02 |  |  |  |
| 22 | -4.4260 | 0.8188 | 0.8371 | 0.02 |  |  |  |
| 23 | -4.6021 | 0.8472 | 0.8390 | 0.02 |  |  |  |
| 24 | -4.9031 | 0.8472 | 0.8410 | 0.02 |  |  |  |

*b) Input parameters for a)*

| Condition For Calculation | | |
|---|---|---|
| Data Cells | Sheet1!$B$5:$B$24 | |
| Fitting Function Column | Sheet1!$C:$C | |
| Parameter Cells | Sheet1!$F$5:$F$7 | |
| √   Error of Data (1/√weight) | Sheet1!$D:$D | |
| Convergence | 0.000001 | Iteration | 100 |
| Clear | Execute | Cancel |

## Table 2. Spreadsheet arrangement of tableau format

*a) An example of the tableau for carbonic acid-acetic acid-NaOH system*

|   | A | B | C | D | E | F | G |
|---|---|---|---|---|---|---|---|
| 1 |   |   |   | H(+) | H$_2$CO$_3$ | CH$_3$COOH | logK |
| 2 |   |   |   | -5.5556779 | -2.0646781 | -2.8601867 |   |
| 3 | H(+) | -5.5557 | 2.78E−06 | 1 |   |   |   |
| 4 | OH(-) | -8.4443 | 3.59E−09 | -1 |   |   | -14 |
| 5 | H$_2$CO$_3$ | -2.0647 | 0.008616 |   | 1 |   |   |
| 6 | HCO$_3$ (-) | -2.859 | 0.001384 | -1 | 1 |   | -6.35 |
| 7 | CO$_3$ (2-) | -7.6333 | 2.33E−08 | -2 | 1 |   | -16.68 |
| 8 | CH$_3$COOH | -2.8602 | 0.00138 |   |   | 1 |   |
| 9 | CH$_3$COO(-) | -2.0645 | 0.00862 | -1 |   | 1 | -4.76 |
| 10 |   |   |   |   |   |   |   |
| 11 | H$_2$CO$_3$ |   | 0.01 |   | 1 |   |   |
| 12 | CH$_3$COOH |   | 0.01 |   |   | 1 |   |
| 13 | NaOH |   | 0.01 | -1 |   |   |   |
| 14 |   |   |   | -5.154E-07 | -8.842E-08 | -5.277E-07 |   |
| 15 |   |   |   | 2.6562E-13 | 7.8188E-15 | 2.7848E-13 | -12.25813 |

*b) Input parameters to repeat calculation of a)*

| Variable Cell | Sheet1!$C$13 | Target Cell | Sheet1!$G$15 |
|---|---|---|---|
| Initial | **0** | Adjusting Cells | Sheet1!$D$2:$F$2 |
| Increment | **0.0005** |   |   |
| Final | **0.035** | Cells to Save | Sheet1!$B$3:$B$9 |
|   |   | Place to save results | Sheet1!$A$18 |
| *Clear* | *Execute* |   | *Cancel* |

# CHEMOMETRIC AND COMPUTER-INTENSIVE TECHNIQUES FOR SPECIATION

**G. Meinrath**
RER Consultants
Schießstattweg 3a, D-94032 Passau, Germany
*and*
Institutes of Geology and Inorganic Chemistry
Technical University Mining Academy Freiberg
D-09596 Freiberg, Germany

**T. Kimura, Y. Kato, Z. Yoshida**
Advanced Science Research Centre
Japan Atomic Energy Research Institute
J-319-1195 Tokai-mura, Japan

**S. Lis**
Department of Rare Earths
Adam-Mickiewicz-University Poznan
Pl-60780 Poznan, Poland

## Abstract

Modern computer-assisted spectroscopic methods are able to collect large data sets from a chemical system under study in comparatively short time. An analysis of these data sets cannot be achieved in a satisfactory way by heuristic analysis but requires suitable numerical data treatment tools. Such tools, e.g. factor analysis, canonical correlation and computer-intensive Monte Carlo resampling techniques, provide detailed statistical informations on the distribution of information in the data set. Application of such techniques also implies that the limitations of each technique are kept in mind. Modern requirements of traceability and comparability in analytical chemistry have given rise to detailed criteria on how analytical data have to be reported. For nuclear science, compliance with these criteria ensures wider public acceptance of nuclear fuel cycle related results outside its own community.

## Introduction

Scientific studies aim at the accurate estimation of true values of parameters that describe physical processes in nature. This activity is distinct from technical studies where focus is on parameters describing technical processes in an often heuristic approach, giving emphasis to precision. As an example, the measurement of pH may be considered. Using standard equipment, like a glass electrode in combination with a voltmeter and a calibration solution, an empirical scale may be established that returns a "pH" value, e.g. for a technical process. Technically, it is sufficient that the calibration of the electrode is done in a reproducible way to establish this "pH" with sufficient precision satisfying the quality requirements of the process. However, such a "pH" value does not return any information on the accuracy of the activity of hydrogen ions in the solution. In scientific context, however, the measurement has to give an indication of the likely difference between measured and true hydrogen ion activity. Establishing standard reference materials and procedures that are fulfilling the metrological requirements is by no means an easy task [1,2]. While for technical process parameters, comparability with other measurement systems is less important, a scientific quantity must be comparable. Hence, traceability via metrological chains of comparison is essential. All uncertainties have to be determined and included.

This example shows that quality assessment and control (QA/QC) in chemistry [3] are interested in improving comparability of data from different measurement processes with respect to each other and with respect to established reference standards and materials. To comply with the ISO QA/QC requirements – also adopted by IUPAC and in the process of adoption by IAEA – for analytical chemistry, it is necessary to evaluate an estimate of measurement uncertainty for each step in the derivation of the final result [4]. Without traceable uncertainty, scientific data can not be compared.

In the following, three speciation examples with relevance to the evaluation of meaningful statistical analysis will be discussed: A solubility study with non-linear model function and non-normal residuals, evaluation of single component spectra from UV-Vis absorption spectra and estimation of fluorescence quantum yields of solution species.

## Solubility study

Extraction of model parameters from experimental data and predicting future events on basis of model parameters are common tasks in science and technology. It is general practice to fit models to experimental data by suitable algorithms that minimise some loss function. The sum of squared residuals (SOR) is most commonly chosen as a loss function:

$$SOR = \Sigma w_i(y_i - f(x_i, p_m))^n, \ (n=2) \tag{1}$$

where $w_i$, weight; $y_i$, measurand; $x_i$, regressor; $p_m$, parameter of the model function f(). The preference for SOR results from linear regression, where under a series of conditions including identically and independently distributed (i.i.d.) residuals the parameter estimates p are the maximum likelihood estimates. However, for non-linear model functions or non-normally distributed residuals or dependent (correlated) residuals, this preference for the loss function Eq. 1 is not any longer valid [5]. The use of SOR as fitting criteria reduces the robustness of an estimation process because an outlying data may govern the result of a numerical analysis via Eq. 1. However, the classical regression technique depends on the loss function Eq. 1 for estimation of variance estimates for each parameter. The variance-covariance matrix is obtained from the slopes of the Jacobian matrix in the minimum of the loss function Eq. 1. By using the more robust sum of residuals (n = 1), however, calculation of the Jacobian is not possible.

If the loss function Eq. 1 is applied to non-linear functions, bias in the parameters will result. Bias estimation in non-linear regression analysis involves rather complicated numerical techniques [5] and, hence, is commonly not applied by chemists. Computer-intensive Monte Carlo resampling techniques overcome these difficulties by replacing sophisticated and often unavailable statistical techniques by brute computing force [6]. They return the estimated probability function of a parameter, thereby eliminating the often incorrect assumption of normally distributed parameter probability densities. A more detailed discussion of the bootstrap is given elsewhere [7-9]. Here, the second-order correct bias-correction capabilities of bootstrap will be demonstrated for the U(VI) solubility data in Figure 1 [9].

**Figure 1. Solubility of U(VI) in equilibrium with $UO_2CO_3(s)$ in 0.1 M perchlorate solution under $CO_2$ atmosphere at 25 °C. The data is interpreted by the single species $UO_2CO_3°$, $UO_2(CO_3)_2^{2-}$ and $UO_2(CO_3)_3^{4-}$. Minor concentration of $(UO_2)_2(OH)_2^{2+}$ detected by spectroscopy are also given (dashed line) but will not be further considered in the text.**

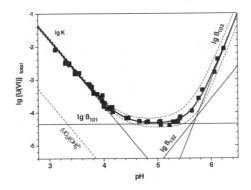

The non-linear model function describing this species model includes terms $K$ $\beta_{10n}$ ($n = 1-3$), where $K$ gives the solubility product of $UO_2CO_3(s)$ and $\beta_{10n}$ the formation constant of species $UO_2(CO_3)_n^{(2-2n)}$. Confidence regions for non-linear model functions can be approximated by Eq. 2:

$$SOR(X) \leq SOR(\overline{X})\left[1 + \frac{P}{n-p}F_{p,n-p,\alpha}\right] \qquad (2)$$

with X, parameter; $\overline{X}$, parameter mean value from least square fit; n, total number of parameters in the model; P, number of parameters for which the confidence region is estimated simultaneously; $F_{(p, n-P, \tau)}$, Fisher's F for n and n-P degrees of freedom and confidence level $\tau$. Eq. 2 is approximate, e.g. due to the comparison with Fisher's F that implies Normally distributed parameters [5].

Bootstrap statistics provides second order correct marginal confidence estimates by bias and acceleration correction ($BC\alpha$) via the cumulative probability distribution obtained for each parameter from a bootstrap analysis [7,9]. A comparison with approximate confidence regions from Eq. 2 with the $BC\alpha$ marginal confidence limits for $\tau = .68$, .90 and .95, respectively, is shown in Figure 2.

A map of the confidence regions according to Eq. 2 has been made by a Monte Carlo routine and three different confidence regions are given in Figure 2. The confidence regions are not always elliptical with the confidence regions of the dicarbonato species even being unbound towards lower parameter values for 99% confidence. A major drawback is the lack of an analytical formula for calculation of the confidence regions (Eq. 2). Hence, transferability of these informations within complex statistical processes is difficult.

**Figure 2. Comparison of non-linear least-squares confidence regions with BC$_\alpha$ bootstrap marginal confidence regions**

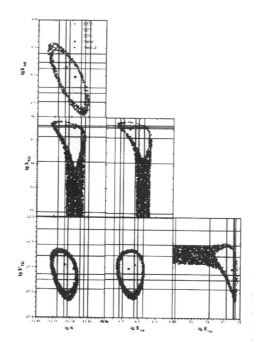

The bootstrap BC$_\alpha$ marginal confidence limits are given by straight lines in Figure 2. The marginal limits are not centred at the regression means because the mean is not a suitable location estimator in case of non-linearity [10]. The median, given by open squares in Figure 2, is more robust because it is less sensitive to outliers. It has been shown, e.g. [11] that BC$_\alpha$ bootstrap confidence regions are second-order correct in contrast to non-linear least-squares confidence region that are – at best – first-order correct. Robustness and optimality properties of the bootstrap add to the confidence put on BC$_\alpha$ bootstrap confidence intervals. As a consequence, non-linear least-square regression estimates of parameters, e.g. of actinide thermodynamic parameters, are most probably biased. Standard deviations reported for these parameters from least-squares procedures (if not even merely guessed) suggest higher confidence than actually justified.

## UV-Vis spectroscopic data analysis

The problem of rank estimation for matrices holding, e.g. UV-Vis spectral data, has challenged analytical chemists for more than a decade. Having a measure for the rank of a matrix, say A, of experimental spectra of a given chemical systems, the number of independently absorbing units in a multi-component mixture can be obtained. Generally these units are individual species. The true rank of A, obtainable from factor analysis, is always obscured by random disturbances (sometimes termed error or noise) that are unavoidable in all experimental science. Therefore, the significance of a rank estimate is always obtained from statistical analysis of the eigenvalues of matrix A. Numerically stable eigen analysis is available by singular value decomposition (SVD) [12]. SVD decomposes the matrix A into three matrices S, V and T such that S holds the column eigenvectors, T holds the row eigenvectors and V gives the square roots of the eigenvalues. The rank of matrix A is estimated from statistical analysis of the eigenvalues. Factor analysis is a method of multivariate linear regression and, hence, assumes the noise in the matrix A being i.i.d. and homoscedastic. It can, however, be shown that these assumptions are not valid for almost all UV-Vis spectra [13].

Table 1 gives some statistical data on the eigen analysis of a nine-spectra matrix A holding UV-Vis absorption spectra of U(VI) in $SO_4^{2-}/ClO_4^-$ medium with sulphate concentration in the range $1\cdot10^{-4}$ to $2\cdot10^{-1}$ M at pH 2.5 and 1 M ionic strength [14].

**Table 1. Eigenvalues and statistical parameters root mean square error (RE) and imbedded error (IE) of a nine-column UV-Vis spectral matrix**

| Eigenvalue | RE | IE |
|---|---|---|
| 12.40 | 0.06275 | 0.02092 |
| 0.1934 | 0.00839 | 0.003955 |
| 0.00132 | 0.00125 | 0.000722 |
| 0.00049 | 0.00059 | 0.000373 |
| 0.00010 | 0.00028 | 0.000215 |
| 0.00002 | 0.00016 | 0.000129 |
| 0.000004 | 0.00010 | 0.000089 |
| 0.000002 | 0.00010 | 0.000090 |

Factor analysis assumes that a certain number of eigenvalues is significant while the remaining smaller eigenvalues represent noise. The strategy is to study the statistical parameters as a function of the number of significant eigenvalues. The root mean square error (RE) estimates the error in the eigenvalue on basis of the noise eigenvalues. The root mean square error indicates the average noise level in the eigenvectors while the imbedded error (IE) is the average error that remains in the data matrix by recalculating matrix A with only the significant eigenvectors and eigenvalues. For less than three significant eigenvalues, the difference in IE in each step is larger than $1\cdot10^{-3}$, but for more than four significant eigenvectors IE becomes less than $2\cdot10^{-4}$ absorption units-a limit to modern UV-Vis spectrometers. The conclusion therefore is that the data matrix has a rank of 3. The rank estimate, however, is build upon the assumption of i.i.d. and homoscedastic residuals. In Figure 3, a Kolmogorov-Smirnov analysis of the residual from the rank 3-data matrix A is given.

**Figure 3. Kolmogorov-Smirnov comparison of residuals from data matrix A with the closest fitting normal distribution**

Mean $= 1.75\cdot10^{-5}$, $\sigma = 4\cdot10^{-4}$ cm$^{-1}$

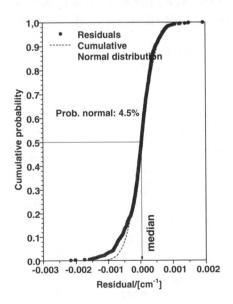

The analysis shows a probability of at best 4.5% that the residuals are a random realisation from a Normal distribution. However, such an analysis of ordered residuals does not account for correlation within the residuals. Their probability therefore may even be less than 4.5%.

The extraction of single component spectra is straightforward via target factor analysis (TFA) The validity of single component spectra of species $UO_2^{2+}$, $UO_2SO_4^{\circ}$ and $UO_2(SO_4)_2^{2-}$ from TFA has to be assessed by an application example. To avoid analysis on the basis of i.i.d. and normally distributed residuals, moving block bootstrap (MBB) analysis of spectral data has been developed [13]. MBB analysis not only returns a variance estimate but the complete probability distribution of the concentration estimator of each species. An application example is shown for a three-component UV-Vis absorption spectrum from the U(VI)-$SO_4^{2-}$ system in Figures 4(a) and 4(b).

**Figure 4(a). Result of a MBB spectral analysis of a mixed component U(VI)-$SO_4^{2-}$ UV-Vis absorption spectrum**

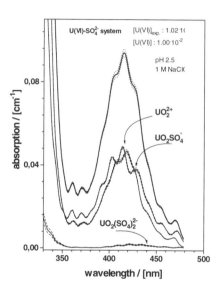

**Figure 4(b). Comparison of LSR 95% confidence ellipses with MBB results from 1 000 resampling cycles for spectrum Figure 4(a). The bias of LSR is obvious. Coverage of LSR ellipses by MBB results is almost negligible.**

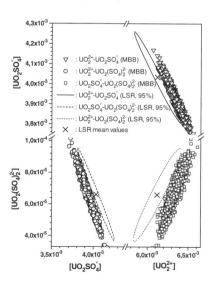

While moving block bootstrap provides second-order correct confidence limits [15], LSR ellipses are deduced from data that does not fulfil the LSR requirements of i.i.d. and Normal distribution. The obvious bias is transferred to thermodynamic data if the concentration ratios from spectral analysis are used for evaluation of thermodynamic data.

## Estimation of fluorescence quantum yields

Derivation of relative fluorescence quantum yields of fluorescing species involves complex experimental and numerical data processing. The main analytical steps involve a) the derivation of single component absorption spectra of the relevant species, b) the derivation of single component emission spectra of the species, c) the assessment of the composition of experimental samples d) the determination of fluorescence intensities from these experimental samples and e) the interpretation of several samples with differing composition to derive a composition dependent fluorescence intensity. It is obvious that the complexity of measurements with their own measurement uncertainty only allows an evaluation of a rough estimate for the relative quantum yield.

Figures 5(a) and (b) give the observed fluorescence lifetime dependence of species $UO_2^{2+}$, $(UO_2)_2(OH)_2^{2+}$ and $(UO_2)_3O\text{-}(OH)_3^+$ as a function of the solution composition expressed by the absorption ratios $r_1 = a((UO_2)_2(OH)_2^{2+})/a(UO_2^{2+})$ and $r_2 = a((UO_2)_3O(OH)_3^+)/a(UO_2^{2+})$ at 420 nm (excitation wave length) [16]. The relative fluorescence yields are $\Phi_r((UO_2)_2(OH)_2^{2+}) = 12 \pm 3$ and $\Phi_r((UO_2)_3O(OH)_3^+) = 10 \pm 3$ relative to $UO_2^{2+}$. The large standard deviations result from the accumulating uncertainties in course of the steps a)-e) given above.

### Figures 5. Variation of relative fluorescence life times
### as a function of the absorption ratio $r_1$ (a) and $r_2$ (b) [16]

*(a)*            *(b)*

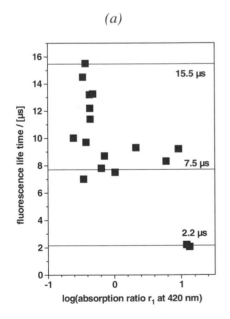 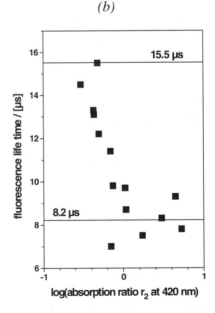

## Concluding remarks

Different experimental methods for determination of thermodynamic data applied to the same system do not usually return the same results. There is few explanation on the sources of this divergence. It is, however, a nuisance factor limiting the value of chemical investigations.

Internationally accepted QA/QC standards stipulate traceability and comparability of measurement results. The examples emphasise the role of bias and inconsistency introduced to the evaluation process by inadequate post-analysis tools where basic assumptions, for example on the data structure, are not met. Robust methods, based on computer-intensive statistical methods, may be helpful to improve this situation.

# REFERENCES

[1]     P. Spitzer, R. Eberhardt, I. Schmidt, U. Sudmeier, *Fresenius J. Anal. Chem.*, 356, 178-181 (1996).

[2]     G. Meinrath, S. Hurst, R. Gatzweiler, *Fresenius J. Anal. Chem.* (2000), in press.

[3]     "ISO Guide to the Expression of Uncertainty in Measurement", ISO, Geneva/CH (1993).

[4]     S. Ellison, W. Wegscheider, A. Williams, *Anal. Chem.*, 69, 607A-611A (1997).

[5]     D.M. Bates, D.G. Watts, *J. Royal Statist. Soc. Ser.*, B, 42, 1-15 (1980).

[6]     B. Efron, R. Tibshirani, *Science*, 253, 390-395 (1991).

[7]     A.C. Davison, D.V. Hinkley, "Bootstrap Methods and their Application", Cambridge University Press, Cambridge/UK (1997), 582 p.

[8]     G. Meinrath, C. Ekberg, A. Landgren, J-O. Liljenzin, *Talanta*, 51, 231-246 (2000).

[9]     G. Meinrath, Y. Kato, T. Kimura, Z. Yoshida, *Radiochim. Acta*, 84, 21-29 (1999).

[10]    M.J. Box, *J. Royal Stat. Soc. Ser.*, B, 33, 171-201 (1971).

[11]    K. Singh, *Ann. Statist.*, 9, 1187-1195 (1981).

[12]    G.H. Golub, C. Reinsch, *Numer. Mathematik*, 14, 403-420 (1970).

[13]    G. Meinrath, *Anal. Chim. Acta*, 415, 105-115 (2000).

[14]    Y. Kato, T.Kimura, Z. Yoshida, G. Meinrath, Report UCRL-135626, Abstr. Migration '99, Lake Tahoe, USA (1999).

[15]    H.R. Künsch, *Ann. Statist.*, 17, 1217 -1241 (1989).

[16]    G. Meinrath, S. Lis, Z. Stryla, C. Noubactep, *J. Alloy Comp.*, 300/301, 107-112 (2000).

# PUREX PROCESS MODELLING – DO WE REALLY NEED SPECIATION DATA?

**Robin J. Taylor and Iain May**
B229, Research and Technology, BNFL Sellafield
Seascale, CA20 1PG, UK

## Abstract

The design of reprocessing flowsheets has become a complex process requiring sophisticated simulation models, containing both chemical and engineering features. Probably the most basic chemical data needed is the distribution of process species between solvent and aqueous phases at equilibrium, which is described by mathematical algorithms. These algorithms have been constructed from experimentally determined distribution coefficients over a wide range of conditions. Distribution algorithms can either be empirical fits of the data or semi-empirical equations, which describe extraction as functions of process variables such as temperature, activity coefficients, uranium loading, etc. Speciation data is not strictly needed in the accumulation of distribution coefficients, which are simple ratios of analyte concentration in the solvent phase to that in the aqueous phase. However, as we construct process models of increasing complexity, speciation data becomes much more important both to raise confidence in the model and to understand the process chemistry at a more fundamental level. UV/vis/NIR spectrophotometry has been our most commonly used speciation method since it is a well-established method for the analysis of actinide ion oxidation states in solution at typical process concentrations. However, with the increasing availability to actinide science of more sophisticated techniques (e.g. NMR; EXAFS) complementary structural information can often be obtained. This paper will, through examples, show how we have used spectrophotometry as a primary tool in distribution and kinetic experiments to obtain data for process models, which are then validated through counter-current flowsheet trials. It will also discuss how spectrophotometry and other speciation methods are allowing us to study the link between molecular structure and extraction behaviour, showing how speciation data really is important in PUREX process modelling.

# SUBGROUP REPORTS

# TRACE CONCENTRATION SPECIATION (<10$^{-6}$)

## Instrumental methods

### *Luminescence*

Chemiluminescense can be used for quantification of certain elements like U(VI), Cm(III) [1] down to 10$^{-12}$ g.

Matrix assisted luminescense (applying UV excitation) can be used for quantification of Np, Pu, U, Am in the range 10$^{-9}$-10$^{-12}$ g (linearity in 5 orders of magnitude) [2,3]. The method can be combined with oxidation specific separation, for example by coprecipitation, extraction [4,5]. Speciation can be achieved in the course of pre-separation by different methods.

### *Advantage*

- High sensitivity.

### *Disadvantage*

- Only few elements can be determined. Sample preparation is time consuming.

### *Future improvement*

Future developments can be made by studying new matrices.

### *Mass spectrometry*

ElectroSpray Mass Spectrometry (ES-MS) can be used for speciation study. It is possible to couple a liquid samples at atmospheric pressure directly to a mass detection system working at reduced pressure with a soft mode of ionisation that allows to give information on chemical species present [6]. Detection limit is ….

Another soft ionisation technique, Matrix Assisted Laser Desorption Ionisation Time of Flight MS (MALDI-TOF MS) [7] can be used for speciation with low detection limits (in some cases down to femto or atto moles). It can be used directly without any separation or it can be combined with separation methods, for example with high performance Capillary Electrophoresis (CE).

CE-MALDI TOF MS is also known on a chip or combined with a multicapillary systems to reach high throughput of the samples.

*Advantages*

On-line CE-MALDI/TOF MS is possible, low sample volume (μL to nano L).

*Disadvantages*

Few applications for inorganic speciation up to now [8], but no experience on actinide speciation.

- Destructive method.

- Further development needed.

*Future*

For actinides speciation further development necessary.

### GC/HPLC separation methods

GC can be applied for iodine-129 speciation with kinetic or MS detection methods [9].

HPLC and Ion Chromatography (IC): with modified detection methods, for example radiometrical or MS detection used as speciation method.

*Advantages*

- Small amount of samples.

- Simple sample pre-treatment.

*Disadvantages*

Future development still needed for radioactive metals application.

### LASER spectrometry

*Photothermal spectroscopy*

Photon-induced thermal process is detected by various sensoric methods for spectroscopy, which is equivalent to UV-VIS absorption spectroscopy but with very low sensitivity ($A = \varepsilon \times c = 10^{-6}$ cm$^{-1}$ is attainable). Speciation of oxidation and complexation states is possible for all 5f elements. Photon absorption of solvent (e.g. of water) is the limitation for the sensitivity but can be compensated to some extents by a dual channel set-up.

*Luminiscence spectroscopy*

Both excitation and emission spectroscopy provide a very high sensitivity. In addition, the relaxation rate of collisions can be used for speciation. Applicability is restricted to some lanthanides (c.f. Eu(III), Gd(III), Tb(III)) and actinides (e.g. Cm(III), Am(III), U(VI)). For the favourable case, the sensitivity can reach down to $10^{-12}$ mol/L.

*Breakdown spectroscopy*

Photon-induced plasma generation can be used for detection of aquatic colloids. The plasma brightness and acoustic signal detections provide a high sensitivity for colloids of smaller size ($< 50$ nm) down to $10^{-12}$ g/mL. Both, particle size and number density can be determined.

## Radiochemical speciation

### Solution

Several radiochemical techniques can be used for speciation of fission products.

- *Oxidation state*. The following methods can be used for the purpose:

  - Extraction [10].

  - Co-precipitation [11].

  - Ion exchangers [12].

  - Sorption [13].

- *Charge-state speciation* (cationic, anionic, neutral) can be achieved applying:

  - Ion exchange [14,15].

  - Electrophoresis [14,15].

  - Capillary diffusion [eventually for the method: 16].

  - Ion focusing and osmosis [17].

- *Complexation state and hydrolysis*. The following methods can he applied for this purpose:

  - Extraction [18].

  - Electrophoresis [19, 20].

  - Ion focusing [21].

  - HPLC [22].

- *Colloidal state*:

    - Membrane filtration [23,24], ultra filtration [14-17].

    - Field Flow Fractionisation (FFF), ultra centrigugation.

    - GPC (Size Exclusion Chromatography or Gel Permeation Chromatography).

### Solid-liquid interface speciation

Mineral surface adsorption of actinides by ion exchange, surface complexation, or surface mineralisation (precipitation) can be speciated by isotherm trace experiment (versus pH, I, etc.) and/ or spectroscopy, for example by TRLFS for Cm (III), Eu(III), U(VI) in the concentration range $10^{-6}$ down to $10^{-9}$ mol/L [25].

Some information about the mechanism of interaction with solid surface can be characterised by sequentional desorption experiments.

### Speciation in solid matrix

There is no scheme (procedures, methods) generally accepted.

- Sample preparation requires careful homogenisation (separation of hot particles, magnetic materials, and others).

- For speciation in soil and sediments consequent leaching as in geochemical studies can be applied [14,26]. Sequention leaching provides phenomenological information for certain cases about ability of different forms of actinides and fission products.

- Redox speciation in different leachates can be done using the same methods as for redox speciation in liquid samples (for example extraction, coprecipitation, and others) [27].

### Gas and aerosol

A few methods have been used. For example, gas chromatography and adsorption technologies are available for iodine speciation [28-30]. Gas chromatography is available for organic iodine compound as described in the section of GC/HPLC separation methods and adsorption technology to Ag filters is available for inorganic iodine compound.

*Subgroup members*

Hikaru Amano

Toshihide Asakura

Blandine Fourest

Yukiko Hanzawa

Josef Havel (Recorder)

Jae-Il Kim

Boris F. Myasoedov (Leader)

Seiya Nagao

Kousaburo Ohashi

Hideyo Takeishi

Shigekazu Usuda

Wenge Zheng

# REFERENCES

[1]    *Unavailable at the time of publication.*

[2]    B.F. Myasoedov, I.A. Lebedev, "Latest Achievements in the Analytical Chemistry of Actinides", *J. Radioanal. Nucl. Chem.,* Art.147, 1, pp. 5-26 (1991).

[3]    P. Thouvenot, S. Hubert, C. Moulin, P. Decambox, P. Mauchien, "Americium Trace Determination in Aqueous and Solid Matrices by Time-Resolved Laser Induced Fluorescence", *Radiochim. Acta*, 6, 15 (1993).

[4]    S.A. Ivanova, M.N. Mikheeva, A.P. Novikov, B.F. Myasoedov, "Preconcentration of Neptunium by Supported Liquid Membranes for Luminescent Analysis of Environmental Samples", *J. Radioanal. Nucl. Chem.*, Art. 186, 4, pp. 341-352 (1994).

[5]    A.P. Novikov, M.N. Mikheeva, O.Iv. Gracheva, B.F. Myasoedov, "Luminescent Determination of Neptunium and Plutonium in the Environment", *J. Radioanal. Nucl. Chem.*, Art. 223, 1-2, pp. 163-166 (1997).

[6]    *Unavailable at the time of publication.*

[7]    *Unavailable at the time of publication.*

[8]    *Unavailable at the time of publication.*

[9]    T. Fukasawa, K. Funabashi, Y. Kondo, "Separation Technology for Radioactive Iodine from Off-Gas Streams of Nuclear Facilities", *J. Nucl. Sci. Technol.*, 31, pp. 1073-1083 (1994).

[10] P.A. Bertrand, G.R. Choppin, "Separation of Actinides in Different Oxidation States by Solvent Extraction", *Radiochemica Acta*, 31, pp. 135-137 (1982).

[11] Y. Inoue, O. Tochiyama, *J. Inorg. Nucl. Chem.*, 39, 1443-1447 (1977).

[12] G.R. Choppin, K.L. Nash, "Actinide Separation Science", *Radiochimica Acta*, 70/71, 225-236 (1995).

[13] A. Kobashi, G.R. Choppin, J.W. Morse, *Radiochemica Acta*, 43, 211 (1988).

[14] B.F. Myasoedov, A.P. Novikov, F.I. Pavlotskaya, "The Study of Radionuclides Behaviour in Aquatic and Terrestrial Ecosystems Near PA "Mayak": Methodology, Objects, Techniques", *Russ. Radiochemistry*, 40, 5, pp. 447-452 (1998).

[15] F.I. Pavlotsky, A.P. Novikov, T.A. Goryachenkova, I.E. Kazinsky, B.F. Myasoedov, "Occurrence Forms of Radionuclides in Water and Bottom Sediments of Some Basins near PA (Mayak)", *Russ. Radiochemistry*, 40, 5, pp. 462-467 (1998).

[16] B. Fourest, J. Duplessis, F. David, "Comparison of Diffusion Coefficents and Hydrated Radii for Some Trivalent Lanthanide and Actinide ions in Aqueous Solutions", *Radiochim. Acta*, 36, pp. 191-195 (1984).

[17] *Unavailable at the time of publication.*

[18] *Unavailable at the time of publication.*

[19] R. Lundqvist, "Scand. Electromigration Method in Tracer Studies of Complex Chemistry. I General", *Acta Chem.*, A 35, 9, pp. 31-41 (1981).

[20] F. Rösch, T. Reinmann, G.V. Buklanov, M. Milanov, V.A. Khalkin, "Electromigration of Carrier-Free Radionuclides 8. Hydrolysis of $^{249}$Cf(III) in Aqueous Solution", *Radiochim. Acta* 47, pp. 187-189 (1989).

[21] *Unavailable at the time of publication.*

[22] *Unavailable at the time of publication.*

[23] H. Amano, T. Matsunaga, S. Nagao, *et al.*, *Organic Geochemistry*, 30, 437-442 (1999).

[24] T. Matsunaga, T. Ueno, H. Amano, *et al.*, *J. Contaminant Hydrology*, 35, 101-113 (1998).

[25] R. Cavellec, C. Lucas, E. Simoni, S. Hubert, N. Edelstein, "Structural Characterisation of Sorption Complexes of Cm(III) at the Phosphate Materials-Solution Interface Using Laser Spectrofluorimetry", *Radiochim. Acta*, 82, 221 (1998).

[26] B.F. Myasoedov, A.P. Novikov, 'Main Sources of Radioactive Contamination in Russia and Methods for their Determination and Speciation", *J. Radioanal. Nucl. Chem.*, Art. 1997, Vol. 229, No. 1-2, pp. 33-38.

[27]  B.F. Myasoedov, A.P. Novikov, "Membrane Methods of Determination and Speciation of Radionuclides in the Environment", In Proc. of Intern. Symp. on Liquid Membrane: Theory and Practice, 21-27 June 1998, Moscow.

[28]  T. Fukasawa, K. Funabashi, Y. Kondo, "Separation Technology for Radioactive Iodine from Off-Gas Streams of Nuclear Facilities", *J. Nucl. Sci. Technol.*, 31, pp. 1073-1083 (1994).

[29]  K. Takeshita, Y. Takashima, S. Matsumoto, S. Inami, "Effects of Grain Size of AgNO$_3$ Loaded in Porous Material on Adsorption of CH~I", *J. Nucl. Sci. Technol.*, 32, pp. 941-943 (1995).

[30]  K. Funabashi, T. Fukasawa, M. Kikuchi, "Investigation of Silver-Impregnated Alumina for removal of Radioactive Methyl Iodine", *Nucl. Technol.*, 109, pp. 366-372 (1995).

# METHODS FOR MACRO CONCENTRATION SPECIATION (>10$^{-6}$ M)

## Introduction

The range of topics covered by the term "speciation" varies from the determination of isotopic composition, through whether an element or nuclide is present and under which form, to the full identification of the molecular compound present or the isolation of species. Therefore, it is quite understandable that a rather large range of techniques may be applicable for speciation purposes and that none may be sufficient by itself to cover all purposes.

We tried in this report to review on technologies available for speciation purposes in nuclear fuel cycle activities and to provide a reference document to guide researchers in choosing the most useful technique for his purposes by summarising their characteristics. Particular attention was given to highlight the advantages and disadvantages/limitations of a given technique as well as, when suitable, expected or necessary future improvements. It was also attempted to indicate a "recommended" application field, though this was not an easy task. In the present part, we concentrate on techniques available for macro concentration speciation with an arbitrary limit fixed around 10$^{-6}$ M. However, one should be aware that some of the techniques discussed in the present may be also applicable for lower concentration (overlap with over Partim. of the whole report) and vice-versa. Then, we consider that the scope of the present can only be achieved considering the whole parts discussed at the OECD/NEA Workshop on evaluation of speciation technology held in October 1999 at JAERI/Tokai-mura.

## Mössbauer spectroscopy [1-14]

The Mössbauer spectroscopy is a γ-resonance spectroscopy. Two great advantages of Mössbauer spectroscopy should be highlighted:

- The absolute selectivity. In a given experiment, a response is registered only for a given isotope of a given element. Furthermore, atoms of that isotope in non-equivalent positions give different responses with magnitudes proportional to the respective populations of these sites.

- The high sensitivity (considering macro concentration samples) determined by the minimum number of resonant atoms needed to get a detectable response. In the conventional transmission Mössbauer spectroscopy for $^{57}$Fe, a response is given by a monolayer with an area of the order of 1 cm$^2$.

It can be used for identification of chemical form of nuclides, which exhibit the Mössbauer effect, by recoilless emission and absorption of γ- radiation. Useful information may be inferred, such as the presence of a given isotope (selectivity and sensitivity), an estimation of its quantity in the sample and may allow to identify and characterise the occupation site(s) of this isotope (charge state, local symmetry, populations ratio in case of different sites, *etc.*) and state on .the chemical bounding (metallic, ionic, covalent), the oxidation state and structural environment of the species investigated.

However, the technique can only be applied to solid samples or frozen solutions and to a limited number of nuclides (detailed information on nuclides may be found in [1]) and only a few of them are suitable for systematic investigations. Several of these are of interest in the nuclear industry i.e. $^{127}$I, $^{129}$I, $^{83}$Kr, $^{129}$Xe, $^{238}$U, $^{237}$Np, $^{243}$Am, some rare earths, etc. The major challenge in experimental works is to produce sources and absorbers that are pure enough and free enough from defects that line broadening from inhomogeneities does not destroy the resonance.

Conventional transmission experiments require a sheet of powered layer of the material to be studied of about 1mm thick over an area of 0.1-1 cm$^2$ (more exact values depending of course of the isotope measured). In general, a few mg to 1g is used (larger quantities needed if the isotopic abundance or concentration of the absorbing element is low). A matrix hosting a Mössbauer absorber should not have an electronic cross section for γ-rays higher than 10 to 100 times that of the Mössbauer cross-section. Finally, low temperature and long time of measurement are often required.

The Mössbauer spectroscopy may have some interesting potential for speciation purposes. Widely used in many chemical applications it is rarely used for speciation purpose in the nuclear field. What is needed is some fundamental work on the real capability of this technique in this field.

If Mössbauer spectroscopy may be suitable for fundamental and modelling applications, it is reserved to high concentration samples and is unlikely to be suitable for environmental materials.


## X-ray techniques

### X-ray diffraction and neutron diffraction [15-18]

X-ray diffraction techniques are some of the most useful in the characterisation of crystalline materials such as metals, intermetallics, ceramics, minerals, polymers, plastics, or other organic and inorganic compounds. X-ray diffraction techniques can be used to identify the phases present in samples from raw starting materials to finished products and to provide information on the physical state of the sample, such as grain size, texture, and crystal perfection. Most of the X-ray techniques are rapid and non-destructive.

The thermal neutron is a unique probe of condensed matter whose utility has been demonstrated in a wide variety of applications. There are numerous aspects that make neutron scattering a valuable tool in materials science.

First, the scattering factor of nuclei for neutrons varied randomly across the periodic table. Consequently, contrast between nearby elements in the periodic chart may be extremely good and "light" atoms are almost always observable even in the presence of "very heavy" atoms, which would dominate in X-ray or electron scattering. Second, the neutron possesses a magnetic moment and interacts with the magnetic fields present within the condensed phase. It is the most direct probe of magnetic ordering in solids. Third, the form factor is essentially constant with diffraction angle, unlike those for X-ray or electron diffraction. Fourth, the neutron is an uncharged particle that interacts weakly with matter. The resulting high penetrability in most materials enables excellent volume sampling, insensitivity to surface condition, the possibility of depth profiling, and ease of measurement in special environments, for example high-temperature furnaces, low temperature cryostats, and high pressure cells.

The application field is applicable whenever you have crystals with sizes greater than about 0.1 micron and the number of defects in the crystals is low.

Advantages are a complete structure analysis if suitable single crystals are available in regard to size and defects, atomic form factors (XRD) or scattering lengths (neutron diffraction) of the ions in the crystal structure can be distinguished, and the phase problem of the structure amplitude can be solved.

The disadvantages for neutron diffraction are that we need:

- A reactor or a spallation source.

- Second containment for transport issues, if elements are involved, which tend to become radioactive (e.g. transition metals).

- The relatively low intensity may require large sample volumes and/or long counting times.

- Strong absorbers (Gd, Cd, Sm, Li, B) must be avoided or kept at low concentrations.

- X-ray studies must generally precede neutron diffraction analysis for structure modelling.

- Substitution of deuterium for hydrogen may be necessary to reduce incoherent scattering contributions to backgrounds.

The improvement should focus on improvement done already on other field that could be applied to nuclear fuel cycle studies.

## X-ray absorption spectroscopy; EXAFS and XANES [19-25]

The application field is wide from fundamental and modelling to reprocessing via waste management.

The advantages are:

- Oxidation state information.

- Co-ordination number information.

- Bond length.

- Neighbouring atoms.

Amorphous state and liquid to solid state are easy to be studied by this technique.

The complex species matrix is difficult to resolve. The limitation access to devoted facilities is an obstacle. The non-unique results if the element of interests exists in multiple non-equivalent sites or valence states are a limitation. Furthermore, the structural results can depend strongly on model systems used in quantitative analysis. The improvement should focus on the availability of the facilities.

## Nuclear magnetic resonance [26-30]

Nuclear magnetic resonance (NMR) is a radio frequency (RF) spectroscopy involving the interaction of the nuclear magnetic dipole or electric quadrupole moments with external or internal magnetic fields or electric-filed gradients. These interactions provide detailed information on the atomic (chemical) environment.

The application field is wide from fundamental and modelling to reprocessing via waste management. Among advantages we have:

- Structural information.

- Conformational information.

A lot of organic information are available compared to other techniques. Dynamic properties are an advantage. Nucleus, para and diamagnetic ions only are disadvantages.

For further improvement, availability should be considered. More devoted instruments for actinide investigations are needed. The NMR field and new pulse sequencing together with the imaging could be improved.

## Vibrational spectroscopy; Raman and FT-IR [31-35]

Infrared spectroscopy is a useful technique for characterising materials and providing information on the molecular structure, dynamics and environment of a compound. It is particularly useful for determining functional groups present in a molecule.

The application is rather wide from fundamental and modelling to reprocessing via waste management. Some advantages are:

- That we have information on chemical forms.

- Easy to handle.

- We have access also to micro Raman.

Recent advances in computerised IR spectroscopy, particularly Fourier transform infrared (FT-IR) spectroscopy, have made it possible to obtain infrared spectra using various sampling techniques.

The disadvantages are:

- The lack of selectivity, it is most useful for "Actinide-$O_2$ systems".

- Little elemental information.

- The molecule must exhibit a change in dipole moment in one of its vibration modes upon exposure to infrared radiation.

The future improvements could be:

- On surface enhanced Raman spectroscopy (SERS).

- On resonance Raman in terms of sensitivity.

The limitations for SERS are on intrusivity.

## X-ray photoelectron spectroscopy: XPS [36-38]

General uses include:

- Elemental analysis of surfaces of all elements except hydrogen.

- Chemical state identification of surface species.

- In-depth profiles of elemental distribution of thin film.

The application field focuses on surface studies in most of nuclear field, applications are on decommissioning...

The advantages are:

- Valence state information.

- Elemental identification.

- Electronic and structural information but the limitations are due to the necessity of high vacuum chamber and they are only few available.

One advantage that was improved in the last 10-15 years is in situ reaction within the apparatus. It is state-of-the art technique today. It is also available for actinides.

The limitations are:

- Poor lateral resolution.

- Charging effects may be a problem.

- The accuracy of quantitative analysis is limited.

## Spectrophotometry [39-52]

UV/Vis absorption spectroscopy, despite the emergence of newer techniques, remains a powerful and important tool. In addition, the instrumentation for UV/Vis spectrophotometry is for the most part relatively inexpensive and widely available. UV/Vis absorption can be used as a qualitative tool to identify and characterise molecular species (including solvated molecular ions) and as a quantitative tool to assess the quantities of inorganic, organic, and biochemical species present in various samples.

The general uses are:

- Quantitative analysis.

- Qualitative analysis, especially of organic compounds.

- Fundamental studies of the electronic structure of atomic and molecular species.

It is applicable in most of the fields except the environmental is received for laser spectroscopy techniques.

The advantages are direct fingerprints in redox, access to complexation phenomena. It is easy to handle, widely applicable to all actinides. We have a fast analysis; it is all in situ measurement. Furthermore, for quantitative analysis of inorganic ions, spectroscopic samples may contain as little as 0.01 mg/l in the case of species that form highly absorbing complexes. For qualitative analysis of organic compounds, concentrations of spectroscopic samples may be as small as 100 nanomolar. The initial capital outlay for ultraviolet/visible spectrophotometric techniques is usually far less than that for related techniques.

The limitations to be considered are a problem of sensitivity in case of a complex matrix. The analyte must absorb radiation from 200 to 800 nm, or be capable of being converted into a species that can absorb radiation in this region.

**Fluorescence spectroscopy** [53-59]

This is a technique applicable to all fields. The advantages are:

- Sensitivity.

- Selectivity.

- Dynamic range.

- *In situ*.

- Rapidity of analysis.

- Access to chemical forms.

- Number of water molecules.

It is possible to perform interface studies.

The disadvantages are the number of possible actinides that can be studied: Am, Cm, U(VI) and eight lanthanides.

The future improvement is microchip laser and real in situ measurement.

**Mass spectrometry; electron-spray mass spectrometry, matrix assisted laser desorption ionisation time of flight mass spectrometry** [60-62]

Mass spectrometer has particularly useful in analysing mixtures of organic compound with the determination of the structure of a compound. With it, information on the molecular weight and the pieces that constitute the molecules in a sample can be obtained.

The application field is wide from fundamental to modelling and to reprocessing via waste management.

The advantages are:

- A multi-elementary analysis.

- Rich isotopy.

- Selectivity.

- Both organic and inorganic speciation.

Disadvantage is the non-intrinsivity has to be verified but recent studies are on their way. Mass spectrometer is useless if two or more compounds are introduced into the samples at a time, because the spectra become unduly complex and difficult to interpret.

Future improvement is a source mechanism comprehension.

## General comments

By considering macro concentration speciation, it is clear that application fields will mostly be devoted to fundamental and modelling aspects ($>10^{-3}$ M) or for stating on species encountered in nuclear plants (reprocessing, storage tanks, waste management, etc.). Most techniques reviewed here are unlikely to find application for environmental materials.

It clearly appears that no technique has a universal character and that all are complementary and suitable for getting information according to their accessibility for researchers and the specificity of the material to be investigated. Best results and understanding are certainly obtained via several technologies including overlapping and cross-correlations.

During our discussion, we also highlighted that efforts should not only be directed towards technologies, but that there were also needs for better molecular modelling, *ab-initio* calculations in order to have a more general approach. Data analysis techniques should also improve, and a better consensus towards methods and procedures should be reached. Reference procedures, materials, techniques and database should also be made available. A spectroscopic database (with special emphasis to raw spectra) may answer to the spectroscopist community expectation involved in nuclear field of applications. A better collaboration between chemists, spectroscopists and modellers may also greatly contribute to a better knowledge and understanding.

Special efforts should be directed to be more attentive to the most recent evolution of techniques developed in other fields (non-nuclear speciation activities) through exchange, collaboration, normalisation, intercomparison and common programmes. On the other hand, care must be given to allow certain accessibility to large facilities for radioactive materials investigations.

Finally, we have to use the extraordinary tools of communication available today to enhance networking and information circulation among the each year smaller community of radiochemists and be aware of maintaining a minimum renewing of the knowledge for the future.

# REFERENCES

*References for Mössbauer spectroscopy*

[1]   "An Introduction to Mössbauer Spectroscopy", L. May, ed., Plenum Press (1971).

[2]   Mössbauer Effect Data Index, J.G. Stevens, V.E. Stevens, eds. (1972).

[3]   "Perspectives in Mössbauer Spectroscopy", S.G. Cohen, M. Pasternak, eds., Plenum Press (1973).

[4]   "Mössbauer Spectroscopy", U. Gonser, ed., Spinger-Verlag (1975).

[5]   "Applications of Mössbauer Spectroscopy", R.L. Cohen, ed., Vol. I & II (1976 & 1980).

[6]   "Mössbauer Isomer Shifts", G.K. Shenoy, E.E. Wagner, eds., North-Holland (1978).

[7]   "Mossbauer Spectroscopy and its Applications", J.G. Stevens, G.K. Shenoy, eds., American Chemical Society, Washington (1981).

[8]   B.D. Dunlap, G.M. Kalvius, "Mössbauer Spectroscopy on Actinides and Their Compounds", in *Handbook on the Physics and Chemistry of Actinides*, A.J. Freeman, G.H. Lander, eds., Chap. 5, Elsevier (1985).

[9]   S.J. Clark, J.D. Donaldson, "Mössbauer Spectroscopy", *Spectrosc. Prop. Inorg. Organomet. Compd.* 31, 364-435 (1998).

[10]  L.J. Swartzendruber, L.H. Bennett, "Nuclear Magnetic Resonance", in *ASM Handbook, Materials Characterisation*, Vol. 10, 287-295 (1998).

[11]  J.P. Sanchez, E. Colineau, V.P. Vulliet, K. Tomala, *J. Alloy and Comp.*, 275-277, 154-160 (1998).

[12]   S. Tsutsui, M. Nakada, M. Saeki, S. Nasu, Y. Haga, E. Yamamoto, Y. Onuki, *Phys. Rev. B-Cond. Matter.*, 60 (1), 37-39 (1999).

[13]   T. Saito, *et al., J. Radioanal. Nucl. Chem.*, 239 (2), 319-323 (1999).

[14]   T. Nakamoto, *et al., J. Radioanal. Nucl. Chem.*, 239 (2), 257-261 (1999).

*References for X-ray diffraction*

[15]   A.D. Krawitz, R. Roberts, J. Faber, *Advances in X-ray Analysis*, Vol. 27, Plenum Press, 1984, pp. 239-249.

[16]   W.B. Yelon, F.K. Ross, A.D. Krawitz, "Neutron Diffraction", in *ASM Handbook, Materials Characterisation*, Vol. 10, 1998, pp. 420-426.

*References for neutron diffraction*

[17]   R.P. Goehner "X-ray Powder Diffraction" in *ASM Handbook, Materials Characterisation*, Vol. 10, 1998, pp. 331-343.

[18]   D.K. Smith, "X-ray Techniques", in *ASM Handbook, Materials Characterisation*, Vol. 10, 1998, pp. 325-332.

*References X-ray absorption spectroscopy; EXAFS and XANES*

[19]   "Spectroscopic Methods in Mineralogy and Geology", *Reviews in Mineralogy*, F.C. Hawthorne, ed., Vol. 18, 1988.

[20]   *X-ray Absorption, Principles, Applications Techniques of EXAFS, SEXAFS and XANES*, D.C. Koningsberger, R. Prins, eds Vol. 92, 1988.

[21]   R.J. Silva, H. Nitsche, *Radiochim. Acta*, 70/71 (1995), 377-396.

[22]   H. Nitsche, *J. Alloys Compd.*, 223(2) (1995), 274-279.

[23]   "Speciation Techniques and Facilities for Radioactive Materials at Synchrotron Light Sources", OECD/NEA Workshop Proceedings, Grenoble France, 4-6 October 1998.

[24]   J. Wong, "Extended X-ray Absorption Fine Structure", *in ASM Handbook, Materials Characterisation*, Vol. 10 (1998), 407-419.

[25]   S.D. Conradson, *Appl. Spectrosc.*, 52(7) (1998), 252A-279A.

*References for nuclear magnetic resonance*

[26]   C.P. Slichter, "Principles of Magnetic Resonance", Harper and Row, 1963.

[27]   R. Richards, K.J. Packer, "Nuclear Magnetic Resonance Spectroscopy in Solids", Cambridge University Press, 1981.

[28]   Y. Minai, G.R. Choppin, D.H. Sisson, *Radiochim. Acta*, 56(4) (1992), 195-199.

[29]  D.L. Clark, S.D. Conradson, S.A. Ekberg, N.J. Hess, D.R. Janecky, M.P. Neu, P.D. Palmer, C.D. Tait, *New J. Chem.*, 20(2) (1996), 211-220.

[30]  L.H. Bennett, L.J. Swartzendruber, "Nuclear Magnetic Resonance", in *ASM Handbook, Materials Characterisation*, Vol. 10, 1998, pp. 277-286.

*References for vibrational spectroscopy; Raman and FT-IR*

[31]  N.B. Colthup, L.H. Daley, S.E. Wiberley, "Introduction Infrared and Raman Spectroscopy", 2nd Ed., Academic Press, 1975.

[32]  J.A. Graham, W.M. Grim III, W.G. Fateley, "Fourier Transform Infrared Spectroscopy, Applications to Chemical Systems", Vol. IV, J.R. Ferraro, L.J. Basile, eds., Academic Press, 1985.

[33]  D.L. Clark, S.D. Conradson, S.A. Ekberg, N.J. Hess, D.R. Janecky, M.P. Neu, P.D. Palmer, C.D. Tait, *New J. Chem.*, 20(2) (1996), 211-220.

[34]  W. Runde, M.P. Neu, S.D. Conradson, D.L. Clark, P.D. Palmer, S.D. Reilly, B.L. Scott, C.D. Tait, Mater. Res. Soc. Symp. Proc. 465 (Scientific Basis for Nuclear Waste Management XX) (1997), 693-703.

[35]  C. Marcott, "Infrared Spectroscopy" in *ASM Handbook, Materials Characterisation*, Vol. 10, 1998, pp. 109-125.

*References for X-ray photoelectron spectroscopy: XPS*

[36]  K. Seigbahn, C. Nordling, A. Fahlman, R. Nordberg, K. Hamrin, J. Hedman, G. Jahansson, T. Bergmark, S. Karlsson, I. Lindgren, B. Lindberg, "Electron Spectroscopy for Chemical Analysis – Atomic, Molecular, And Solid State Structure Studies by Means of Electron Spectroscopy", Almqvist, Wiksells, Stockholm, 1967.

[37]  C.J. Dodge, A.J. Francis, C.R. Clayton, Synchrotron Radiat. Tech. Ind., Chem., Mater. Sci., [Proc. Comb.Symp. Appl. Synchrotron Res. Mater. Sci. Appl. Synchrotron Radiat. Chem. Relat. Fields] (1996), Meeting Date 1994-1995, 159-168.

[38]  J.B. Lumsden, "X-Ray Photoelectron Spectroscopy" in *ASM Handbook, Materials Characterisation*, Vol. 10, 1998, pp. 568-580.

*References for spectrophotometry*

[39]  D.J. Swinehart, *J. Chem. Educ.*, Vol. 39, 1962, p. 333.

[40]  D.P. Karim, D.J. Lam, A.M. Friedman, N. Susak, P. Rickert, J.C. Sullivan, Mater. Res. Soc. Symp. Proc. 6 (Sci. Basis Nucl. Waste Manage.) (1982), 661-666.

[41]  "Practical Absorption Spectrometry, Ultraviolet Spectrometry Group", C. Knowles, C. Burgess, ed., Chapman and Hall, 1984.

[42]  H. Nitsche, N.M. Edelstein, *Radiochim. Acta*, 39(1) (1985), 23-33.

[43]  I. Grenthe, B. Lagerman, *Acta Chem. Scand.*, 45(3) (1991), 231-238.

[44] H. Capdevila, P. Vitorge, E. Giffaut, L. Delmau, *Radiochim. Acta*, 74 (1996), 93-98.

[45] M.P. Neu, S.D. Reilly, W.H. Runde, Mater. Res. Soc. Symp. Proc. 465 (Scientific Basis for Nuclear Waste Management XX) (1997), 759-765.

[46] G.R. Choppin, P.J. Wong, *Aquat. Geochem.*, 4(1) (1998), 77-101.

[47] G.D. Brabson, "Ultraviolet/Visible Absorption Spectroscopy", in *ASM Handbook, Materials Characterisation*, Vol. 10, 1998, pp. 60-71.

*References for PAS and TL*

[48] T. Kimura, J.G. Serrano, S. Nakayama, K. Takahashi, H. Takeishi, *Radiochim. Acta*, 58-59 (Pt. 1) (1992), 173-178.

[49] J.M. Berg, C.D. Tait, D.E. Morris, W.H. Woodruff, Mater. Res. Soc. Symp. Proc. 212 (Sci. Basis Nucl. Waste Manage. 14) (1991), 531-538.

[50] R.A. Torres, C.E.A. Palmer, P.A. Baisden, R.E. Russo, R.J. Silva, *Anal. Chem.* 62(3) (1990), 298-303.

[51] C. Moulin, N. Delorme, T. Berthoud, P. Mauchien, *Radiochim. Acta*, 44-45 (Pt. 1) (1988), 103-106.

[52] G. Bidoglio, G. Tanet, P. Cavalli, N. Omenetto, *Inorg. Chim. Acta*, 140(1-2) (1987), 293-296.

*References for fluorescence spectroscopy*

[53] Th. Fanghanel, J.I. Kim, *J. Alloys Compd.*, 271-273 (1998), 728-737.

[54] C. Moulin, P. Decambox, P. Mauchien, *J. Radioanal. Nucl. Chem.*, 226(1-2) (1997), 135-138.

[55] T. Kimura, Y. Kato, G. Meinrath, Z. Yoshida, G.R. Choppin, JAERI-Conf 95-005 (Vol. 2), (1995), 473-486.

[56] C. Moulin, P. Decambox, V. Moulin, J.G. Decaillon, *Anal. Chem.*, 67(2) (1995), 348-353.

[57] R. Klenze, J.I. Kim, H. Wimmer, *Radiochim. Acta*, 52-53 (Pt. 1) (1991), 97-103.

[58] "Analytical Applications of Lasers", E. Piepmeier, ed., Vol. 87, 1986.

[59] "Lanthanide and Actinide Chemistry and Spectroscopy", N. Edelstein, ed., ACS Symposium Series 131, 1980.

*References for mass spectrometry*

[60] Fugerio, "Essential Aspects of Mass Spectrometry", Spectrum Publications, 1974.

[61] L. Raphaelian, "Mass Spectrometry" in *ASM Handbook, Materials Characterisation*, Vol. 10, 1998, pp. 639-648.

[62] C. Moulin, *et al.,* "Speciation of Uranium by ES-MS: Comparison with TRLIF", *Applied Spectroscopy* (1999), submitted.

# METHODS FOR EMPIRICAL FORMULA, MOLECULAR STRUCTURE DETERMINATION AND COLLOID CHARACTERISATION

## Introduction

Determination of empirical formula and of molecular structures in issues pertinent to the fuel cycle, waste management and remediation, as well as in risk assessment associated with radionuclide release and its migration to the near and far field of a repository present a particular challenge. Speciation techniques in these areas are needed for extraction chemistry of actinides, aquatic reactions of actinide ions (e.g. tetravalent actinide hydrolysis, formation of ternary complexes and redox chemistry), reactions at the water/mineral interface, formation of secondary phases and the formation of and interaction with colloids. The range of concentrations of interest is extreme, from relatively pure substances such as the fuels themselves, to tracer levels in far-field release scenarios. Actinide speciation in extractant solutions used for separation techniques for partitioning technologies involves investigation of solutions with often acidic pH, whereas speciation in cement pore waters, for example, can involve very basic solutions. Finally, the amount of sample available may be limited, e.g. pore water samples can be as small as $\mu L$ quantities.

There exist a variety of techniques for determining empirical formula and molecular structure of radionuclide species. The actual contents and components of the sample at hand in addition to the degree of complexity required dictates the method that is chosen.

The speciation of radionuclides is often governed by their association with or formation to colloids. The characterisation of such colloids is imperative and their quantification and characterisation is, therefore, included as a separate treatise.

## Empirical formula

Determination of the empirical formula involves quantification of the elements in a sample of known mass. The analysis requires that the substance be pure. Classical analytical and radioanalytical techniques can be used for such determinations [1, 2]. The choice of the method of determination is dictated by the type of elements in the sample and the quantity of sample available. Care must be given the fact that, in order to define the stoichiometry, the precision of the analysis for each element must be better than its atomic weight.

If the sample cannot be isolated in pure form, alternative methods must be implemented. For example, a number of titration techniques can be used for determining the ratio of metal cations to the number of ligands. This provides indirect evidence for the empirical formula. Further specialised techniques can be used for determining empirical formulas of solution species. TRLFS is also used for determining the number of water molecules in the first hydration sphere of certain metal cations. Co-ordination numbers are extractable from both EXAFS and NMR spectra, thereby allowing empirical formula to be determined. Secondary mass spectrometry provides information of molecular weight to charge ratios. From this, the empirical formula can also be determined.

## Molecular structure

Molecular structure determination in a strict sense is the determination of the metrical parameters, i.e. atomic distances and bond angles, of a species structure. The number of techniques that are available to achieve this, however, is limited. These include structural analysis with diffraction techniques, X-ray absorption spectroscopy and in part NMR. However, a number of other spectroscopies provide information, which is related to molecular structure such as, for example, co-ordination symmetry, crystal field splitting and/or nature and number of chemical functional groups in the species. This information can be used to develop models for the molecular structure of the system under study.

The following discusses available instrumental spectroscopic techniques for the determination of either molecular structure or the determination of parameters related to molecular structure. According to the energy of the particles used for excitation (photons, electrons, neutrons etc.), different parts of the molecule will interact and different structural information will be obtained. Depending on the relaxation process, each method has a characteristic time scale. Especially for NMR and fluorescence of f-element ions, the relaxation rate may be faster than the rate constant of the reaction under study. The applicability of each method, its particular advantages, as well as its limitations are presented below. The discussion is restricted to speciation methods, which find broader application. It is not an exhaustive list. The spectroscopic methods are discussed in order of increasing excitation energy. Table 1 summarises these methods and pertinent information for convenient reference.

The application of theoretical tools for predicting molecular structure such as, for example, *ab initio* and density functional methods calculations, have not been included in the discussion. Such tools provide only a first approximation to the molecular structure. This is because ionic systems and the relativistic parameters for the very heavy actinides present problems in the analysis. There is much room for further development of theoretical molecular structure calculations applied to radionuclide speciation before it becomes a reliable predictive tool.

## *NMR*

The potentials of NMR are reported in detail in one of the annexed papers [3]. NMR finds its main application in the analysis of solutions, using $^1H$ as the most sensitive nucleus; $^{19}F$ and $^{31}P$ nuclei are also used for speciation of fluoride and phosphorous containing complexes. NMR yields information on chemical functional groups of organic ligands. The entire arsenal of two-dimensional techniques can be used for finding spatial connections between nuclei. Speciation of metal complexes and determination of stability constants are possible, provided separate peaks are observed for each species. This often requires performing experiments at low temperature and/or adding an organic solvent to an aqueous solution. If the metal complex under investigation is sufficiently rigid so that resonances do not arise from rapid exchanges between different conformations, complete geometrical models can be determined. Symmetry information assists the determination of geometry and NMR is thus particularly useful for the study of macrocyclic extractants. The NMR relaxation times are in the msec range, or longer, so that speciation of fast equilibrium reactions is not possible.

Structural analyses of complexes of diamagnetic metals are often limited by the small spectral shifts induced by complexation with ligands, However, the relative magnitude of the very large paramagnetic shifts of the $Yb^{3+}$ complexes can be fruitfully used to build geometrical models. A full determination of the complex structure is then possible. Analyses of paramagnetic shifts can be complemented by relaxation time measurements as these yield quantitative information concerning metal nuclei distances.

The frequency dependence of the longitudinal relaxation time of the solvent peak of a $Gd^{3+}$ solution (relaxity) is directly related to the tumbling time of the metal complex. Quantitative formation on the oligomerisation of $Gd^{3+}$ complexes is then derived from the Solomon-Bloembergen equations.

The major limitations of NMR are its limited sensitivity ($10^{-4}$ M in $^{1}H$, $^{19}F$, $^{31}P$ NMR spectroscopy and $10^{-5}$-$10^{-6}$ M in relaxity measurements; greater concentrations are necessary for other nuclei) and the lack of spectrometers dedicated to radionuclide investigations. It is expected that higher sensitivities will be reached in the future as more powerful magnets and improved pulse sequences become available.

## Vibrational spectroscopies

Infra-red and Raman vibrational spectroscopies provide complementary information concerning the type of functional groups present in a species as well as bond strength. A specialised application of vibrational spectroscopies for actinide research is the study of the actinyl asymmetric stretching vibration band near 920 cm$^{-1}$. The bond distance to the actinyl oxygen atoms can be estimated from the band position. Also, the presence of more than one actinyl stretching band can be used to identify the number of actinyl species present in a mixture. The functional groups bound to a radionuclide can often be determined from band energy shifts following complexation. Special techniques such as measuring isomer shifts following isotopic exchange to identify band associations are also useful in molecular structure studies. For investigations of relatively concentrated systems, such as An(Ln)-extractant interactions for partitioning, conventional instrumentation may suffice. In solutions, the sensitivity is limited by the background absorption of the solvent. Conventional instrumentation for IR and Raman vibrational spectroscopies lack the sensitivity required for many radionuclide speciation studies. An increase in sensitivity down to μM levels, is possible using a variety of specialised techniques, including resonant Raman and coherent Raman spectroscopies. This technique may be especially interesting for the speciation of actinyl ions.

## Absorption spectroscopy

Absorption spectroscopy on the transitions within the 5f electron shell has been a very powerful tool for the characterisation of oxidation states and complexation form since the beginning of actinide chemistry. Study of the abundance of the sharp features of lanthanide and actinide ions spectra also provided basis information for the theoretical understanding of the electronic structure of 4f and 5f elements [4]. The parity forbidden transitions are rather weak in intensity (molar absorptivity $\varepsilon < 10$ L mol$^{-1}$ cm$^{-1}$ for lanthanide and $\varepsilon < 500$ L mol$^{-1}$ cm$^{-1}$ for actinide ions). Thus, for application to speciation of actinide ions under natural conditions, e.g. in the neutral pH range, where concentrations are usually far below $10^{-6}$ mol/L, classical UV/Vis (transmission) spectroscopy is not sensitive enough.

For this reason, special detection methods have been developed that make use of the high light intensities of laser radiation sources. These photothermal absorption techniques are based on the detection of the heat generated by the light absorbed within the sample. The heat may be measured by the pressure wave generated, i.e. photo acoustic effect Laser-Induced Photo Acoustic Spectroscopy (LPAS) or by the change of the refractive index with temperature sensed by a probe laser beam Thermal Lensing (TLS) and Photothermal Deflection spectroscopy (PDS). The sensitivity of these methods is about 2-3 orders of magnitude higher than transmission spectroscopy. LPAS has been applied to various actinide oxidation states, including U(IV), U(VI), Np(IV), Np(V), Pu(III), Pu(IV), Pu(V), Pu(VI), Am(III) and Am(V) [5]. The detection limit found for Am(III) (in 0.1 M HClO$_4$, absorption maximum at 503.2 nm, $\varepsilon = 410$ L mol$^{-1}$ cm$^{-1}$) is $7 \times 10^{-9}$ mol/L, allowing speciation of this cation in the sub-micro-molar concentration range [5].

The detection limit of these methods is largely limited by the absorption of the solvent, e.g. water, the extinction of which increases strongly from 500 to 1 000 nm. Thus, an increase in sensitivity necessitates improved compensation techniques for background absorption. The use of deuterated solvents (e.g. heavy water) will reduce strongly the absorption in the near IR compared to protonated solvents.

## TRLFS

Detection of emitted light from excited states of actinide ions provides most sensitive speciation techniques. However, sufficient quantum yield of radiative decay is observed for only few actinide oxidation states, including U(VI), Cm(III) and Am(III) [4]. Sensitivity for Time-resolved Laser Fluorescence Spectroscopy (TRLFS) is extremely high, detection limits for U(VI) and Cm(III) are about $10^{-12}$ mol/L. Speciation of Cm(III) is possible in natural aquatic systems down to $10^{-9}$ mol/L.

Spectroscopic information on the species is derived from excitation spectra, emission spectra and relaxation dynamics of the excited state. Excitation spectra are obtained by scanning the excitation wavelength, and recording the emission within a fixed wavelength range, yielding similar information as absorption spectra. However, discrimination of light scattering in the medium and absorption by fluorescing organic compounds by time-resolved detection is possible. Spectral shifts and eventually splitting of emission bands are the main sources for fingerprinting the speciation of the actinide ion. For the same reasons discussed for the absorption spectra, changes in the peak maxima of excitation and emission spectra are indicators for ligand field the strength and symmetry, but give no direct structural information.

With respect to sensitivity and selectivity, Cm(III) is the most appropriate element to be studied by TRLFS [4,6]. The complexation of Cm(III) with various ligands, e.g. $OH^-$, $F^-$, $Cl^-$, $CO_3^{2-}$, $SO_4^{2-}$ [7], humic and fulvic acid as well as reactions at the mineral-water interface have been studied by TRLFS. The method has also been used to speciate Cm(III) in ground waters under natural conditions for the validation of geochemical models and databases. Similar studies have been performed on U(VI) [8]. Absorption and emission spectra of the uranyl ion show a characteristic vibronic structure. However, the fluorescence emission of U(VI) is strongly quenched by many ligands, e.g. $Cl^-$ and humic substances, somewhat limiting the applicability to natural aquatic systems.

Compared to Cm(III), the shift of the emission bands observed for complexation of trivalent lanthanides such as Eu(III) and Tb(III) on is rather small and usually not appropriate for fingerprinting. The $^5D_0 \rightarrow {}^7F_2$ transition for Eu(III) is hypersensitive. That means its intensity strongly varies with ligand environment, whereas the intensity of the $^5D_0 \rightarrow {}^7F_1$ transition is very insensitive to co-ordination changes. Thus, the intensity ratio of both of these emission bands may be used for Eu(III) speciation in a binary system. Direct structural information may be deduced from the $^5D_0 \rightarrow {}^7F_0$ transition in Eu(III). The multiplicity of this single electronic transition indicates the number of different Eu(III) species present in the system. Due to the weak transition probability, Eu(III) concentrations higher than $10^{-4}$ mol/L are necessary.

The lifetimes of f-element ions discussed above are in the μsec to msec range. The purely radiative lifetime of the excited states, as well as the intensity of the absorption bands of f-elements, may be calculated by the semi-empirical Judd-Ofelt theory. The observed lifetimes result from both radiative and non-radiative decay. The non-radiative decay rate depends directly on the energy gap between the emitting state and the next lower lying state. High fluorescence yield is found only for f-configurations having a large energy gap, such as $f^7$ (Gd(III), Cm(III)), $f^6$ (Eu(III), Am(III)) and $f^8$ (Tb(III), Bk(III)). In aqueous solution, the dominating non-radiative decay occurs via loss of energy to OH vibrations of co-ordinated $H_2O$ and $OH^-$ ligands. From the fluorescence lifetime measured in

$H_2O$ and $D_2O$, the number of water molecules directly co-ordinated to the metal ion, may be deduced by an empirical linear relationship. This method was derived by Horrocks [9] for Eu(III) and Tb(III) and has also been applied to Cm(III), Am(III), Sm(III) and Dy(III) [10].

## Diffraction techniques

X-ray, neutron and electron diffraction techniques are used to determine crystal structures and, hence, can be used for molecular structure determinations. Due to its high resolution, it is the method of choice for structure determinations in crystalline compounds, either single crystals or crystalline powders. It is not applicable to amorphous phases. One must keep in mind that the average bulk crystalline structure determined from diffraction studies is expressed as simple, small, symmetric arrangements of atoms in a unit cell. However, local deviations from this average structure are often the driving force behind the collective behaviour of a crystalline compound [11].

For radionuclide speciation, applications of XRD have been used for structure determinations of crystalline waste forms, structural characterisation of An(Ln)-extractant molecules and characterisation of equilibrium phases [12,13,14].

## XANES/EXAFS

EXAFS data contain the metrical parameters, co-ordination number, bond distances, and mean-square displacements for atoms comprising the first few co-ordination spheres surrounding an absorbing element of interest. This information is extracted from the EXAFS oscillations, previously isolated from background and atomic portion of the absorption, using non-linear least-square fit procedures. It is important in such analyses to compare metrical parameters obtained from experiments on model or reference compounds to those for samples of unknown structure, in order to avoid ambiguity in the interpretation of results. The absorption spectra in the XANES region contain information concerning co-ordination geometry and metal cation valence. A reliable theoretical *ab initio* code based on multiple-scattering theory is often used for calculating XANES spectra [15]. This program has been extended to include relativistic effects for high Z elements [16]. By using these codes to calculate XANES of atomic clusters of proposed structures and comparing these to experiment, structural models can be developed and improved [17].

The fundamental advantage of XAFS for speciation studies is that it is elemental specific and applicable for investigating radionuclides embedded in complex matrices. Separation steps are not necessary, thereby avoiding speciation changes and shifts in equilibria that might occur during such steps. Furthermore, XAFS is a short-range order method so that, unlike XRD investigations, amorphous systems and the molecular structure of solutions species can be studies. No vacuum is necessary in the hard X-ray regime, so that wet samples can be investigated. The application of XAFS to radionuclide speciation investigations is manifold [18]. An example of the application of the EXAFS technique for molecular structure elucidation is found in this proceedings for the characterisation of the Er(III)/Cm(III)-diglycolamide molecule in an study of the extractant properties used for Ln-An separation [19]. An overview article of the XAFS technique and its application to radionuclide speciation in waste forms, geological media and geochemical model systems, as well as a discussion of some of the method's limitations, is also found in this proceedings [20].

## Other methods

A number of other specialised methods exist that can also be used for determination of parameters related to molecular structure. These include (and this listing is in no way exhaustive) the following:

By analysing the chemical shifts and quadrupole splitting in Mössbauer spectra of [237]Np containing samples, information on oxidation state and local structure can be obtained. No actinide nuclides exist, which can serve as a probe in PAC investigations. However, [181]Hf is a well-known, suitable PAC probe nuclide. PAC studies using Hf(IV) as a chemical homologue for the tetravalent actinides yield parameters defining the components of the electric field gradient tensor and its symmetry associated with the Hf-probe nuclei and, thereby, images the Hf(IV) chemical (structural) environment. Theoretically, EPR spectroscopy provides information on the paramagnetic ion under investigation and its near-neighbour environment. However, the application of EPR to the study of actinides is very limited due to extensive line broadening even at low temperatures. XPS is an excellent technique for determining an atom's electron binding energies. By studying the chemical shift of the electron binding energy and the associated peak heights of an unknown, relative to standards or known reference materials, some predictions of the unknown structure can be made. In optimal cases, lattice imaging from HREM micrographs yields direct information on the crystal structure. Thermal analysis using DTA/TG can be used for determining the number of hydrating water molecules in solid compounds from the weight loss in the TG diagram where dehydration occurs, corresponding to an endorthermic peak in the DTA diagram.

## Colloid quantification and characterisation

Natural colloids present in any groundwater to more or less amounts may interact with polyvalent metal ions and thus stabilise them in the mobile aqueous phase. Therefore, colloid facilitated transport is regarded as a key uncertainty for the migration of radionuclides in the geosphere. One of the main problems in assessing the colloid influence on radionuclide transport is the lack of information on particle number and size distribution in the groundwater.

Direct methods for the quantification of aquatic colloids are based on static and dynamic light scattering. However, intensity of scattered light decreases with sixth power of inverse particle diameter (Rayleigh scattering), restricting a sensitive detection to particles with >100 nm diameter. Detection of smaller particles by other methods, e.g. separation by ultrafiltration and counting by SEM, may be hampered by artifacts produced during sample preparation. Within the past few years a very sensitive method to detect colloids, even in ultra pure waters, based on the laser induced breakdown effect has been developed [21,22]. This method is highly sensitive to small particle diameters down to 10 nm, with a concentration detection limit in the sub-ppt range.

The laser induced breakdown effect results from the ionisation of matter in the electric field generated in the focus area of a pulsed laser beam. The critical power density to produce a plasma, the breakdown threshold, is much lower for solids compared to liquids [23]. If the laser pulse power density is below the breakdown threshold of the solvent, a plasma is generated almost exclusively by colloidal particles being present within the focus volume. Particle concentrations are determined from the breakdown probability. Additional, particle size distribution may be derived from the spatial distribution of the breakdown events within the focus volume. Based on the atomic emission spectra from the plasma emission, the elementary composition of the particles may be derived, discriminating against dissolved metal ions.

While atomic force microscopy (AFM) can be used for studying colloid morphology even under aqueous conditions, scanning transmission X-ray microscopes (STXM) for X-ray spectro-microscopy can be used to study both the morphology as well as the composition of colloids. The spatial resolution for STXM using available micro-zone focusing plates is less than that attained by AFM. However, by varying the incident radiation wavelength used for creating the images, a material contrast can be made. Wavelengths within the "water window" are often used, i.e. energies between the carbon and oxygen K absorption edges (284 eV and 537 eV, respectively). By creating images, for example, at an energy below and again at the carbon K edge, a contrast is produced that can differentiate organic colloids from inorganic colloids. Furthermore, by analysing a stack of images recorded stepwise at energies ranging from, e.g. about 275 to 310 eV, the carbon K edge XANES of an organic colloid can be registered. Changes in the carbon K edge XANES features can be used, e.g. to determine which functional groups are on the colloid surface as a function of pH or identify functional groups involved in metal complexation.

## Recommendations

Future development of spectroscopic speciation methods will depend on the availability of more powerful photon sources, e.g. tunable lasers in the visible and IR range, synchrotron light sources, X-ray lasers and free electron lasers. In addition, in techniques involving synchrotron light sources, the development of detector technology is lagging far behind the development of machine characteristics. New detector concepts for fluorescent detection to improve XAFS sensitivity and detectors able to withstand high photon fluxes are needed.

*Subgroup members*

Gregory R. Choppin
Melissa A. Denecke (Recorder)
Jean F. Desreux
Reinhardt Klenze (Leader)
Toshihiko Ohnuki
Satoshi Skurai
Ulla S. Vuorinen
Tsuyoshi Yaita

## REFERENCES

[1]    Skoog and Leary, *Instrumentelle Analytik*, Springer-Verlag, Heidelberg (1996).

[2]    J.P. Adloff, R. Guillaumont, "Fundamentals of Radiochemistry", CRC Press, Inc., Tokyo, 1993.

[3]    J.F. Desreux, *et al.*, these proceedings.

[4]    J.V. Beitz, in *Handbook on the Physics and Chemistry of Rare Earth*, K.A. Gschneider Jr., L. Eyring, G.R. Choppin, G.H. Lander, eds., Vol. 18, North-Holland, Amsterdam, 1994, pp. 159-195.

[5]    J.I. Kim, R. Stumpe, R. Klenze, "Chemical Applications of Nuclear Probes", in *Topics in*

*Current Chemistry.* No. 157, K. Yoshihara, K., ed., Springer (Berlin) (1990) S. 129-183.

[6]  J.I. Kim, R. Klenze, H. Wimmer, *Eur. J. Solid State Inorg. Chem.*, 28 (Suppl.), 347-56 (1991).

[7]  Th. Fanghänel, J.I. Kim, *J. Alloys and Compounds*, 271-273, 728-737 (1998).

[8]  G. Bernhard, G. Geipel, V. Brendler, H. Nitsche, *J. Alloys and Compounds*, 271-273, 201-5 (1998).

[9]  W.D. Horrocks Jr., D.R. Sudnick, *J. Am. Chem. Soc.,* 101, 334, (1979).

[10]  T. Kimura, these proceedings.

[11]  S.D. Conradson, *Appl. Spectrosc.*, 52 (1998) 252A-79A.

[12]  E.R. Vance, A. Jostsons, R.A. Day, C.J. Ball, B.D. Begg, P.J. Angel, "Scientific Basis for Nuclear Waste Management XIX", W.M. Murphy, D.A. Knecht, eds., Mater. Res. Soc., Philadephia (1996), pp.41-7.

[13]  P. Byers, M.G.B. Drew, M.J. Hudson, N.S. Isaacs, C. Madic, *Polyhedron,* 13, 349-52 (1994).

[14]  A.G. Sowder, S.B. Clark, R.A. Fjeld, *Radiochim. Acta,* 74, 45-49 (1996).

[15]  J. Mustre de Leon, J.J. Rehr, S.I. Zabinsky, R.C. Alber, *Phys. Rev.*, B69, 4146 (1991).

[16]  A.L. Ankoudinov, PhD Thesis, "Relativistic Spin-Dependent X-ray Absorption Theory", University of Washington (1996).

[17]  E.A. Hudson, P.G. Allen, L.J. Terminello, M.A. Denecke, T. Reich, *Phys. Rev.*, B 54, 156-65 (1996).

[18]  Speciation, Techniques and Facilities for Radioactive Materials at Synchrotron Light Sources, Workshop Proceedings, Grenoble, France, 4-6 October 1998, OECD/NEA (1999).

[19]  Yaita, *et al.*, these proceedings.

[20]  M.A. Denecke, *et al.*, these proceedings.

[21]  Kitamori, *et al.* (1988).

[22]  Scherbaum, *et al.* (1996).

[23]  Radziemski, *et al.* (1989).

**Table 1**

| METHOD | STRUCTURAL INFORMATION CONTENT | STATE* |
|---|---|---|
| **NMR** | Functional groups, molecular structure, exchange reaction rates | S, L |
| **IR/Raman** | Functional groups, bond energies | S, L, G |
| **UV-Vis/LPAS** | Crystal field splitting | S, L |
| **TRLFS** | Crystal field splitting, number of hydrating water molecules, number of species | S, L |
| **XAFS** | Oxidation state, co-ordination geometry, type, number and distances of neighbouring atoms | S, L, G |
| **XRD, neutron and electron diffraction** | Space group, crystal structure, crystallite size | S |

* Required sample state: S = solid; L = liquid, G = gas

# AN OVERVIEW OF ACTINIDE REDOX SPECIATION METHODS: THEIR APPLICABILITY, ADVANTAGES AND LIMITATIONS

## Objectives

The determination of oxidation states is a first step towards characterising aqueous species. The methods available can be divided into three different categories: chemical/radiochemical separation methods, spectroscopic methods, and electrochemical methods. In this paper, we discuss methods of oxidation state analysis of actinide elements mainly based on the reviews presented by Rai and Hess [1] and by Saito [2] in OECD/NEA Workshop on Evaluation of Speciation Technology held in Tokai-mura, Japan, 26-28 October 1999. Among the actinides that may be of environmental concern, only U, Np, and Pu show multiple oxidation states that are stable under ordinary environmental conditions. Therefore, the applicability, advantages, and limitations of some of the most commonly used or available methods for oxidation state analyses of U, Np, and Pu are briefly discussed in this paper.

## Radiochemical/chemical separation

The radiochemical/chemical separation is the only method capable of redox speciation of these elements at extremely low concentrations. Redox speciation by chemical separation methods means the determination of species distribution of an element among multiple oxidation states that exist simultaneously in the system, using procedures that do not alter the oxidation state of the element during the measurement. Since the chemical separation uses the difference in chemical behaviour among the species of different oxidation states, each method can be applied to U, Np, Pu, and Am, in principle. The commonly used techniques are solvent extraction, precipitation/co-precipitation, adsorption, ion exchange, and electrophoresis. Selective extraction chromatographic materials are also developed recently. Since there is significant difference in redox potentials among the redox couples, the basic method has to be modified for each element. Table 1 [3-21] summarises the chemical separation methods for redox speciation with the conditions, key analytical parameters, advantages and limitations as well as their references. Common advantages of the chemical separation methods are:

- The speciation by chemical separation can be applied to any oxidation state.

- The methods can be applied to a wide range of concentration. Determination of nuclides by radioactivity measurement enables to detect extremely small amount of nuclides. The lower limit of the concentration depends on the half-life and the decay mode of the nuclide, and on the detection instruments available. Recovery of enough amount of activity is possible by choosing an appropriate sample volume.

- The method can be combined with any other analytical method. Usually, the amount of the fractionated nuclide is determined by its radioactivity. However, any method can be applied for the determination, if it is more favourable as to the detection limit, handling, etc.

- A number of different separation techniques are available for the same information.

- Processes are simple without the use of complicated mechanical systems.

- Principles of the methods are well established.

Limitations common to the chemical separation methods are:

- Procedures may shift the redox equilibrium. The condition of the sample solution, such as pH and the concentrations of the reagents, must be adjusted so as to facilitate the separation. This may perturb and shift the steady state oxidation state distribution.

- Experiences and skills are required.

- Methods are only applicable to solution samples. Species adsorbed on solid surfaces have to be extracted, or the solid samples are to be totally decomposed. But through these pre-treatment of samples, redox states will not be same as those at the original state.

## Spectroscopic methods

There are a large number of spectroscopic techniques that can be used to determine redox speciation. The most commonly used techniques are the electronic absorption (ultraviolet/visible/near-infrared [UV-Vis-NIR]) spectroscopy, fluorescence spectroscopy, laser-induced photothermal spectroscopy (LPS), vibrational (Raman) spectroscopy, X-ray absorption spectroscopy (XAS), and nuclear magnetic resonance (NMR) spectroscopy. Several methods are available for quantifying oxidation state distributions in aqueous solutions and solids, but only a few of these methods can be applied at low concentrations, such as sorbed species or dilute solutions and solids. UV-Vis-NIR, LPS, XAS are capable of simultaneous determination of multiple oxidation states, while fluorescence, Raman and NMR are the methods for single oxidation state determination.

### Electronic absorption and fluorescence spectroscopy

Electronic absorption (UV-Vis-NIR) spectroscopy is the main method for simultaneous determination of U, Np, and Pu oxidation states. In general, absorption spectroscopy has two advantages:

- The absorption spectra is dependent on the oxidation state and the nature of complexes, thus allowing direct determination of species.

- It is non-invasive and non-perturbing.

- Oxidation states of a given actinide element show a distinct absorption spectrum for each of the oxidation states (see Table 2 [22-29]), enabling identification and quantification of different oxidation states of the actinide. The limitations of UV-Vis-NIR include:

- The method is suitable for analyses of relatively concentrated samples ($>\sim0.0001$ M).

- The method is not appropriate for direct study of environmental samples.

- Information about the dominant species and their molar extinction coefficients is required for reliable quantification.

Related to this technique is complexation spectrophotometry where the oxidation state of interest is treated with a reagent, such as Arsenazo III for M(IV) to form a coloured complex [29]. The same equipment as for standard UV-Vis-NIR spectroscopy is used. Detection limits may be improved by one to two orders of magnitude; these are still likely to be insufficient for samples of environmental interest. A further disadvantage is that it is an invasive technique.

Fluorescence is fairly sensitive ($10^{-6}$ to $10^{-8}$ M or even lower) allowing speciation of sorbed species [30]. In addition, the fluorescence spectrum and lifetime are sensitive to the chemical environment, and by using time-resolved techniques individual species can be discriminated [31]. However, the major limitation of this method is that only a few elements and oxidation states fluoresce. Among U, Np, and Pu, only U(VI) fluoresces. Although these selective methods can only be used to quantify actinyl species, in the case of U and Np, where only two oxidation states are expected to be important under most environmental conditions, it should in principle be possible to estimate the tetravalent states by subtracting the actinyl concentrations from the total.

## Laser-induced photothermal spectroscopy

Laser-induced photothermal spectroscopy using photoacoustic (PAS), photothermal deflection (PDS), and thermal lensing (TLS) techniques is generally 100 to 1000 times more sensitive than the conventional UV-Vis-NIR[32]. A major limitation is the difficulties in the experimental set-up, such as the poor reproducibility in PAS for coupling of a piezoelectric detector to a sample cell, and the difficulty in maintaining the experimental arrangement of TLS and PDS.

## Raman spectroscopy

The vibrational spectra of the actinyl species of U, Np, and Pu show specific Raman shift (see Table 3 [33-37]) making it possible to identify the presence of the actinyl species of these elements. The advantages of this method are:

- It is non-invasive and non-perturbing.

- The magnitude of the Raman shift is sensitive to the chemical environment.

The limitations of this method include:

- Only actinyl species can be detected.

- It requires solid samples or high concentrations of solutions ($>\sim10^{-4}$ M) are required.

## X-ray absorption spectroscopy

With the availability of third generation synchrotrons, X-ray absorption spectroscopy (XAS) is becoming a powerful technique for the speciation study. X-ray absorption near edge spectra (XANES) of the different oxidation states of the actinide elements occur at specific energies (Figure 1), allowing differentiation between the actinide elements and their oxidation states. Major advantages of XANES over UV-Vis-NIR spectroscopy are:

- X-ray absorption edge energies for a given actinide fall in a narrow energy range that is well separated from the other actinides (Figure 2), allowing discrimination of contributions from the other actinide elements.

- It is about a factor of 100 more sensitive than the UV-Vis-NIR.

The major limitations of this method are:

- The shift in absorption edge energy is non-linear with oxidation state.

- It requires the transportation of samples off site to a synchrotron facility, thereby prohibiting analysis of time-sensitive complexes.

### *Nuclear magnetic resonance spectroscopy*

The vast majority of NMR studies of actinide complexes are conducted with the ligand atoms as the NMR active nuclide. However, when the actinide metal is selected as the NMR active nuclide, this method is isotope and oxidation-state specific (see Table 4 [38]). Only a few of the isotopes are NMR sensitive and, in general, only the actinyl species can be identified by this method. The major advantage of this method is that the chemical environment of the metal can be determined. The major limitations of this method are:

- Very high concentrations are required (>~0.01 M).

- Only actinyl species can be detected.

Although this method can potentially be applied to both solid and solution studies, thus far no studies utilising the actinide metal as the NMR active nuclide have been conducted in solution.

## Electrochemical methods

An electrochemical method is applicable, in principle, to the speciation of the oxidation state of an objective species, since the method is based on a differentiation of electrochemical behaviour among the species of different oxidation states. The commonly employed techniques are divided into two groups; one is voltammetry including polarography or coulometry on the basis of the measurement of a current or a coulomb number at a polarised electrode and the other is potentiometry on the basis of the measurement of a potential generating at a depolarised electrode.

A current-potential curve obtained by cyclic voltammetry or polarography gives both qualitative information on the species from the potential and quantitative information of its concentration from the current simultaneously. The data of the redox potential at which the reduction or oxidation of a given species in the solutions of various compositions takes place have been accumulated. Standard redox potentials, $E^0$, for the reversible electrode processes of U, Np and Pu ions are evaluated and recommended [39]. Coulometry is also feasible for the redox speciation of the ions. In particular, flow-coulometry using multi-step column-electrodes system is powerful for the oxidation state analysis of U, Np, and Pu ions [40], and this technique was applied [41] even to the oxidation state analysis of U in the solid sample of $UO_2 + UO_3$ mixture.

Main advantages of the electrochemical method are:

- The method is essentially non-destructive and non-contaminating.

- The method has characteristics as a rapid method, both of which allows an *in situ* measurement and monitoring applications.

- Usually high precision can be achieved, e.g. by using highly precise controlled potential coulometry.

Main disadvantage or limitation of the electrochemical method is:

- A sensitivity is not enough high for the environmental application; the detection limit for a conventional voltammetry is around $10^{-6}$ M.

- The interference from the electroactive coexisting ions should be precisely evaluated and eliminated.

Potentiometry using ion-selective electrode (ISE) is promising in future. One of the most unique characteristics of ISE is its sensitivity to "activity" rather than "concentration" of the objective species. Despite many advantages of an analytical application of an ISE, there have been very limited studies on the ISE for actinide ions other than several $UO_2^{2+}$-ISE proposed [42]. Further studies are required for the development of new actinides-specific ISE and its application for the speciation.

*Subgroup members*

Hisao Aoyagi
Hisanori Imura
Kwang-Wook Kim
Yoshihiro Kitatsuji
Xin Li
Dhanpat Rai (Recorder)
Peter Rance
Akira Saito
Mohammad Samadfan
Osamu Tochiyama (Leader)
Hongxian Wang
Steve J. Williams
Zenko Yoshida

# REFERENCES

[1]    D. Rai, N.J. Hess, Proceedings of OECD/NEA Workshop on Evaluation of Speciation Technology, Tokai-mura, Japan, 26-28 Oct. 1999, OECD/NEA (2000).

[2]    A. Saito, Proceedings of OECD/NEA Workshop on Evaluation of Speciation Technology, Tokai-mura, Japan, 26-28 Oct. 1999, OECD/NEA (2000).

*References for radiochemical/chemical separation methods*

[3]    S.C. Foti, E.C. Freiling, "The Determination of the Oxidation States of Tracer Uranium, Neptunium and Plutonium in Aqueous Media", *Talanta*, 11, 385- 392 (1964).

*TTA extraction and LaF$_3$ precipitation methods*

[4]    E.A. Bondieeti, S.A. Reynolds, "Field and Laboratory Observations on Plutonium Oxidation States", BNWL-2117, Proceedings of the Actinide-Sediment Reactions Working Meeting, Seattle, Washington, USA, 1976, p. 505-530 (1976).

*Solvent extraction with TTA, Hexone, Diethylether; precipitation by Zr(IO$_4$)$_4$ , PrF$_3$ , UO$_2^{2+}$ acetate*

[5]    P.A. Bertland, G.R. Choppin, "Separation of Actinides in Different Oxidation States by Solvent Extraction", *Radiochim. Acta*, 31, 135-137 (1982).

*TTA extraction method*

[6]    A. Saito, G.R. Choppin, "Separation of Actinides in Different Oxidation States from Neutral Solutions by Solvent Extraction" *Anal. Chem.*, 55, 2454-2457 (1983).

*Extraction method with use of DBM*

[7]    H. Nitsche, S.C. Lee, R.C. Gatti, "Determination of Plutonium Oxidation States at Tracer Levels Pertinent to Nuclear Waste Disposal", *J. Radioanal. Nucl. Chem., Articles*, 124, 171-185 (1988).

*Extraction method with use of TTA, HDEHP, and Hexone; LaF$_3$ precipitation*

[8]    M.P. Neu, D.C. Hoffman, K. Roberts, H. Nitsche, R.J. Silva, "Comparison of Chemical Extractions and Laser Photoacoustic Spectroscopy for the Determination of Plutonium Species in Near-Neutral Carbonate Solutions", *Radiochim. Acta*, 66/67, 251-258 (1994).

*Extraction method with use of PMBP, 4-benzoyl-3-methyl-1-phenyl-2-pyrazolin-5-one*

[9]   C.F. Novak, H. Nitsche, H.B. Silber, K. Roberts, Ph.C. Torretto, T. Prussin, K. Becraft, S.A., D. Carpenter, E. Hobart, I. AlMahamid, "Neptunium(V) and Neptunium(VI) "Solubilities in Synthetic Brines of Interest to the Waste Isolation Pilot Plant (WIPP)", *Radiochim. Acta*, 74, 31-36 (1996).

*Demonstrate that the method works on Np as well as Pu*

[10]  D.M. Nelson and M.B. Lovett, "Determination of Some Oxidation States of Plutonium in Sea Water and Associated Particulate Matter" in *Techniques for Identifying Transuranic Speciation in Aquatic Environments*, Proc. Joint CEC/IAEA Technical Meeting, Ispra, 1980 (1981).

*LaF₃ precipitation method*

[11]  D.M. Nelson and M.B. Lovett, "Oxidation State of Plutonium in the Irish Sea", *Nature*, 276, 599-601 (1978).

*LaF₃ precipitation method*

[12]  E.A. Bondietti and F.H. Sweeton, "Transuranic Speciation in the Environment", *NVO-178* (Nevada Applied Ecology Group, USDOE), 1977, 449-476 (Oak Ridge National Laboratory) (1977).

*BiPO₄ precipitation*

[13]  S.J. Malcolm, P.J. Kershaw, M.B. Lovett, B.R. Harvey, "The Interstitial Water Chemistry of $^{239,240}$Pu and $^{241}$Am in the Sediments of the Northeast Irish Sea", *Geochim. Cosmochim. Acta*, 54, 29-35 (1990).

*Precipitation method with use of BiPO₄*

[14]  T.G. Scott, S.A. Reynolds, *Radiochem. Radioanal. Letters*, 23, 275-281 (1975).

*Conditions for BiPO₄ precipitation*

[15]  Y. Inoue, O. Tochiyama, "Determination of the Oxidation States of Neptunium at a Tracer Concentrations by Adsorption on Silica Gel and Barium Sulfate", *J. Inorg. Nucl. Chem.*, 39, 1443-1447 (1977).

*Adsorption method with use of silica gel and BaSO₄*

[16]  E.A. Bondietti, J.R. Trabalka, "Evidence for Plutonium(V) in an Alkaline, Freshwater Pond", *Radiochem. Radioanal. Lett.*, 42, 169-176 (1980).

*TiO₂ adsorption*

[17]  J.M. Cleveland , *The Chemistry of Plutonium*, Gordon and Breach Science Pub., 1970, p. 147.

[18]  A. Kobashi, G.R. Choppin, J.W. Morse, "A Study of Techniques for Separating Plutonium in Different Oxidation States", *Radiochim. Acta*, 43, 211-215 (1988).

*Sorption methods with use of silica gel and CaCO₃*

[19]   H. Gehmecker, N. Trautmann, G. Herrmann, "Separation of Plutonium Oxidation States by Ion Exchange Chromatography", *Radiochim. Acta*, 40, 81-88 (1986).

*Ion exchange with silica gel based inorganic exchanger*

[20]   E.P. Horwitz, M.L. Dietz, R. Chiarizia, H. Diamond, S.L. Maxwell III, D.R. Nelson, *Anal. Chim. Acta*, 310, 63 (1995).

*Extraction chromatography based on well-known solvent extraction methods*

[21]   M. Mang, H. Gehmecker, N. Trautmann, G. Herrmann, "Separation of Oxidation States of Neptunium and Plutonium by Continuous Electrophoretic Ion Focusing", *Radiochim. Acta*, 62, 49-54 (1993).

*References for spectroscopic methods*

[22]   J.M. Cleveland, "The Chemistry of Plutonium", American Nuclear Society, p. 653 (1979).

[23]   J.J. Katz, G.T. Seaborg, L.R. Morss, "The Chemistry of the Actinides", Chapman Hall, New York, Vol. 1, 465-469, 485-487 & 784-787 (1986).

[24]   J.J. Katz, E. Rabinowitch, "The Chemistry of Uranium", Atomic Energy Commission, p. 769 (1958).

[25]   D. Cohen, W.T. Carnall, *J. Phys. Chem.*, 64, 1944 (1960).

[26]   W. Runde, M.P. Neu, S.D. Conradson, D.L. Clark, P.D. Palmer, S.D. Reilly, B.L. Scott, C.D. Tait, *Sci. Basis Nucl. Waste Man.* XX. W.J. Gray, I.R. Triay, eds., MRS, 693 (1997).

[27]   G.A. Burney, R.M. Harbour, "Radiochemistry of Neptunium", NAS-NS-3060 (1974).

[28]   H. Nitsche, E.M. Standifer, R.J. Silva, "Neptunium(V) Complexation with Carbonate", *Lanthanide and Actinide Research* 3, 203 (1990).

[29]   J.L. Swanson, D. Rai, *Radiochem. Radioanal. Letters* 50, 89-98 (1981).

[30]   D.E. Morris, C.J. Chisholm-Brause, M.E. Barr, *Geochim. Cosmochim. Acta* 58, 3613-3619 (1994).

[31]   C. Moulin, I. Laszak, V. Moulin, C. Tondre, *Appl. Spectrosc.* 52, 528-535 (1998).

[32]   M.K. Richmann, D.T. Reed, Mater. Res. Soc. Proceedings, 412, 623-630 (1996).

[33]   L.J. Basile, J.C. Sullivan, J.R. Ferraro, P. LaBonville, *Appl. Spectrosc.* 28, 144-148 (1974).

[34]   L.J. Basile, J.R. Ferraro, M.L. Mitchell, J.C. Sullivan, *Appl. Spectrosc.* 32, 535-537 (1978).

[35]   C. Madic, D.E. Hobart, G.M. Begun, *Inorg. Chem.*, 22, 1494-1503 (1983).

[36]   L. Maya, G.M. Begun *J. Inorg. Nucl. Chem.*, 43, 2827 (1981).

[37]  W. Runde, M.P. Neu, S.D. Conradson, D.L. Clark, P.D. Palmer, S.D. Reilly, B.L. Scott, C.D. Tait, *Sci. Basis Nucl. Waste Man*. XX. W.J. Gray, I.R. Triay, eds., MRS, 693 (1997).

*References for nuclear magnetic resonance spectroscopy*

[38]  R.D. Fischer, in *NMR of Paramagnetic Molecules: Principles and Applications*, G.D. La Mar, W.D. Horrocks Jr., R.H. Holm, eds., Academic, New York, pp. 521-553 (1973).

*References for electrochemical methods*

[39]  S. Kihara, Z. Yoshida, H. Aoyagi, K. Maeda, O. Shirai, Y. Kitatsuji, Y. Yoshida, *Pure Appl. Chem.*, in press (2000).

[40]  H. Aoyagi, Z. Yoshida, S. Kihara, *Anal. Chem.*, 59, 400-405 (1987).

[41]  S. Kihara, Z. Yoshida, H. Muto, H. Aoyagi, Y. Baba, H. Hashitani, *Anal. Chem.*, 52, 1601-1606 (1980).

[42]  I. Goldberg, D. Meyerstein, *Anal. Chem.*, 52, 2105 (1980).

## Table 1. Chemical separation methods for redox speciation

### Table 1a. Solvent extraction

| Reagent | Conditions, key anal. parameters | Actinides (III) | (IV) | (V) | (VI) | Advantages/disadvantages/limitations, other comments | Ref. |
|---|---|---|---|---|---|---|---|
| TTA | pH-dependent D  pH 0-1  pH 4.5 | X  O | O  O | X  X | X  O | Rapid, quantitative separation  TTA is photosensitive | 3, 4, 5, 7, 9 |
| PMBP | pH-dependent D  pH 0-1  pH 4.5 | X  O | O  O | X  X | X  O | Used to replace TTA  Stable reagent  Very small pKa (4.11) | 8, 9 |
| DBM | pH-dependent D  pH 8  pH 5.0-5.5 | O  X | X  O | X  X | O  O | Stable reagent  pKa = 9.2  Slow extn. kinetics | 6 |
| HDEHP | pH-dependent D  pH 0  pH 1 | X  O | O  O | X  X | O  O | Liq. cation exchanger  Low selectivity for redox state | 7 |
| Hexone | 4.0 M Ca(NO₃)₂  1 M HNO₃, D ≤ 30 | X | O | X | O | High salt concn. required | 4 |
| Diethylether | 10 M NH₄NO₃  1 M HNO₃ | X | X | X | O | High salt concn. required  Selective for An(VI), but not quantitative  Difficult manipulation of the solvent | 4 |

O: extracted, X: not extracted

### Table 1b. Precipitation

| Reagent | Conditions, key anal. parameters | Actinides (III) | (IV) | (V) | (VI) | Advantages/disadvantages/limitations, other comments | Ref. |
|---|---|---|---|---|---|---|---|
| LnF₃ | 0.8 M HNO₃,  0.25 M H₂SO₄,  0.5mM K₂Cr₂O₇ | O | O | X | X | Drastic treatment is required to redissolve.  Incapable of distinguishing V and VI | 3, 4, 7, 9, 10, 11 |
| BiPO₄ | 0.25 M HNO₃  Bi³⁺ 25 mg/L  H₃PO₄ ≤ 0.1 M | O | O | X | X | Dence, crystalline ppt. Readily separable.  Soluble in concn. HNO₃  Incapable of distinguishing V and VI | 12, 13, 14 |
| Zr(IO₃)₄ | 2 M HNO₃ | – | O | X | X | Probable involvement of IO₃⁻ in redox reaction (III ↔ IV) | 4 |
| NaUO₂(AcO)₃ | 1 M HNO₃  5 M NaNO₃  3 M CH₃COONa | X | X | X | O | Specific for AnO₂²⁺  Introduce α-activity to the sample.  Time consuming | 4 |

O: precipitated, X: not precipitated

## Table 1c. Adsorption

| Reagent | Conditions, key anal. parameters | Actinides (III) | (IV) | (V) | (VI) | Advantages/disadvantages/ limitations, other comments | Ref. |
|---------|----------------------------------|-----------------|------|-----|------|-------------------------------------------------------|------|
| $BaSO_4$ | 0.2 M $HClO_4$ <br> Slight excess of $SO_4^{2-}$ over $Ba^{2+}$ necessary | – | O | X | X | Adsorption of An(V) (~10%) | 15 |
| Silica gel | pH 8 (seawater condition) <br> pH 9-10.5 <br> 10 g adsorbent/L <br> Contact time 2-4 hrs | – <br> – | – <br> – | X <br> O | O <br> O | Careful pre-treatment of adsorbent is necessary <br> Long contact time | 15, 18 |
| $CaCO_3$ | pH 8 (seawater condition) <br> 50g adsorbent/L, <br> Contact time 5-10 min | – | – | O | X | Short contact time <br> Poor separation efficiency <br> Provide direct evidence of An(V) | 16 |
| $TiO_2$ | pH 9 (natural water) | – | O | X | X | Basic data are lacking. | 16, 17 |

O: adsorbed, X: not adsorbed

## Table 1d. Ion exchange

| Exchanger | Conditions, key anal. parameters | Actinides (III) | (IV) | (V) | (VI) | Advantages/disadvantages/ limitations, other comments | Ref. |
|-----------|----------------------------------|-----------------|------|-----|------|-------------------------------------------------------|------|
| Li Chrospher Si 100 | 0.01 M $HClO_4$ | X | O | X | X | (Merck) <br> Functional group-SiOH | 19 |
| Li Chrosorb KAT | 0.01 M $HClO_4$ | O | O | X | O | (Merck) <br> Functional group-$SO_3^-$ | 19 |

O: adsorbed, X: not adsorbed

## Table 1e. Extraction chromatography

| Resin | Conditions, key anal. parameters | Actinides (III) | (IV) | (V) | (VI) | Advantages/disadvantages/ limitations, other comments | Ref. |
|-------|----------------------------------|-----------------|------|-----|------|-------------------------------------------------------|------|
| TRU™ | 4M HCl, 2M HCl, 0.1 M $NH_4C_2O_4$ | O | O | X | O | 0.75 M CMPO inTBP | 20 |
| TEVA™ | | X | O | X | X | Aliquat 336 | 20 |
| U/TEVA™ | | X | O | X | O | $(C_5H_{11}O_2)_2(C_5H_{11})P=O$ | 20 |

O: retained, X: not retained

## Table 1f. Electrophoresis

| Methods | Conditions, key anal. parameters | Actinides (III) | (IV) | (V) | (VI) | Advantages/disadvantages/ limitations, other comments | Ref. |
|---------|----------------------------------|-----------------|------|-----|------|-------------------------------------------------------|------|
| Electro-phoresis | Net charges on the complexed species <br> Applied voltage <br> Complexing reagents | O | O | O | O | Flow method <br> Chromatographic separation <br> Time consuming | 21 |

O: separated each other

**Table 2. Electronic absorption methods for redox speciation**

| Oxidation state | Estimated detection limit /M | | Maximum wavelength and molar absorptivity | Ref. |
|---|---|---|---|---|
| | UV-Vis-NIR | LPAS | | |
| U(IV) | $2 \times 10^{-3}$ | $5 \times 10^{-6}$ | 650 nm, $\varepsilon = 58$ | 25 |
| U(VI) | $1 \times 10^{-2}$ | $6 \times 10^{-6}$ | 413 nm, $\varepsilon = 8$ | 26 |
| Np(IV) | $8 \times 10^{-4}$ | $7 \times 10^{-6}$ | 723 nm, $\varepsilon = 127$ | 23, 26, 27 |
| | | | 959 nm, $\varepsilon = 162$ | |
| Np(V) | $2 \times 10^{-4}$ | $2 \times 10^{-6}$ | 980 nm, $\varepsilon = 395$ | 23, 26, 27, 28 |
| Np(VI) | $2 \times 10^{-3}$ | $5 \times 10^{-5}$ | 1230 nm, $\varepsilon = 45$ | 23, 26, 27 |
| Pu(III) | $3 \times 10^{-3}$ | $5 \times 10^{-6}$ | 600 nm, $\varepsilon = 35$ | 23, 29 |
| Pu(IV) | $2 \times 10^{-3}$ | $4 \times 10^{-7}$ | 475 nm, $\varepsilon = 50$ | 23, 29 |
| Pu(V) | $5 \times 10^{-3}$ | $5 \times 10^{-6}$ | 569 nm, $\varepsilon = 17$ | 23 |
| Pu(VI) | $2 \times 10^{-4}$ | $2 \times 10^{-6}$ | 830 nm, $\varepsilon = 550$ | 23, 26, 29 |

Molar absorptivities are given in $M^{-1}cm^{-1}$

**Table 3. Raman spectroscopy for redox speciation**

| Oxidation state | Approximate concentration range/M | Solution conditions | Vibrational frequencies/cm$^{-1}$ | Ref. |
|---|---|---|---|---|
| U(VI) | $>10^{-1} - 10^{-3}$ | 0.01 M HClO$_4$ | 872 | 34 |
| | | | 885-851 | 37 |
| | | | 870 | 36 |
| Np(V) s | $>10^{-1} - 10^{-3}$ | 1.3 M HClO$_4$ | 767 | 34 |
| | | | 767 | 37 |
| Np(VI) | $>10^{-1} - 10^{-3}$ | 0.01 M HClO$_4$ | 863 | 34 |
| | | | 854 | 36 |
| Pu(V) | $>10^{-1} - 10^{-3}$ | Na$_2$CO$_3$ | 755 | 35 |
| Pu(VI) | $>10^{-1} - 10^{-3}$ | 0.01 M HClO$_4$ | 835 | 34 |
| | | | 830-840 | 36 |

**Table 4. NMR spectroscopy for redox speciation**

| Isotope | Observable oxidation state | Estimated threshold concentrations/M | Ref. |
|---|---|---|---|
| $^{229}$Th | 4 | $2.3 \times 10^{-2}$ | 38 |
| $^{231}$Pa | 5 | $2.3 \times 10^{-4}$ | 38 |
| $^{233}$U | 6 | $1.1 \times 10^{-2}$ | 38 |
| $^{235}$U | 6 | $6.4 \times 10^{-2}$ | 38 |
| $^{237}$Np | 7 | $1.2 \times 10^{-4}$ | 38 |

## Figure 1. X-ray absorption edge energies for U, Np, and Pu

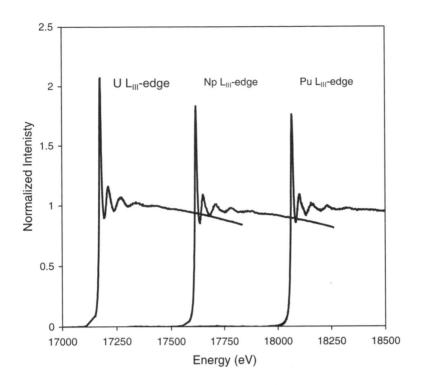

## Figure 2. Shift in absorption edge energy for the different oxidation states of Pu

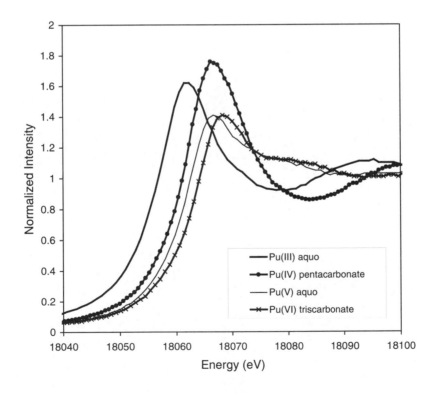

# PREDICTIVE APPROACH TO SPECIATION

## Introduction

In this section, the term "speciation" will be taken to mean the chemical form of the element or substance, including the oxidation state, charge and complexation properties. The behaviour of trace elements in natural systems, especially those displaying multiple valence states, is known to be critically dependent on speciation, hence the attention devoted to this subject in the context of environmental protection and waste management.

However, consideration of speciation is not limited to these particular fields. The principal situations to which this document pertains are:

- Performance assessment of a repository for radioactive waste.

- Decommissioning of nuclear facilities.

- Land remediation activities following unintentional and/or unregulated release of radioactive materials.

Implicit in the above is the likelihood that chemical separations will be required and these, in turn, require adequate knowledge of the species present.

Extant thermodynamic databases and related information are frequently inadequate to meet the need for addressing speciation in the principal situations outlined above. Where gaps in the database are apparent, the time frames on which information is needed are much shorter than those required for measuring the desired data experimentally. The result is often the abandonment of scientifically rigorous methods in favour of empirical approaches, which though simple to implement, are generally difficult to defend in adversarial fora. For these reasons, methods for estimating thermodynamic data have been developed over the years and are reviewed briefly in the current report. To date, these have tended to concentrate on aqueous systems. Predictive methods for complex solids are less advanced, though such systems are important in controlling the release and transport of toxic/radioactive materials.

Methods for determining speciation are numerous and greatly exceed those considered in the present discussion. The scope of the report will be restricted to the actinide elements, although the principles could be adapted to include other radioactive materials and toxic metals. Organic compounds are considered only in regard to their role in actinide chemistry, notably metal-organic complexation and the formation of organic colloids. Three main categories are recognised:

- Metallo-organic materials that have been disposed of or accidentally released.

- Naturally occurring high molecular weight organics (humic and fulvic acids).

- Degradation products of, for example, cellulose that have the potential to bind actinide ions.

The following sections address the various approaches available for acquiring speciation information. This does not constitute an exhaustive review but rather the intention is to provide a working outline for addressing the prediction of speciation.

## Requirements and state of the art

The term speciation depends upon the framework of application. The present state of the art may be highly variable for different conditions encountered, depending on whether they relate to well-controlled laboratory experiments or highly complex field conditions.

When estimating data for thermodynamic calculations it is essential that the system be defined adequately as few data are truly generic. Examples of systems for which knowledge of the speciation is required include high-level waste repositories, polluted soils or waters and facilities contaminated by previous industrial activities or accidental releases of toxic/radioactive materials. The core information needed includes the mass of contaminant and related materials present, the flux and composition (pH, Eh etc.) of water, contact with or absence of air and the temperature. The amount and nature of solid phases present are also needed to determine source terms and possible substrates for fixation. Not surprisingly, this exhaustive list of requirements is rarely (if ever) met.

A firm theoretical framework is also a pre-requisite for extrapolation. However, the applicability of scientific knowledge is inextricably coupled to the study system. Thus, although a large amount of information may be available for a given element or species, it may not be directly applicable to the situation of interest. Care must be exercised to ensure that impossible predictions do not arise from the exercise. Natural systems where predictive speciation calculations are required are usually far more complex than those in which the data were originally determined. Thus, observational level information must be factored into the prediction as a control.

In the event that thermodynamic data are completely lacking, then estimates of elemental solubility or quantities derived from species distribution, mainly "sorption" values or "bio availability" necessarily rely on a more general data correlation, "raw" observations of natural systems, or "expert judgements". The further the framework of prediction is from relevant data and/or a theoretical basis, the less reliable the predicted speciation will be and the greater the likelihood that the results will not be accepted by scientific and regulatory bodies.

At some point, a balance must be struck between the sophistication sought in predictive modelling and the limitations of the experimental information available. There will always be a desire for experimental verification of theoretical predictions. However, time and resource constraints will impose practical limits for performing laboratory and field experiments. Criteria for determining the acceptability or otherwise of predictions need to be defined beforehand, recognising the inherent uncertainty in thermodynamic calculations.

The adoption of ISO/BIPM's *Guide for Expression of Uncertainty in Measurement* by IUPAC and IAEA, means that strict criteria are now stipulated for evaluating and reporting thermodynamic data. This should enhance quality control in the future but does little to address problems of 'legacy' data, which comprise the bulk of current thermodynamic compilations and also the absence of measurements in a number of important areas.

Efforts made to overcome the gap that exists between the current situation and the level of knowledge needed can be sub-divided into:

- Parametric interpretation/extrapolation, based on chemical systematics and including consideration of fundamental thermodynamic properties (enthalpy, entropy, heat capacity, molar volume etc.).

- The use of chemical, mineralogical or structural analogies.

- More subjective approaches, for example, Delphi methods where data are selected on the basis of recommendations made by expert review panels.

Emphasis here is focused on the first category, together with the often-valuable use of analogy. Elicitation of expert opinion is outside the scope of this report though it is noted that compilation of a "standard" database itself incorporates a degree of subjective judgement.

One of the most important aspects of understanding actinide behaviour is defining how speciation responds to changing environmental conditions. Differences in temperature, ionic strength and major element composition will determine the species present and alter their relative proportions. By the simple use of analogy, valuable information can be acquired from existing thermodynamic databases if the data included were obtained under conditions similar to those of interest. Caution is required, of course, if the thermodynamically stable form is not the actual species encountered, owing to kinetic factors. The key question is always, how good is the analogy?

A comparison between the actinides and lanthanide elements immediately reveals the distinction between the early actinide series elements from the other series in the periodic table. There is a relatively regular change of properties between, for example, the homologous transition metals whereas the redox chemistry in particular of the light actinides (Pa-Pu) is variable and does not change smoothly. In order to predict the structure of even simple aqueous species it is important to recognise this distinction, which can be attributed to different factors.

First, non-relativistic calculations show that the 5f electrons are less shielded than 4f electrons, due to the relatively larger extension of the 5f orbitals. Hence, the lighter actinides have a greater tendency for variable oxidation states and normally favour higher oxidation states, as exemplified by comparing homologs in the lanthanide series.

The second factor results from relativity considerations. Owing to the high velocity of the inner shells of a heavier atom, the mass of an inner electron increases and, thus, the electron is stabilised; an effect that is largest for s and p electrons. The effective potential of the nucleus on the d and f electrons is correspondingly decreased. Relativistic shielding by d and f orbitals increases with $Z^2$, where Z is the atomic number. Hence, relativity exerts a greater influence on the actinides as compared to the lanthanides and this must be recognised in modelling.

## Types of models and their fitness for purpose

The need to evaluate and/or predict actinide behaviour for very long periods of time (up to $10^6$ years) will always exceed our ability to acquire the necessary experimental data/information. Predictive approaches for assessing speciation start in each case with a critical evaluation of the experimental evidence and development of a methodology for extrapolation. In assigning confidence to a model, directly measured experimental data from the laboratory are normally afforded first priority and are supplemented by documented information, experience of related work and, where feasible, field studies. In practice, established information of the latter type is normally employed to evaluate proposed models and to refine them.

Modelling approaches fall into different categories, ranging from those built solely on empirical information to those established purely on theory or thermodynamic principles. In the latter, the limitation of the thermodynamic method must be considered; that is, this approach looks at the initial and final states. It is well recognised that kinetic factors may greatly influence the actual results.

Another important aspect of modelling is the physical state of the material being assessed. An engineered repository may offer the most controlled (i.e. predictable) parameters, as the inventory is often reasonably well characterised. Other situations such as those involving accidental releases into the environment entail much wider, variable or even unknown conditions, which may severely test the best of models in making realistic assessments. Designated restoration sites may be considered an intermediate situation with regard to defining parameters but still provide a strong challenge with respect to predicting speciation. The main point here is that modelling results need to be considered within the limits of their validity/applicability and should not be afforded undue credence in a performance assessment context.

The modelling of purely aqueous systems, though far from straightforward, has been studied most extensively and is the area where fewer difficulties in estimating data are likely to be encountered. Modelling of solid phase interactions is more difficult. The degree of crystallinity and variable (non-ideal) stoichiometry will have a significant effect on solubility. Adequate enthalpy and free energy of formation data are important for establishing the appropriate phase(s) to be considered in any calculation of precipitation-dissolution processes. For mineral solids, models often rely on observations from natural systems and interpolation between ideal "end-member" compositions. The error introduced by such simplifications may be substantial.

In performance assessments, it is common to use solubility ($K_{sp}$) and "equilibrium sorption" ($K_d$) values to quantify the source term and migration of contaminant species, respectively. Depending on the situation, significant variation in $K_{sp}$ values may not generate a serious modelling error, even though such an uncertainty may not be desirable from a scientific standpoint. This needs to be assessed on a case by case basis. The use of directly calculated elemental $K_d$ "constants", however, is extremely difficult to justify. The problems are sufficiently well known that they need not be repeated here. Thermodynamic sorption models have been developed to address choice of distribution coefficients for radionuclides in bentonite systems [1] and for simulating transport of U(VI) species through bentonite [2].

Simple organic complexes in aqueous systems have been modelled successfully using existing experimental data or estimates based on the substitution of reasonable counterparts. The vast array of experimental data and established equilibrium constants for simple ligands may also be used with care to provide estimates for more complex moieties containing the same or similar configurations of functional groups. A comparison of approaches for modelling metal-humate complexes is given by, for example, Schussler, *et al.* [3]. It must be recognised, however, that humics are rather ill-defined multi-component materials) encompassing carboxylic groups, proteins and nucleic acids. Efforts to describe such complex material by formation constants can be considered only on the basis of a "first approximation".

The use of natural analogues offers another means of fitting or predicting behaviour [4]. As in performance assessment, application of natural analogues proceeds by:

- Conceptual modelling.

- Provision/verification of results.

- Final validation of the models.

Modelling of colloidal systems offers a different challenge. In the nuclear field they have been classified as "pure" colloids, usually consisting of agglomerations of sparingly soluble hydroxides or "pseudo-colloids" formed by the attachment of contaminant species to a substrate (e.g. clay minerals) in the colloidal size range. The latter are more difficult to simulate, as the mode of attachment may not be known. Attachment may come about via surface adsorption, ion exchange, complexation or (co-) precipitation and may not be reversible. Further, the conditions under which the substrate may be stable can be very different from the conditions under which ionic species (or a pure colloid of the material) would be stable. Therefore, it becomes very difficult to assign phase behaviour and transport properties to these materials in natural waters. Typically one is forced to rely on experimental observations for assessment purposes, though invariably these cannot fully replicate conditions in the field. Modelling the transport behaviour of colloids in fractured natural media has been discussed by for example, [5].

There have been numerous attempts over the years to model such systems. Kim and co-workers describe an electrostatic method for predicting actinide complexes with inorganic ligands [6], a semi-empirical approach based on energy criteria describing inter-ligand electrostatic repulsion. This model has been implemented using a number of different codes and the INE PITZER database [7]. Other options are the use of linear Gibb's free energies of formation correlated to known data for aqueous cations and crystalline solids [8] or the use of a triple-layer surface complexation model for the calculation of mobility [9]. Radiotracer techniques in conjunction with modelling of surface attachment, including consideration of pH, ionic strength, solid to solution ratios, contact times, etc., have also been used to verify actinide speciation and sorption behaviour [10].

Conditions for formation of colloids are not well understood. Certainly pH, ionic strength and composition of the aqueous phase are important parameters. Colloids are often regarded as transient features since a change of experimental conditions may render them unstable. However, there is some evidence that removal of colloids from a system by, for example, filtration, results merely in the generation of new colloids to restore the status quo. Attempts have been made to model the formation of colloids during corrosion using electric double layer theory to provide some prediction of the colloid stability [11]. This approach may help quantify the potential impact of colloids on assessment calculations.

Modelling of colloid migration is also complicated but tends not to be included in performance assessment calculations, even though it is recognised as a potentially critical factor [12]. Deposition and re-suspension may occur and given the small size of colloids, a deposited colloid may become mobile merely by a mechanical disturbance, a situation difficult to model using existing methods.

It is obvious that many different situations must be considered in developing predictive assessments and that not all of these are amenable to thermodynamic treatment. Ideally, thermodynamics and empirical knowledge should be used together in arriving at a defensible case. Where modelling is largely or wholly reliant on estimated data it is imperative that the limitations of the models be stated explicitly. The limitations of the purely thermodynamic approach when applied to repository systems has been addressed, albeit somewhat superficially, by Berner [13]. The treatment of parameter uncertainty can be explored in an automated fashion using Monte Carlo sampling in the HARPROB computer program [14].

## Examples of the estimation methods available

Notwithstanding the limitations of thermodynamic models for describing dynamic natural systems, establishing confidence in the thermodynamic database is of primary importance as this is the starting point for subsequent kinetic or transport calculations. There are still problems of consistency among 'standard' compilations and extensive efforts, for example in the NEA-TDB project, are being devoted to the evaluation of available thermodynamic data [15]. This applies also to the estimation and extrapolation of data [16] in cases where the database is incomplete. It is important to determine the scientific basis, range of applicability and the accuracy of such methods together with their limitations. Some examples are shown below.

### Aqueous species

Numerous predictive approaches have been developed over the past three decades predominantly for aqueous species [15]. These are based on correlation between ions of similar charge-size characteristics and the same or similar configuration of valence electrons. They are distinguished by the empirical relationships employed and the relative weighting attached to complexation affinities with various ligands.

In simple electrostatic models, the stability of a species in solution is discussed in terms of the charge and ionic radius of the central metal ion. For example, the hydrolysis constants of metal-hydroxide complexes correlate well with the ratio z/d, where z is the formal cationic charge and d is the interatomic distance [17]. The theoretical premise behind such a correlation is the ionic model of chemical bonding, whereby the chemical forces between atoms are largely of electrostatic type. It is expected, therefore, that correlations of this type will work better for chemical systems where the electrostatic component in the chemical bonding is significant as in the bonding between hard electron acceptors and hard donors.

The actinide ions are regarded as hard acids in the framework of the Pearson hard and soft acid and bases concept. Co-ordination occurs preferably with hard ligands, such as hydroxide, carbonate, phosphate and carboxylic acids indicating preference for mainly electrostatic interactions. It seems feasible to describe the interactions of such ligands with the metal ion by a model that adds the energy changes resulting from successive replacement of water molecules in the hydration sphere by the ligand. An example of this type of approach is that proposed by Münze [18]. The thermodynamic functions, $\Delta G$, $\Delta H$ and $\Delta S$, of the species' formation reactions are derived from electrostatic considerations of the ligand-central ion interactions. The model includes a term that allows calculation of the thermodynamic functions over a range of temperatures. Comparison of experimental and calculated formation constants for lanthanide and actinide acetates was found to be satisfactory, but for other ligands the agreement was not as good.

Another example is an extended hard sphere model in which the effective rather than the nominal charge of a metal ion is considered [19,20]. This model has also been applied to the analysis of the step-wise hydrolysis constants of actinide ions and the effective charges of the central actinide ions have been determined. Though the effective charges obtained suggest a contribution from non-electrostatic interactions, it is expected that similar concepts may be applied with caution to the analysis of stability constants for actinide solid phases.

The so-called unified theory of metal-ion complex formation constants is an example of an extended electrostatic model that considers the electronic structure as well as charge and ionic radius [21]. The concept of "electronicity", which was developed in this model, describes the ability of a ligand to bind a metal ion and allows prediction of the hydrolysis and complexation behaviour of most cations in the periodic table. Thus, one can predict, with variable degrees of confidence, stability constants for a large number of aqueous species composed of metal ions and ligands that are considered important in safety assessments [22].

The so-called unified theory of metal-ion complex formation constants is an example of an extended electrostatic model that considers the electronic structure as well as charge and ionic radius [21]. The concept of "electronicity", which was developed in this model, describes the ability of a ligand to bind a metal ion and allows prediction of the hydrolysis and complexation behaviour of most cations in the periodic table. Thus, one can predict, with variable degrees of confidence, stability constants for a large number of aqueous species composed of metal ions and ligands that are considered important in safety assessments [22].

Since few direct measurements are available for the temperatures expected in a high level waste repository, extrapolation of standard state thermodynamic data for the relevant aqueous species is needed. Established methods include the correspondence principle [23], the isocoulombic principle [24] and the use of density functions [25]. These methods now need to be extended to include radiologically critical elements.

It is important to note that the models employed are not purely theoretical but contains empirical or, at least, semi-empirical functions and/or parameter values. Thus, in determining their "fitness for purpose" the accuracy of the auxiliary thermodynamic data used for the estimation of the functions, needs to be checked. For the majority of the actinides, for example, the species present are very sensitive to the redox conditions and the data for oxidation and reduction reactions are far from being fully satisfactory.

### Solid phase and colloidal species

In comparison to aqueous species, there is much more uncertainty in the prediction of thermodynamic quantities for solid phase species, since the stability of solids is very much dependent on their composition and structure. One approach is to use correlations between the thermodynamic stability and certain measurable crystal parameters. For multi-component complex structures, however, the individual bond energies are not easily calculated and one must attempt to find some crystal-chemical regularity in the available thermodynamic data. In the case of silicate minerals, for example, an approach has been proposed in which similar energies are assumed in the individual polyhedral units to permit the proportional addition of these unit energies to obtain the overall crystal energy [16,26].

With reference to entropy, it has been shown that approximately linear relationships exist between molar volumes and entropy for many isostructural compounds. Departures from this idealised behaviour have been attributed to crystal field effects for which corrections can be applied. More recent hybrid approaches combine it with statistical theory. Thus, many of the thermodynamic data required could be derived from basic principles and 'internally consistent' databases of this type have been published. The different philosophies behind the published compilations are explored in [27]. Comparing the performance of the alternative approaches (Table 1) emphasises the importance of assumptions made regarding the structure of the solid.

**Table 1: Comparison of •G°$_f$ for selected minerals at 1 013 kPa and 25°C [28-30] (kJ mol$^{-1}$)**

| Mineral | Formula | [28][a] | [29] | [30][b] |
|---|---|---|---|---|
| calcite | $CaCO_3$ | -1129.7 | -1128.0 ± 0.8 | |
| aragonite | $CaCO_3$ | -1129.3 | -1127.7 ± 1.0 | |
| rhodochrosite | $MnCO_3$ | -816.7 | -816.5 ± 1.6 | |
| magnesite | $MgCO_3$ | -1030.1 | -1029.0 ± 1.0 | |
| | $UO_2CO_3$ | -1513.8 | | -1562.2 ± 3.3 |
| | $UO_2(OH)_2$ | -1285.3 | | -1633.3 ± 3.5 |

a) Re-calculated with 1 cal = 4.184 J, b) Experimental data

The level of agreement with the more common carbonate minerals is very good; discrepancies are within an order of magnitude of the estimated uncertainties given by La Iglesia and Felix [29]. In comparison, the estimates for the uranium (VI) solids show much larger differences, for two main reasons. First, the linear regression approach requires data for the free energy of formation of $UO_2^{2+}$(aq) and even though this is known to within about ±1.0 kcal mol$^{-1}$ [28], the uncertainty is sufficient to significantly influence the final estimate. Second, it was assumed that $UO_2CO_3$ has a calcite-type structure. However, $UO_2CO_3$ forms layer structures with two distinct stacking patterns. Similarly, $UO_2(OH)_2$ does not possess a $CdCl_2$ structure as was assumed. Rather, $UO_3 \cdot 2H_2O$ forms with a complex network of pentagonally co-ordinated $UO_2^{2+}$ units, μ-bridging oxo groups and both bridging and fully co-ordinated hydroxides [31]. The much larger differences between predicted and experimentally determined data in the last row of Table 1 are, therefore, not surprising. This example serves to illustrate that any prediction of thermodynamic data for chemically analogous solid phases may be misleading if underlying assumptions regarding the predicted state are not valid.

Molecular modelling techniques are useful in many cases and can bridge the gap between analytical data and extant understanding of the chemistry of a system. Vibrational spectroscopy and theoretical calculations of infrared and/or Raman data can be used to assess the accuracy of hypothetical species. Nuclear magnetic resonance shifts can also be used to determine speciation. These approaches have been very valuable in establishing the stability of complexes as a function of experimental conditions.

Colloids present a different perspective. It is well known that the stability of colloidal species is sensitive to solution conditions, particularly pH and ionic strength. As a result, colloids of many elements form a relatively minor component in comparison to the total concentration in the aqueous phase. In the case of actinide ions, however the formation of colloids is often observed, due to their strong tendency to hydrolyse, even at pH values below 3. The formation of so-called pseudo colloids is often encountered as actinide species sorb readily on mineral or other surfaces.

It is important to establish the composition, structure and the size distribution of these colloids and to devise a means of predicting their properties theoretically. For colloids having comparable structures to well-established silicate minerals, similar methods to the polyhedral approach may be applied to assess the stability and nature of the colloidal forms. A polymer model applied to the size distribution of colloidal species of iron is an example of this approach [32]. However, other approaches may be needed for organic species, such as complex humic substances, which exhibit much more complicated behaviour and can also form pseudo-colloids. A fully satisfactory method for addressing colloids has not been devised to date.

# Concluding remarks

Prediction is always based on existing knowledge and all experimental data are affected by uncertainty. The predictive power, for example, of linear free energy relationships, is limited by the uncertainties in those data employed in the numerical equations used for the estimation. The advances made in computer-based technology do not alter these basic facts but they do assist in the numerical treatment of complex systems. In particular, the development of statistical methods that are able to deal with uncertainties by distributional approximations or re-sampling methods [33] are a significant aid. Similarly, chemometrics has become a recognised field of analytical chemistry [34] and now the assignment of a meaningful standard deviation to any experimentally derived parameter has tended to become the norm [35]. Correlated or mutually 'induced' uncertainties caused by a large number of potential species in a given solution could be considered by stratified sampling methods, for example, the technique of Latin Hypercube Sampling [36].

General rules, e.g. the Gillespie-Nyholm valence shell repulsion model may be applied to actinide ions in aqueous solutions, especially in cases where the valence shell is empty. In this context, it is unfortunate that experimental data on the hydration of actinides in dilute aqueous solutions is rather scarce. The increased application of EXAFS to investigating actinide solution chemistry may generate knowledge regarding these important fundamental data. In some cases, simple geometric considerations may provide sufficient information. For U(VI) carbonato species, structures have been proposed by simple geometrical and space-filling considerations [37]. A parallel study of the respective Np(V) species by EXAFS [38] produced comparable results. These findings have proved helpful in explaining the solution behaviour of U(VI) [39] and Np(VI) [40] carbonato species.

Determining the speciation of radioelements in environmental systems entails consideration of the entire spectrum from simple aqueous ions through colloid sized material to meta-stable and stable solids. A firm theoretical basis is required to justify the predictions made though formal thermodynamic principles may not apply in all cases. Having defined the system a decision needs to be made at the outset with regard to the type and rigour of modelling approach commensurate with the level of understanding and data available. Thus, kinetic or thermodynamic methods, possibly with caveats, are preferable followed by quasi-thermodynamic bespoke models for less well-defined systems, e.g. cementitious waste forms or natural organic materials. Purely empirical approaches should be viewed as a last resort rather than the first port of call, as is frequently the case at present.

Where thermodynamic methods can be applied with an acceptable degree of confidence (ionic aqueous species, stoichiometric solids exhibiting congruent dissolution), a variety of techniques are available for estimating the necessary parameters, including charge-size relationships and use of isostructural solids, as discussed above. All involve the use of analogies either explicitly or implicitly. In those cases where classical theory is not applicable, operational models may be used to scope potential outcomes. These should acknowledge the role of expert scientific judgement by clearly documenting assumptions employed.

In addition to establishing a theoretical basis for prediction at least some reference data are required as a starting point for extrapolation and for verification of the results. This requirement is best met for simple, binary, aqueous complexes. Relatively little is known of the mechanisms via which the transuranics may be incorporated into mineral solids in groundwater environments and this remains an area of concern. A broad picture of their likely fate may be obtained from observations of natural U, Th and rare earth minerals as a guide to prioritising both experimental and modelling tasks. Predictive methods widely applied in petrology [26] offer considerable potential in this regard.

The full range of environmental parameters likely to be encountered must also be accommodated: temperatures in excess of 100°C; pH values ranging from 0 to over 13 (e.g. cemented waste forms); and for situations ranging from strongly reducing to strongly oxidising.

Finally, a dialogue is needed urgently between workers engaged in speciation research and those carrying out performance assessments to ensure that the latter conform to the "best available practice". The role of speciation modelling in building a defensible safety case is not self-evident and decision-makers need to be convinced of the merits of adopting a more phenomenological approach. An outline methodology for addressing the issue in a performance assessment framework is suggested in Figure 1.

*Subgroup members*

| |
|---|
| John A. Berry |
| Richard G. Haire (Recorder) |
| Pertti Koukkari |
| Günter Meinrath |
| Hirotake Moriyama |
| David Read (Chairman) |
| Fabian Rorif |
| Seiichiro Uno |
| Raymond G. Wymer |

## REFERENCES

[1]    M. Ochs, B. Lothenbach, M. Yui, "Application of Thermodynamic Sorption Models to Distribution Coefficients of Radionuclides in Bentonite", *Mat. Res. Soc. Symp. Proc.* 506, 765 (1998).

[2]    K. Stamberg, D. Vopalka, J. Skrkal, P. Benes, K. Chalupska, "Simulation of Transport of Uranium (VI) Species through the Bed of Bentonite", *Mat. Res. Soc. Symp. Proc.* 506, 485 (1998).

[3]    W. Schussler, R. Artinger, J.I. Kim, N.D. Bryan, D. Griffin, "Modelling of Humic Colloid Borne Americium (III) Migration in Column Experiments using the Transport/Speciation Code K1-D and the KICAM Model", *Abstract Migration-99*, 26 Sept.-1 Oct. 1999, Lake Tahoe, Nevada, USA.

[4]    B. Miller, R. Alexander, N. Chapman, I. McKinley, J. Smellie, "Natural Analogues Revisited", *Radiochimica Acta*, 66/67, 545(1994).

[5]    B. Harmand, M. Sardin, "Modelling the Coupled Transport of Colloids and Radionuclides in a Fractured Natural Medium", *Radiochimica Acta*, 66/67, 691(1994).

[6]   V. Neck and J.I. Kim, *An Electrostatic Approach for the Prediction of Actinide Complexation Constants*, UCRL-135626, p. 241, *Abstract Migration-99*, 26 Sept.-1 Oct., 1999, Lake Tahoe, Nevada, CA.

[7]   B. Kienzler, B. Luckscheiter, S. Wilhelm, "Waste Form Corrosion Modelling Comparison with Experimental Results", UCRL-135626, p. 243, *Abstract of Migration-99*, 26 Sept.-1 Oct. 1999, Lake Tahoe, Nevada, USA.

[8]   H. Xu, Y. Wang, L. Barton, "A Linear Free Energy Relationship for Aqueous Ions and Crystalline Solids of MO2, M(OH)4, Garnet and MzrTi$_2$O$_7$", UCRL-135626, p. 244, *Abstract of Migration-99*, 26 Sept.-1 Oct. 1999, Lake Tahoe, Nevada, USA.

[9]   S. Carroll, S. Roberts, P. O'Day, N. Sahai, "Sorption and Precipitation of Strontium at Kaolinite, Gibbsite, Silica Gel and Goethite", UCRL-135626, p. 245, *Abstract of Migration-99*, 26 Sept.-1 Oct. 1999, Lake Tahoe, Nevada, USA.

[10]  P. Benes, K. Stamberg, L. Siroky, J. Mizera, "Radiotracer Study and Modelling of Sorption of Europium on Gorleben Sand From Aqueous Solutions Containing Humic Acid Substances", UCRL-135626, p. 246, *Abstract of Migration-99*, 26 Sept.-1 Oct. 1999, Lake Tahoe, Nevada, USA.

[11]  E.C. Buck, "Modelling the Formation of Colloids During Nuclear Waste From Corrosion", UCRL-135626, p. 242, *Abstract of Migration-99*, 26 Sept.-1 Oct. 1999, Lake Tahoe, Nevada, USA.

[12]  J.S. Contardi, D.R. Turner, T.M. Ahn, "Modelling Colloid Transport for Performance Assessment", UCRL-135626, p. 254, *Abstract of Migration-99*, 26 Sept.-1 Oct. 1999, Lake Tahoe, Nevada, USA.

[13]  U. Berner, "Geochemical Modelling of Repository Systems: Limitations of the Thermodynamic Approach", *Radiochim. Acta*, 82, 423 (1998).

[14]  A. Haworth, A.M. Thompson, C.J. Tweed, "Use of the HARPROB Program to Evaluate the Effect of Parameter Uncertainty on Chemical Modelling Predictions", *Radiochim. Acta*, 82, (1998) 429.

[15]  I. Grenthe, J. Fuger, R. Konings, R. Lemire, A. Muller, C. Nguyen-Trung, H. Wanner, "The Chemical Thermodynamics of Uranium", North-Holland, Amsterdam (1992).

[16]  I. Grenthe, I. Puigdomenech, eds., "Modelling in Aquatic Chemistry", OECD/NEA, Paris (1997).

[17]  C.F. Baes Jr., R.E. Mesmer, "The Hydrolysis of Cations", John Wiley and Sons, New York, (1976).

[18]  R.J. Münze, *Inorg. Nucl. Chem.*, 34, 661 (1972).

[19]  H. Moriyama, M.I. Pratopo, K. Higashi, "Systematics of Hydrolysis and Carbonate Complexation Constants of Ions of Transuranium Elements", *Radiochim. Acta.*, 66/67, 899 (1994).

[20]  H. Moriyama, A. Kitamura, K. Fujiwara, H. Yamana, "Analysis of Mononuclear Hydrolysis Constants of Actinide Ions by Hard Sphere Model", *Radiochim. Acta* (in press).

[21] P.L. Brown, R.N. Sylva, "Unified Theory of Metal-Ion-Complex Formation Constants", *J. Chem. Res.*, (S)4-5, (M)0110 (1987).

[22] P.L. Brown, H. Wanner, "Predicted Formation Constants using the Unified Theory of Metal Ion Complexation", OECD-NEA, Paris, (1987).

[23] C.M. Criss, J.W. Cobble, "The Thermodynamic Properties of High Temperature Aqueous Solutions. IV. Entropies of the Ions up to 200°C and the Correspondence Principle", *Amer. Chem. Soc.*, 86 5385 (1964).

[24] Y. Gu, C.H. Gammons, M.S. Bloom, "A One-term Extrapolation Method for Estimating Equilibrium Constants of Aqueous Reactions at Elevated Temperatures", *Geochim. Cosmochim. Acta*, 58, 3545 (1994).

[25] G.M. Anderson, S. Castet, J. Schott, R.E. Mesmer, "The Density Model for Estimation of Thermodynamic Parameters of Reactions at High Temperatures and Pressures", *Geochim. Cosmochim. Acta*, 55, 1769 (1991).

[26] J.A. Chemak, J.D. Rimstidt, "Estimating the Thermodynamic Properties of Silicates Minerals at 298 K from the Sum of Polyhedral Contributions", *Am. Mineral.*, 74, 1023 (1989).

[27] M. Gottschalk, "Internally Consistent Thermodynamic Data for Rock-Forming Minerals", *Europ. J. Min.*, 9, 175 (1997).

[28] D.A. Sverjensky, P.A. Molling, *Nature*, 356, 231 (1992).

[29] A. La Iglesia, J.F. Félix, *Geochim. Cosochim. Acta*, 58, 3883 (1994).

[30] G. Meinrath, T. Kimura, *Inorg. Chim. Acta*, 204, 79 (1993).

[31] R.J. Finch, M.A. Cooper, F.C. Hawthorne, R.C. Ewing, *Can. Mineral*, 34, 1071 (1996).

[32] S. Nakayama, H. Moriyama, H. Arimoto, K. Higashi, "Behaviors of Americium in Aqueous Solutions Containing Iron", *J. Nucl. Sci. Technol.*, 23, 731 (1986).

[33] T. Roy, *J. Chemometrics*, 8, 37 (1994).

[34] ISO/IUPAC/BIPM, "Guide to the Expression of Uncertainty in Measurement", 1993, International Standards Organization, Geneva, Switzerland.

[35] S. Ellison, W. Wegscheider, A. Williams, *Anal. Chem.*, 69, 607A (1997).

[36] G. Meinrath, C. Ekberg, A. Lanfgren, J.-O. Liljenzin, *Talanta*, 51, 231-246 (2000).

[37] G. Meinrath, *J. Radioanal. Nucl. Chem*, 211, 349 (1996).

[38] D.J. Clark, S. Conradson, S.A. Ekberg, N.J. Hess, M.P. Neu, P.D. Palmer, W. Runde, C.D. Tait, *J. Am. Chem. Soc.*, 118, 2089 (1996).

[39] G. Meinrath, Y. Kato, T. Kimura, Z. Yoshida, *Radiochim. Acta*, 84, 21 (1999).

[40] Y. Kato, T. Kimura, Z. Yoshida, N. Nitani, *Radiochim. Acta,* 74, 21 (1996).

**Figure 1**

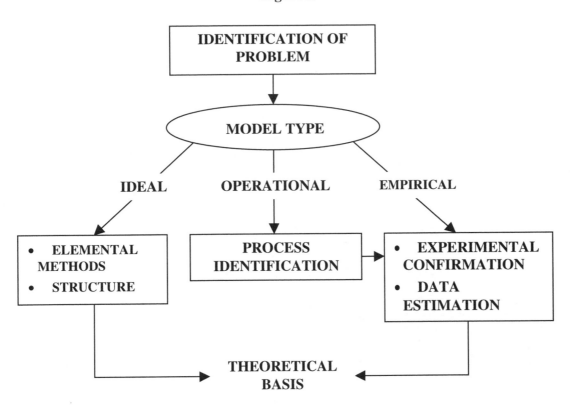

IDENTIFICATION OF PROBLEM

MODEL TYPE

IDEAL          OPERATIONAL          EMPIRICAL

- ELEMENTAL METHODS
- STRUCTURE

PROCESS IDENTIFICATION

- EXPERIMENTAL CONFIRMATION
- DATA ESTIMATION

THEORETICAL BASIS

# RECOMMENDATION

It was recommended that a working group be set up in order to review the reports and to develop them into a "Speciation Web Site" within the NEA web site. The web site would provide the latest information on speciation technology and would serve to guide researchers in choosing the most useful technique. The following proposal was drafted by Dr. Wymer, including the comments given at the open discussion of the closing session.

## Proposal for a speciation web site

A web site to make information on speciation of radioactive and toxic materials easily available would be useful and productive.

It is both appropriate and desirable for the OECD Nuclear Energy Agency to consider sponsoring such a site. The benefits to OECD Member countries would include reduction in duplication of effort, reduction in the time required to solve speciation problems (and consequently a reduction in the time required to solve remediation problems), and unnecessary dissipation of the efforts of an ever-decreasing number of trained and knowledgeable scientists. There would result a significant decrease in the net cost to member countries in carrying out their respective speciation activities.

The very large number of areas in which speciation is useful dictates that a selection be made of areas especially well suited to OECD Member country needs. Important areas for consideration include:

- Contaminants in water.

- Contaminants in soil.

- Contaminants on surfaces.

- Materials associated with high-level waste repositories.

- Interim storage areas.

There are two important phases involved in setting up and operating the web site:

- Deciding the scope and specific content of the site.

- Maintaining the site over the long term.

These phases are elaborated on in the following sections.

**Scope and content of the site**

It will be necessary to convene a working group to define in detail what the web site will include. Members of the group should include individuals that possess, at a minimum, the following skills:

- Speciation specialists.

- Chemical thermodynamicists.

- Separations chemists.

- Theoretical chemists.

- Actinide chemists.

- Waste management specialists.

- Environmental pollution experts.

- Computer database experts.

This group will be needed during site establishment and periodically thereafter to ensure that the site is adequately maintained and relevant.

**Maintaining the site**

It will be necessary to maintain, improve and update the site on a scheduled basis. This will require availability of a database expert on an as-needed basis. The need will be determined by the periodic meetings of the working group or an appropriate representation of that group.

It is desirable to establish two databases in the web site. The first database would contain existing information on speciation equipment and techniques. The second database would contain indirect, predictive, information and tools. These are discussed below.

*Existing speciation equipment and techniques database*

This database would contain, at a minimum, the following fields:

- Technique name.

- Sensitivity range.

- Limitations of usefulness, e.g. in the laboratory, in the field, etc.

- Likely areas of applicability.

- Relative cost (need for sophisticated equipment, tedious sample preparation, etc.).

- References to the literature.

There should be allowance for database user input on speciation needs and utility of the database and for improvements in it.

## *Predictive information and tools database*

The extraordinary complexity of the natural environment and of some other contamination situations makes it a certainty that speciation methods needed to aid in solving clean-up and remediation problems will not always be available. In these instances resort may be made to information that is not derived directly from results of speciation. Instead, useful information and insights that may be obtained from more general theoretical bases, from studies of systematic behaviours of related materials, from analogies, and from expert judgement. Some of the kinds of information that addresses speciation problems at this predictive level are listed below.

- Basic theoretical constructs about the equilibria of interest.

- Studies of the systematic behaviours of elements, compounds, complexes, valence states, etc., that are expected to encompass the behaviour of the material of interest.

- Analogies drawn from experience or from nature, e.g. from geologic analogues.

Just as for the other database, it will be necessary for this database to include an appropriate suite of database fields, and it will be desirable to accommodate use input.

## Conclusion

The OECD Nuclear Energy Agency has already provided scientific reports, or has work in progress, on most of the scientific areas relevant to the databases discussed above. The suggested web site would provide an excellent manner of organising and making easily available the large amount of work done to date. In addition, through user input new areas for needed additional work emerge.

# CLOSING REMARKS

## Raymond G. Wymer

Thank you for this opportunity to have the last word in this Workshop on Evaluation of Evaluation of Speciation Technology. I think we have had a productive and successful meeting. We began with a discussion of the broad areas of need for speciation of radioactive and toxic materials found in waste management and environmental pollution research, development and, finally, in remediation activities.

Over the past two days, we have heard presentations by authorities in a very wide range of speciation specialities. Of necessities, the presentations and posters were brief and provided only a glimpse of the expertise of the scientists attending this workshop. I look forward to receiving the publication of the full papers and group reports.

A comparison of the areas of need for speciation with the areas discussed here shows that while a great deal has been accomplished in speciation techniques and methodologies, much still remains to be done. This is not a criticism of the work to date, but instead is a testimony to the difficulty and complexity of the task before us as we face the ever-increasing speciation challenges. One example is extending the temperature range.

As we close this workshop, I would like to suggest that we convey our thanks to the OECD/NEA Nuclear Science committee for organising this excellent workshop and to the individuals who have worked so hard to ensure its success. Not only has the technical side of the meeting been handled in fine fashion, but the human aspects have been taken care of as well. From long contact with our Japanese friends, this is what I have come to expect.

We need to acknowledge and thank our host, Japan Atomic Energy Research Institute, for the excellent care they have provided for us while we were here.

If the recommendation for a web site on speciation sponsored and maintained by the NEA is approved I will see all of you on the Internet.

Thank you. I hope to see you again.

# LIST OF PARTICIPANTS

## BELGIUM

DESREUX, Jean F.
University of Liège
Sart Tilman (B16), Liège B-4000

Tel: +32 4 366 35 01
Fax: +32 4 366 47 36
Eml: jf.desreux@ulg.ac.be

* FUGER, Jean
Laboratory of Radiochemistry
University of Liège,
Sart Tilman (B16), Liège B-4000

Tel: +32 4 366 3470
Fax: +32 4 366 4736
Eml: jfuger@ulg.ac.be

RORIF, Fabian Francois
Radiochemistry, University of Liège
Sart Tilman (B16), Liège B-4000

Tel: +32 4 366 3476
Fax: +32 4 366 2927
Eml: Fabian.Rorif@ulg.ac.be

## CZECH REPUBLIC

HAVEL, Josef
Dept. of Anal. Chem., Faculty of Science
Masaryk University
Kotlarska 2, 611 37 BRNO

Tel: +420 5 41 129 568
Fax: +420 5 41 211 214
Eml: havel@chemi.muni.cz

HAVLOVA, Michaela
(c/o Prof. Josef HAVEL )
Molecular Biology and Genetics
Masaryk University
Kotlarska 2, 611 37 BRNO

Tel:
Fax:
Eml:

## FINLAND

KOUKKARI Pertti
Process Physics
VTT Chemical Technology
Otakaari 3 A, Espoo, P.O. Box 1404
FIN-02044 VTT

Tel: +358 9 456 6366
Fax:
Eml: Pertti.Koukkari@vtt.fi

VUORINEN, Ulla Sinikka
VTT Chemical Technology
Otakaari 3 A, Espoo, P.O. Box 1404
FIN-02044 VTT

Tel: +358 0 456 6389
Fax: +358 0 456 6390
Eml: Ulla.Vuorinen@vtt.fi

# FRANCE

FOUREST, Blandine
Institut de Physique Nucléaire
Université Paris-Sud
91406 Orsay-Cedex

Tel: +33 1 6915 7485
Fax: +33 1 6915 6470
Eml: fourest@ipno.in2p3.fr

MOULIN, Christophe
French Atomic Energy Commission (CEA)
DCC/DPE/SPCP/LASO
91191 Gif-sur-Yvette

Tel: +33 1 6908 7751
Fax: +33 1 6908 5411
Eml: cmoulin1@cea.fr

# GERMANY

DENECKE, Melissa A.
Forschungszentrum Karlsruhe (INE)
P.O.Box 3640 Karlsruhe D-76021

Tel: +49 7247 825536
Fax: +49 7247 823927
Eml: denecke@ine.fzk.de

KIM, Jae-Il
Forschungszentrum Karlsruhe (INE)
P.O.Box 3640 Karlsruhe D-76021

Tel: +49 7247822230/31
Fax: +49 7247824308
Eml: kim@ine.fzk.de

KLENZE, Reinhardt
Institut fur Nukleare Entsorgungstechnik
Forschungszentrum, Karlsruhe D-76021

Tel: +49 7247 824602
Fax: +49 7247 824306
Eml: klenze@ine.fzk.de

MEINRATH, Günter
TU Bergakademie Freiberg,
Institute of Geology
G.-Zeuner-Str. 12, Freiberg D-09596

Tel: +49 851 70372
Fax: +49 851 70372
Eml: meinrath@geo.tu-freiberg.de

# JAPAN

ADACHI, Takeo
Dept. of Environmental Science
Japan Atomic Energy Research Institute
Shirakata 2-4, Tokai, Ibaraki 319-1195

Tel: +81 29 282 5530
Fax: +81 29 282 5820
Eml: adachi@analchem.tokai.jaeri.go.jp

AMANO, Hikaru
Dept. of Environmental Science
Japan Atomic Energy Research Institute
Shirakata 2-4, Tokai, Ibaraki 319-1195

Tel: 81 29 282 5090
Fax: 81 29 282 6757
Eml: amano@popsvr.tokai.jaeri.go.jp

AOYAGI, Hisao
Advanced Science Research Centre
Japan Atomic Energy Research Institute
Shirakata 2-4, Tokai, Ibaraki 319-1195

Tel: +81 29 282 5915
Fax: +81 29 282 5927
Eml: h_aoyagi@popsvr.tokai.jaeri.go.jp

ARISAKA, Makoto
Advanced Science Research Centre
Japan Atomic Energy Research Institute
Shirakata 2-4, Tokai, Ibaraki 319-1195

Tel: +81 29 282 5493
Fax: +81 29 282 5935
Eml:

ASAKURA, Toshihide
Dept. of Fuel Cycle Safety Research
Japan Atomic Energy Research Institute
Shirakata 2-4, Tokai, Ibaraki 319-1195

Tel: +81 29 282 6660
Fax: +81 29 282 6677
Eml: asakura@procsafe.tokai.jaeri.go.jp

FUJITA, Reiko
TOSHIBA CORP.
Nuclear Engineering Lab.
4-1,Ukishima-cho, Kawasaki-ku
Kawasaki 210

Tel: +81 44 288 8153
Fax: +81 44 270 1807
Eml: fujita@cpg.nel.rdc.toshiba.co.jp

FUKASAWA, Tetsuo
Nuclear Systems Div., Hitachi, Ltd.
3-1-1 Saiwai, Hitachi, Ibaraki 317-8511

Tel: +81 294 23 5458
Fax: +81 294 23 6626
Eml: tetsuo_fukasawa@cm.hitachi.hitachi.co.jp

HANZAWA, Yukiko
Dept. of Environmental Science
Japan Atomic Energy Research Institute
Tokai, Naka, Ibaraki, 319-1195 Japan

Tel: +81 29 282 5544
Fax: +81 29 282 6950
Eml: hanzawa@popsvr.tokai.jaeri.go.jp

HOTOKEZAKA, Hiroyasu
Dept of Quantum Engineering and Systems Science
The University of Tokyo
7-3-1 Hongo, Bunkyo-ku, Tokyo 113-8156

Tel: +81 3 5841 6970
Fax: +81 3 3818 3455
Eml: hotoke@flanker.q.t.u-tokyo.ac.jp

IGARASHI, H.
Japan Nuclear Cycle Development Institute
4-33 muramatu, Tokai, Ibaraki 319-1194

Tel:
Fax:
Eml:

IKEDA, Yasuhisa
Insitute of Research and Innovation
1201, Takada, Kashiwa, Chiba 277-0861

Tel: +81 471 44 8837
Fax: +81 471 44 7602
Eml: ikeda@iri.or.jp

IMURA, Hisanori
Ibaraki University
Bunkyo, Mito 310-8512

Tel: +81 29 228 8364
Fax: +81 29 228 8403
Eml: imura@mito.ipc.ibaraki.ac.jp

ISO, Shuichi
Advanced Science Research Centre
Japan Atomic Energy Research Institute
Shirakata 2-4, Tokai, Ibaraki 319-1195

Tel: +81 29 282 5532
Fax: +81 29 282 5927
Eml:

KAMEO, Yutaka
Dept. of Decommissioning and Waste Management
Japan Atomic Energy Research Institute
Shirakata 2-4, Tokai, Ibaraki 319-1195

Tel: +81 29 282 5562
Fax: +81 29 282 5998
Eml: kameo@lynx.tokai.jaeri.go.jp

KATO Yoshiharu
Advanced Science Research Centre
Japan Atomic Energy Research Institute
Shirakata 2-4, Tokai, Ibaraki 319-1195

Tel: +81 29 282 5532
Fax: +81 29 282 6898
Eml: ykato@analchem.tokai.jaeri.go.jp

* KIHARA Sorin
Kyoto Institute of Technology
Matsugasaki, Sakyou-ku, Kyoto 606-8585

Tel: +81 75 724 7518
Fax: +81 75 724 7518
Eml: kiharas@ipc.kit.ac.jp

KIMURA, Takaumi
Advanced Science Research Centre
Japan Atomic Energy Research Institute
Shirakata 2-4, Tokai, Ibaraki 319-1195

Tel: +81 29 282 5493
Fax: +81 29 282 5935
Eml: kimura@analchem.tokai.jaeri.go.jp

KITAMURA, Akira
Waste Management and Fuel Cycle Research Centre
Japan Nuclear Cycle Development Institute
4-33 muramatu, Tokai, Ibaraki 319-1194

Tel: +81 29 287 3695
Fax: +81 29 287 3258
Eml: ak@tokai.jnc.go.jp

KITATSUJI, Yoshihiro
Advanced Science Research Centre
Japan Atomic Energy Research Institute
Shirakata 2-4, Tokai, Ibaraki 319-1195, Japan

Tel: +81 29 282 5537
Fax: +81 29 282 6898
Eml: kita@popsvr.tokai.jaeri.go.jp

LI, Xin
Advanced Science Research Centre
Japan Atomic Energy Research Institute
Shirakata 2-4, Tokai, Ibaraki 319-1195

Tel: +81 29 282 5532
Fax: +81 29 282 5927
Eml:

MAEDA, Mitsuru
Japan Atomic Energy Research Institute
Fukoku Seimei Bldg., 2-2-2 Uchisaiwai-cho
Chiyoda-ku, Tokyo 100-0011

Tel: +81 3 3592 2111
Fax: +81 3 3580 6107
Eml:

MEGURO, Yoshihiro
Advanced Science Research Centre
Japan Atomic Energy Research Institute
Shirakata 2-4, Tokai, Ibaraki 319-1195

Tel: +81 29 282 5532
Fax: +81 29 282 6898
Eml: meguro@analchem.tokai.jaeri.go.jp

MINAI, Yoshitaka
Musashi University
1-26-1 Toyotama-kami, Nerima-ku,
Tokyo 176-8534

Tel: +81 3 5984 3845
Fax: +81 3 3991 1198
Eml: minai@cc.musashi.ac.jp

MORIYAMA Hirotake
Kyoto University
Kumatori-cho, Sennan-gun, Osaka 590-0494

Tel: +81 724 51 2424
Fax: +81 724 51 2634
Eml: moriyama@rri.kyoto-u.ac.jp

NAGAISHI, Ryuji
Dept. of Environmental Science
Japan Atomic Energy Research Institute
Shirakata 2-4, Tokai, Ibaraki 319-1195

Tel: +81 29 282 5493
Fax: +81 29 282 5935
Eml: nagaishi@analchem.tokai.jaeri.go.jp

NAGAO, Seiya
Dept. of Fuel Cycle Safety Research
Japan Atomic Energy Research Institute
Shirakata 2-4, Tokai, Ibaraki 319-1195

Tel: +81 29 282 5085
Fax: +81 29 282 5934
Eml: nagao@sparclt.tokai.jaeri.go.jp

NAGASAKI, Shinya
The University of Tokyo
7-3-1 Hongo, Bunkyo-ku, Tokyo 113-8156

Tel: +81 3 3812 2111
Fax: +81 3 5800 6856
Eml: nagasaki@q.t.u-tokyo.ac.jp

NAKAYAMA, Shinichi
Dept. of Fuel Cycle Safety Research
Japan Atomic Energy Research Institute
Shirakata 2-4, Tokai, Ibaraki 319-1195

Tel: +81 29 282 6157
Fax: +81 29 282 5958
Eml: nakayama@sparclt.tokai.jaeri.go.jp

OGAWA, Toru
Dept. of Materials Science
Japan Atomic Energy Research Institute
Shirakata 2-4, Tokai, Ibaraki 319-1195

Tel: +81 29 282 5382
Fax: +81 29 282 5922
Eml: ogawa@molten.tokai.jaeri.go.jp

OHASHI, Kousaburo
Ibaraki University
Bunkyo 2-1-1 Mito 310-8512

Tel: +81 29 228 8363
Fax: +81 29 228 8406
Eml: ohasi@mito.ipc.ibaraki.ac.jp

OHNUKI, Toshihiko
Dept. of Environmental Science
Japan Atomic Energy Research Institute
Shirakata 2-4, Tokai, Ibaraki 319-1195

Tel: +81 29 282 6008
Fax: +81 29 282 5958
Eml: ohnuki@sparclt.tokai.jaeri.go.jp

OKADA, Sohei
Advanced Science Research Centre
Japan Atomic Energy Research Institute
Shirakata 2-4, Tokai, Ibaraki 319-1195

Tel: +81 29 282 5416
Fax: +81 29 282 5927
Eml:

SAITO, Akira
Department of Chemistry
Tokyo Gakugei University
4-1-1 Nukii Kita-machi, Koganei, Tokyo 184-8501

Tel: +81 42 329 7507
Fax: +81 42 329 7508
Eml: akira@u-gakugei.ac.jp

SAMADFAN, Mohammad
Dept. of Environmental Science
Japan Atomic Energy Research Institute
Shirakata 2-4, Tokai, Ibaraki 319-1195

Tel: +81 29 282 6180
Fax: +81 29 282 5958
Eml: samaal@sparclt.tokai.jaeri.go.jp

SHIBUTANI, Tomoki
Waste Management and Fuel Cycle Research Centre
Japan Nuclear Cycle Development Institute
4-33 Muramatu, Tokai, Ibaraki 319-1194

Tel: +81 29 287 3695
Fax: +81 29 287 3258
Eml: tomoki@tokai.jnc.go.jp

SHIOKAWA, Yoshinobu
Institute for Materials Science
Tohoku University
Katahira 2-1-1, aoba-ku, Sendai 980-8577

Tel: +81 22 215 2122
Fax: +81 22 215 2121
Eml: shiokawa@imr.edu

TACHIKAWA, Enzo
Japan Chemical Analsys Centre
295-3 Sannoh-cho, Inage-ku
Chiba-city 263-0002

Tel: +81 43 423 5325
Fax: +81 43 423 5372
Eml: e-tachikawa@jcac.or.jp

TACHIMORI, Shoichi
Dept. of Materials Science
Japan Atomic Energy Research Institute

Tel: +81 29 282 5198
Fax: +81 29 282 6723
Eml: tachi@mummy.tokai.jaeri.go.jp

TAKEISHI, Hideyo
Advanced Science Research Centre
Japan Atomic Energy Research Institute
Shirakata 2-4, Tokai, Ibaraki 319-1195

Tel: +81 29 282 5532
Fax: +81 29 282 5927
Eml:

TANAKA, Satoru
The University of Tokyo
7-3-1 Hongo, Bunkyo-ku, Tokyo 113-8156

Tel: +81 3 5841 6968
Fax: +81 3 5841 8625
Eml: chitanak@q.t.u-tokyo.ac.jp

TOCHIYAMA, Osamu
Tohoku Univ. Graduate School of Engineering
Dept. of Quantum Sci. & Eng.
Aramaki-Aza-Aoba, Sendai 01 980-8579

Tel: +81 22 217 7918
Fax: +81 22 217 7918
Eml: osamu.tochiyama@qse.tohoku.ac.jp

UNO, Seiichiro
Advanced Science Research Centre
Japan Atomic Energy Research INSTITUTE
Shirakata 2-4, Tokai, Ibaraki 319-1195

Tel: +81 29 282 5532
Fax: +81 29 282 5927
Eml:

USUDA, Shigekazu
Dept. of Environmental Science
Japan Atomic Energy Research Institute
Shirakata 2-4, Tokai, Ibaraki 319-1195

Tel: +81 29 282 6814
Fax: +81 29 282 6798
Eml: usuda@sglsun.tokai.jaeri.go.jp

YAITA, Tsuyoshi
Dept. of Materials Science
Japan Atomic Energy Research Institute
Shirakata 2-4, Tokai, Ibaraki 319-1195

Tel: +81 29 282 6162
Fax: +81 29 282 6723
Eml: yaita@mummy.tokai.jaeri.go.jp

YASUOKA, Hiroshi
Advanced Science Research Centre
Japan Atomic Energy Research Institute
Shirakata 2-4, Tokai, Ibaraki 319-1195

Tel: +81 29 282 5093
Fax: +81 29 282 5927
Eml:

YOSHIDA, Zenko
Advanced Science Research Centre
Japan Atomic Energy Research Institute
Shirakata 2-4, Tokai, Ibaraki 319-1195

Tel: +81 29 282 5535
Fax: +81 29 282 5927
Eml: zyoshida@popsvr.tokai.jaeri.go.jp

WANG, Hongxian
Advanced Science Research Centre
Japan Atomic Energy Research Institute
Shirakata 2-4, Tokai, Ibaraki 319-1195

Tel: +81 29 282 5915
Fax: +81 29 282 5927
Eml: hx_wang@popsvr.tokai.jaeri.go.jp

YUI, Mikazu
Waste Management and Fuel Cycle Research Centre
Japan Nuclear Cycle Development Institute
4-33 muramatu, Tokai, Ibaraki 319-1194

Tel: +81 29-287-1540
Fax: +81 29-287-3704
Eml: yui@tokai.jnc.go.jp

ZHENG, Wenge
Dept. of Environmental Science
Japan Atomic Energy Research Institute
Shirakata 2-4, Tokai, Ibaraki 319-1195

Tel: +81 29 282 5533
Fax: +81 29 282 6097
Eml: zhengwenge@hotmail.com

## KOREA

KIM, Kwang-Wook
Korea Atomic Energy Research Institute
P.O. Box 105, Yusong, Taejon 305-600

Tel: +82 42 868 2044
Fax: +82 42 863 2042
Eml: nkwkim@nanum.kaeri.re.kr

* YOO, Jae-Hyung
Korea Atomic Energy Research Institute
P.O. Box 105, Yusong, Taejon 305-600

Tel: +82 42 868 2043
Fax: +82 42 863 1236
Eml: njhyou@nanum.kaeri.re.kr

## RUSSIA

MYASOEDOV, Boris Fedorovich
Vernadsky Institute of Geochemistry
and Analytical Chemistry
Kosygin St., 19 117975 Moscow, Russia

Tel: +7 095 237 8081
Fax: +7 095 954 2228
Eml: bf@ncrc.msk.ru

## UKRAINE

* DRYAPACHENKO, Ihor Pavel
Institute for Nuclear Researches
47 Nauka Avenue 252028 Kyiv, UKRAINE

Tel: +38 044 2656072
Fax: +38 044 2651402
Eml: dryapach@marion.iop.kiev.ua

## UNITED KINGDOM

BERRY, John A.
AEA Technology plc.
220 32 Harwell, Didcot, Oxon OX11 ORA

Tel:
Fax: +44 1235 43 4385
Eml: john.berry@aeat.co.uk

\* DUFFIELD, John Ralph
Chemical and Physical Sciences
Faculty of Applied Sciences
University of the West of England, Bristol
Frenchay Campus, Coldharbour Lane
Bristol BS16 1QY

Tel: +44 0117 965 6261 (ext. 3815)
Fax: +44 0117 976 3871
Eml: John.Duffield@uwe.ac.uk

RANCE, Peter
British Nuclear Fuels Limited
Sellafield, Seascale, Cumbria CA20 1PG

Tel: +44 19467 75763
Fax: +44 19467 85740
Eml: PJWR1@bnfl.com

READ, David
University of Reading
Whiteknights, READING RG6 6PX BERKSHIRE

Tel: +44 118 986 8250
Fax: +44 137 245 1520
Eml: enterpri@readinguni.u-net.com

\* TAYLOR, Robin J.
British Nuclear Fuels Limited
Sellafield, Seascale, Cumbria CA20 1PG

Tel: +44 19467 74596
Fax: +44 19467 85740
Eml: rjt2@bnfl.com

WILLIAMS, Steve J.
AEA Technology plc.
220 32 Harwell, Didcot, Oxon OX11 ORA

Tel:
Fax: +44 1235 43 4385
Eml: steve.j.williams@aeat.co.uk

## UNITED STATES OF AMERICA

BERG, John M.
Los Alamos National Laboratory
MS E510, NMT-6 Los Alamos NM 87545

Tel: +1 505 665-8262
Fax: +1 505 665-4459
Eml: jberg@lanl.gov

CHOPPIN, Gregory R.
Dept. of Chemistry, Florida State University
B-164 Tallahassee, Florida 32306-4390

Tel: +1 904 644 3875
Fax: +1 904 644 8281
Eml: choppin@chemmail.chem.fsu.edu

HAIRE, Richard G.
Head of Actinide Research
Oak Ridge National Laboratory,
P. O. Box 2008 MS 6375, Oak Ridge
TN 37831-6375

Tel: +1 423 574 5007
Fax: +1 423 574 4987
Eml: hairerg@ornl.gov

PAVIET-HARTMANN, Patricia
Los Alamos National Laboratory
Chemical and Science Technology Division
CST-7, MS J514 Los Alamos NM 87545

Tel: +1 505 667-5711
Fax: +1 505 665-4955
Eml: ppaviet-hartmann@lanl.gov

RAI, Dhanpat
Pacific Northwest Laboratory
P.O. Box 999/MS P7-50
Richland WA 99352

Tel: +1 509 373 5988
Fax: +1 509 372 1632
Eml: dhan.rai@pnl.gov

\* REED, Donald Timothy
Actrinide Speciation Group
Argonne National Laboratory
Chemical Technology Division, Bld. 205, 9700 S.
Cass Ave Argonne IL 60439

Tel: +1 630-252-7964
Fax: +1 630-972-4431
Eml: reedd@cmt.anl.gov

\* RUNDE, Wolfgang Harald
Los Alamos National Laboratory
CST-7, Mail Stop J514, Los Alamos NM 87544

Tel: +1 505 667 3350
Fax: +1 505 665 4955
Eml: runde@lanl.gov

WYMER, Raymond George
Consultant
188-A Outer Drive, Oak Ridge TN 37830

Tel: +1 423 483 5013
Fax: +1 423 483 9309
Eml: rgwymer@prodigy.net

\* XU, Jide
Department of Chemistry,
Univ. of California, Berkeley
Berkeley, CA 94720

Tel: +1 510 642 7219
Fax: +1 510 486 5283
Eml: jide@uclink4.Berkeley.EDU

## INTERNATIONAL ORGANIZATIONS

SAKURAI, Satoshi
OECD/NEA
12, Bd des Iles 92130 Issy-les-Moulineaux, France

Tel: +33 1 4524 1152
Fax: +33 1 4524 1110
Eml: sakurai@nea.fr

WASTIN, Franck
European Commission
JRC, Inst. Transuranium Elements
Postfach 2340 Karlsruhe D-76125, Germany

Tel: +49 7247 951 387
Fax: +49 7247 951 593
Eml: wastin@itu.fzk.de

\* Regrets to have been unable to attend

*Annex 2*

## WORKSHOP ORGANISATION

## General Chairperson

        M. Maeda        (JAERI, Japan)

## Co-chairpersons

        G.R. Choppin        (Florida StateUniv., USA)
        J. Fuger        (Univ. of Liège, Belgium)
        Z. Yoshida        (JAERI, Japan)

## International Scientific Members

        T. Adachi        (JAERI, Japan)
        J.F. Desreux        (Univ. of Liège, Belgium)
        J.I. Kim        (FZK, Germany)
        T. Kimura        (JAERI, Japan)
        F. Livens        (Univ. of Manchester, UK)
        C. Madic        (CEA, France)
        H. Moriyama        (Kyoto Univ., Japan)
        C. Moulin        (CEA, France)
        B. Myasoedov        (V.I. of GeoChem. & Anal. Chem., Russia)
        H. Nitsche        (Univ. of California, Berkeley, USA)
        D. Rai        (PNL, USA)
        D. Read        (Univ. of Reading, UK)
        A. Saito        (Tokyo Gakugei Univ., Japan)
        S. Sakurai        (OECD/NEA, France)
        S. Tanaka        (Tokyo Univ., Japan)
        O. Tochiyama        (Tohoku Univ., Japan)

## Local Organising Committee

        Y. Kato        (JAERI, Japan)
        T. Kimura        (JAERI, Japan)
        Y. Meguro        (JAERI, Japan)
        T. Ohnuki        (JAERI, Japan)
        S. Tachimori        (JAERI, Japan)
        Z. Yoshida        (JAERI, Japan)

*Participants of the Workshop on Evaluation of Speciation Technology*

*ALSO AVAILABLE*

# NEA Publications of General Interest

*1999 Annual Report* (2000)                                    *Free: available on Web.*

*NEA News*
ISSN 1605-9581                          Yearly subscription: FF 240  US$ 45  DM 75  £ 26  ¥ 4 800

*Geologic Disposal of Radioactive Waste in Perspective* (2000)
ISBN 92-64-18425-2                          Price: FF 130  US$ 20  DM 39  £ 12  ¥ 2 050

*Radiation in Perspective – Applications, Risks and Protection (1997)*
ISBN 92-64-15483-3                          Price: FF 135 US$ 27  DM 40  £ 17  ¥ 2 850

*Radioactive Waste Management in Perspective* (1996)
ISBN 92-64-14692-X                          Price: FF 310  US$ 63  DM 89  £ 44

# Nuclear Science

*Pyrochemical Separations* (2001)
ISBN 92-64-18443-0                          Price: FF 500  US$ 66  DM 149  £ 46  ¥ 7 230
*Core Monitoring for Commercial Reactors: Improvements in Systems and Methods* (2000)
ISBN 92-64-17659-4                          Price: FF 460  US$ 71  DM 137  £ 44  ¥ 7 450
*Physics and Fuel Performance of Reactor-Based Plutonium Disposition* (1999)
ISBN 92-64-17050-2                          Price: FF 400  US$ 70  DM 119  £ 43  ¥ 8 200
*Shielding Aspects of Accelerators, Targets and Irradiation Facilities (SATIF-4)*(1999)
ISBN 92-64-17044-8                          Price: FF 500  US$ 88  DM 149  £ 53  ¥ 10 300

*3-D Radiation Transport Benchmarks for Simple Geometries with Void Regions*
(2000)                                                        *Free on request.*
*Benchmark Calculations of Power Distribution Within Fuel Assemblies*
Phase II: Comparison of Data Reduction and Power Reconstruction Methods in Production Codes
(2000)                                                        *Free on request.*
*Benchmark on the VENUS-2 MOX Core Measurements*
(2000)                                                        *Free on request.*
*Calculations of Different Transmutation Concepts: An International Benchmark Exercise*
(2000)                                                        *Free on request.*
*Prediction of Neutron Embrittlement in the Reactor Pressure Vessel: VENUS-1 and VENUS-3 Benchmarks*
(2000)                                                        *Free on request.*
*Pressurised Water Reactor Main Steam Line Break (MSLB) Benchmark*
(2000)                                                        *Free on request.*

**International Evaluation Co-operation** *(Free on request)*

Volume 1:  *Comparison of Evaluated Data for Chromium-58, Iron-56 and Nickel-58* (1996)
Volume 2:  *Generation of Covariance Files for Iron-56 and Natural Iron* (1996)
Volume 3:  *Actinide Data in the Thermal Energy Range* (1996)
Volume 4:  $^{238}U$ *Capture and Inelastic Cross-Sections* (1999)
Volume 5:  *Plutonium-239 Fission Cross-Section between 1 and 100 keV* (1996)
Volume 8:  *Present Status of Minor Actinide Data* (1999)
Volume 12: *Nuclear Model to 200 MeV for High-Energy Data Evaluations* (1998)
Volume 13: *Intermediate Energy Data* (1998)
Volume 14: *Processing and Validation of Intermediate Energy Evaluated Data Files* (2000)
Volume 15: *Cross-Section Fluctuations and Shelf-Shielding Effects in the Unresolved Resonance Region* (1996)
Volume16: *Effects of Shape Differences in the Level Densities of Three Formalisms on Calculated Cross-Sections* (1998)
Volume 17: *Status of Pseudo-Fission Product Cross-Sections for Fast Reactors* (1998)
Volume 18: *Epithermal Capture Cross-Section of* $^{235}U$ (1999)

*Order form on reverse side.*

# ORDER FORM

**OECD Nuclear Energy Agency, 12 boulevard des Iles, F-92130 Issy-les-Moulineaux, France**
**Tel. 33 (0)1 45 24 10 10, Fax 33 (0)1 45 24 11 10, E-mail: nea@nea.fr, Internet: www.nea.fr**

| Qty | Title | ISBN | Price | Amount |
|-----|-------|------|-------|--------|
|     |       |      |       |        |
|     |       |      |       |        |
|     |       |      |       |        |
|     |       |      |       |        |
|     |       |      |       |        |
|     |       |      |       |        |
|     |       |      |       |        |
|     |       |      |       |        |
|     |       | **Postage fees\*** | | |
|     |       | **Total** | | |

*European Union: FF 15 – Other countries: FF 20

❏ Payment enclosed (cheque or money order payable to OECD Publications).

Charge my credit card  ❏ VISA  ❏ Mastercard  ❏ Eurocard  ❏ American Express

*(N.B.: You will be charged in French francs).*

| Card No. | Expiration date | Signature |
|----------|-----------------|-----------|
| Name | | |
| Address | Country | |
| Telephone | Fax | |
| E-mail | | |

OECD PUBLICATIONS, 2, rue André-Pascal, 75775 PARIS CEDEX 16
PRINTED IN FRANCE
(66 2001 04 1 P) ISBN 92-64-18667-0 – No. 51863 2001